LINEAR ALGEBRA

LINEAR ALGEBRA

By

A.K. Sharma

DISCOVERY PUBLISHING HOUSE

NEW DELHI-110002

First Published – 2007

Reprinted – 2025

ISBN: 978-81-8356-222-5

Linear Algebra

Published by:

DISCOVERY PUBLISHING HOUSE PVT. LTD.
4383/4B, Ansari Road, Darya Ganj
New Delhi-110 002 (India)
Phone: +91-11-23279245; 23253475; 43596065
Mobile: +91 9811179893 / +91 9871656464
E-mail: discoverybooksindia@gmail.com
orderdphbooks@gmail.com
namitwasan9@gmail.com
web: www.discoverypublishinggroup.com

Printed at:
Infinity Imaging Systems
Delhi

Preface

This book Linear Algebra has been written for the use of students of Degree, Degree Honours and Postgraduate classes of all Indian Universities.

All the examples have been completely solved. The subject matter has been discussed in such a simple way that the students will find no difficulty to understand it. The students should first try to understand the theorems and then they should try to solve the questions independently.

Author

CONTENTS

Convergence and Baire's Theorem, Continuous Mappings, Uniform Convergence and Pointwise Convergence, Cauchy Schwarz Inequality, Minkowski Inequality, Minkowski's Inequality of Norms, Holder's Inequality, Definition Cantor's Tennary Sets, Definition of Neighbourhood, Interior, Closure, Space of Continuous Functions, Embedding, Embedding Theorems and Metriazability, Quotient Topology Space, Tychonoff Topological Space, Product Invariant Properties, Tychonoff Theorem, Tychonoff Cubes, Projection Mappings, Quotient Map, Diagonal Map, Quotient Space, Upper Semi-continuous Decomposition, Evaluation Map, Solved Examples, Exercises.

1

VECTOR SPACES

INTRODUCTION

Vector Space can in large part, trace its origins to topics in Geometry and Physics. Vector Space our their importance to the fact that so many models arising in the solution of specific problems turn out to be vector spaces for this reason the basic concept introduced in them have a certain university and are ones we encounter, and keep encountering, in so many diverse contexts. Among these fundamental notation are those of linear dependence, basis and dimension which will be developed in this chapter.

BINARY OPERATION ON A SET

Let G be a non-empty set.

Then $G \times G = \{(a, b) : a \in G, b \in G\}$. If $f : G \times G \to G$, then f is said to be a binary operation on the set G. The image of the ordered pair (a, b) under the function f is denoted by f(a, b) or by a f b. Often we use the symbols +, ×, ., o, * etc. to denote binary operations on a set. Thus '+' will be a binary operation on G iff

$$a + b \in G \;\forall\; a, b \in G \text{ and } a * b \text{ is unique.}$$

Similarly '*' will be a binary operation on G iff

$$a * b \in G \;\forall\; a, b \in G \text{ and } a * b \text{ is unique.}$$

A binary operation on a set G is sometimes also called a *binary composition* in the set G. If '*' is a binary composition in G, then $\forall$ a, b $\in$ G, a * b is a unique clement of G. If $a * b \in G \;\forall\; a, b \in G$, then we also say that G is closed with respect to the composition denoted by *.

If there is a binary composition in a set G, the most convenient notation to denote this composition is the multiplicative notation. In this notation if a, b $\in$ G, then ab represents the element obtained on multiplying a and b.

Thus ab $\in$ G $\forall$ a, b $\in$ G if the binary composition in G has been denoted multiplicatively.

Examples:

Addition is a binary operation on the set N of natural numbers. The sum of two natural numbers is also a natural number. Therefore N is closed with respect to addition

i.e., $a + b \in N \ \forall \ a, b \in N$.

Subtraction is not a binary operation on N. We have $4-7 = -3 \notin N$ whereas $4 \in N$, $7 \in N$. Thus N is not closed with respect to subtraction.

But subtraction is a binary operation in the set of integers I.

We have $a - b \in I \ \forall \ a, b \in I$.

Division is not a binary operation in the set R of all real numbers. We have $0 \in R$, $5 \in R$ but $5 \div 0$ is not an element of R.

ALGEBRAIC STRUCTURE

Definition : A non-empty set G equipped with one or more binary operations is called an algebraic structure. Suppose * is a binary operation on G.

Then (G, *) is an algebraic structure. (N, +), (I, +), (I, -), (R, +, .) are all algebraic structures.

Obviously addition and multiplication are both binary operations on the set R of real numoers. Therefore (R, +,.) is an algebraic structure equipped with two operations.

GROUP

Definition : Let G be a non-empty set equipped with a binary operation denoted by the symbol * *i.e.*, $a * b \in G \ \forall \ a, b \in G$. Then this algebraic structure is a group if the binary operation * satisfies the following postulates:

1. *Associativity i.e.*, $(a * b) * c = a * (b * c) \ \forall \ a, b, c \in G$.
2. *Existence of right Identity :* There exists an element $e \in G$ such that $a = a \ \forall \ a \in G$.

   The element e is called the *right identity*.
3. *Existence of right inverse :* Each element of G possesses right inverse *i.e.*, for each elemental $\in$ G, there exists an element

   $a^{-1} \in G$ such that $a * a^{-1} = \in$.

   The element a^{-1} is then called the right inverse of a.

Abelian Group or Commutative Group

Definition : A group G is said to be abelian or commutative if in addition to the above three postulates the following postulate is also satisfied:

4. *Commutativity i.e.*, $a * b = b * a \ \forall \ a, b \in G$.

Note 1 : In our definition of .a group we have denoted the composition in G by the symbol : However we can use any symbol like ×, o, ., etc. to denote the composition. If we use the additive notation '+' to denote the composition in G, then the right inverse of an element $a \in G$ is denoted by the symbol $-a$ *i.e.*, we have

$$a + (-a) = e.$$

Note 2 : If we use multiplicative notation to denote the composition in G, then often we denote the right identity by the symbol '1'. Thus 1 is an element of G, such that $a1 = a \ \forall \ a \in G$.

There should be no confusion about 1. It is not the number 1 but it is an element of the set G whatever it may be.

In multiplicative notation the right inverse a^{-1} of a is often denoted by 1/a.

In additive notation, we often denote the right identity by the symbol '0'. Thus 0 is an element of G such that

$$a + Q = a \ \forall \ a \in G.$$

Note 3 : In additive notation the element $a + (-b) \in G$ is denoted by $a - b$. In multiplicative notation the element $ab^{-1} \in G$ is denoted by a/b.

Example:

Show that the set I of all integers

$$\ldots, -4, -3, -2, -1, 0, 1, 2, 3, 4, \ldots$$

is an abelian group with respect to the operation of addition of integers.

Some General Properties of Groups

Suppose G is a group with binary operation being denoted multiplicatively. Then we have the following properties in G,

Theorem 1:

Existence of right cancellation law i.e., if a, b, c are in G,

then $$ba = ca \Rightarrow b = c.$$

Proof:

Since $a \in G$ therefore $\exists \ a^{-1} \in G$ such that $aa^{-1} = e$ where e is the right identity.

Now $ba = ca$

$\Rightarrow (ba)\, a^{-1} = (ca)\, a^{-1}$

$\Rightarrow b\, (aa^{-1})\, c\, (aa^{-1})$ [by associativity]

$\Rightarrow be = ce$ [$\because a^{-1}$ is right inverse of a]

$\Rightarrow b = c$ [$\because$ e is right identity 1

Theorem 2:

The right identity is also the left identity i.e., if e is the right identity then $ea = a\ \forall\ a \in G$.

Proof:

Let $a \in G$ and e be the right identity. Since a possesses right inverse, therefore there exists $a^{-1} \in G$ such that

$aa^{-1} = e.$

Now $(ea)\, a^{-1} = e\, (aa^{-1})$ [by associativity]

$= ee$ [$\because aa^{-1} = e$]

$= e$ [$\because$ e is right identity]

$= aa^{-1}$ [$\because aa^{-1} = e$]

$\Rightarrow ea = a.$ [by right cancellation law]

$\therefore$ e is also the left identity.

Hence e is the-identity *i.e.*, $ea = a = ae\ \forall\ a \in G$.

Theorem 3:

The right inverse of an element is also its left inverse i e., if a^{-1} is the right inverse of a then $a^{-1} a = e$.

Proof:

Let a e G and e be the identity element. Let a^{-1} be the right inverse of a *i.e.*, $aa^{-1} = e$. To prove that $a^{-1} a = e$.

We have

$(a^{-1} a)\, a^{-1} = a^{-1} (aa^{-1})$ [by associativity]

$= a^{-1} e$ [$\because aa^{-1} = e$]

$= a^{-1}$ [$\because$ e is right identity]

$= ea^{-1}$ [$\because$ e is also left identity]

Now $(a^{-1} a)\, a^{-1} = ea^{-1}$

$\Rightarrow a^{-1} a = e.$ [by right cancellation law]

$\therefore$ a^{-1} is also the left inverse of a.

$\therefore$ a^{-1} is the inverse of a *i.e.*, $a^{-1}a = e = aa^{-1}$.

Theorem 4:

***Uniqueness of Identity** : The identity element in a group is unique.*

Proof:

Suppose e and e' are two identity elements of a group G. We have

$ee' = e$ if e' is identity

and $ee' = e'$ if e is identity.

But ee' is a unique element of G.

$ee' = e$ and $ee' = e' \Rightarrow e = e'$.

Hence the identity element is unique.

Theorem 5:

***Uniqueness of Inverse** : The inverse of each element of a group is unique.*

Proof:

Let a be any element of a group G and let e be the identity element. Suppose b and c are two inverses of a

i.e., $ba = e = ab$

and $co = e = ac$.

We have $b(ac) = be$

$= b$ $[\because$ e is identity$]$

Also $(ba)c = ec$ $[\because ba = e]$

$= c$ $[\because$ e is identity$]$

But in a group the composition is associative.

$\therefore$ $b(ac) = (ba)c$

$\Rightarrow$ $b = c$

Theorem 6:

Existence of left cancellation law i.e., if a, b, c are in G, then

$$ab = ac \Rightarrow b = c.$$

Proof:

We have $ab = ac$

$\Rightarrow a^{-1} (ab) = a^{-1} (ac)$

$\Rightarrow (a^{-1} a) b = (a^{-1} a) c$

$\Rightarrow eb = ec$ [$\because$ a^{-1} is left inverse of a]

$\Rightarrow b = c$ [$\because$ e is left identity]

THE RELATION OF ISOMORPHISM IN THE SET OF ALL GROUPS

Theorem:

The relation of isomorphism in the set of all groups is an equivalence relation.

Proof:

We shall prove that the relation of isomorphism denoted by $\cong$ in the set of all groups is reflexive, symmetric and transitive.

Reflexive: If G is any group, then $G \cong G$. Let f b the identity mapping on G *i.e.*, let f: G $\rightarrow$ *i.e.*, let f: $G \rightarrow G$ such that $f(x) = x \ \forall \ x \in x \ G$. Obviously f is one-one onto. Also if x, y are any elements of G, then $f(x) = x$ and $f(y) = y$.

Also $f(xy) = xy$ [$\because$ f is identity mapping]

$= f(x) f(y)$.

$\therefore$ f is composition preserving also. Thus f is an isomorphism of G onto G.

Hence $G \cong G$.

Symmetric *i.e.*, $G \cong G' \Rightarrow G' \Rightarrow G$. Suppose a group G is isomorphic to another group G'. Let f be an isomorphism of G onto G'. Then f is one-one onto and preserves compositions in G and G'. Since f is one-one onto, therefore it is inversible *i.e.*, f^{-1} exists. Also we know that the mapping f^{-1} is also one-one onto.

Now, we shall show that $f^{-1}: G' \rightarrow G$ is also composition preserving. Let x', y' be any elements of G'. Then there exist elements $x, y \in G$ such that

$f^{-1}(x') = x, \ f^{-1}(y') = y$ ...(1)

and $f(x) = x', \ f(y) = y'$ [From (2)]

Now $f^{-1}(x' y') = f^{-1}[f(x') f[(y)]$

$= f^{-1}[f(xy)]$, since $f(xy) = f(x) f(y)$

$= xy$ [by def. of f^{-1}]

$= f^{-1}(x') f^{-1}(y')$. [From (1)]

$\therefore f^{-1}$ preserves compositions in G' and G.

Hence $G' \cong G$.

Transitive *i.e.*, $G \cong G', G' \cong G'' \Rightarrow G \cong G''$. Suppose G is isomorphic to G' and G' is isomorphic to G". Further suppose that $f: G \to G'$ and $g: G' \to G''$ are the respective isomorphic mappings. We know that the composite mapping $g \circ f: G \to G''$ defined by

$$(g \circ f)(x) = g[f(x)] \ \forall \ x \in g$$

is also one-one onto if both f and g are one-one, onto.

Further if x, y are any elements of G, then

$(g \circ f)(xy) = g[f(xy)]$ [by definition of g o f]

$= g[f(x)] f(y)]$ [$\because$ f is composition preserving]

$= g[f(x)] g[f(y)]$ [$\because$ g is also an isomorphism]

$= [(g \circ f)(x)] [(g \circ f)(y)]$.

Hence the relation of isomorphism in the of G onto G" and $G \cong G''$.

Hence the relation of isomorphism in the set of all groups is an equivalence relation.

Note: The relation of isomorphism in the set of all groups will partition the set of all groups into disjoint equivalence classes. If G_1 is any group, then all the groups isomorphic to G_1 will form one equivalence class. If G_2 is another group not isomorphic to G_1, then all the groups isomorphic to G_2 will form another equivalence class, and so on.

TRANSFERENCE OF GROUP STRUCTURES

Theorem:

If G is a group and G' is a set equipped with a composition (supposed denoted multiplicatively) and if there exists a one-one mapping f of G onto G' such that

$$f(ab) = f(a) f(b) \ \forall \ a, b \in G,$$

then G' is also a group isomorphic to G for the composition in question.

Proof:

To show that G' is a group we are to show that the composition in G' satisfies the group postulates.

Associatively: Suppose a', b', c' are any three elements of G'. Since f is an onto function, there exist three elements a, b, c of G such that f (a) =

a', f (b) = b', f (c) = c'.

Now (a'b') c' = [f (a) f (b)] f (c)

= [f (ab)] f (c) [∵ f preserves compositions]

= f [(ab) c] [∵ f is composition preserving]

= f [a (bc)] [by associativity in the group G]

= f (a) [f (bc)] = f (a) [f (b) f (c)]

= a' (b' c').

Therefore the composition in G' is associative.

Existence of Identity: Suppose e is the identity of G. Then f (e) is an element of G'. If a' is any element of G', there exists an element a Î G such that f (a) = a'.

Now [f (e)] a' = [f (e)] [f (a)] = f (ea) = f (a) = a'.

Also a' [f (e)] = [f (a)] [f (e)] = f (ae) = f (a) = a'.

∴ f (e) which is the f-image of the identity e in G is the identity in G'.

Existence of Inverse: If a' is any element of G', there exists an element a ∈ G such that f (a) = a'.

Now a ∈ G ⇒ a^{-1} ∈ G where a^{-1} a = e = aa^{-1}.

Also f(a^{-1}) is an element of G'.

Now [f (a^{-1}) a' = [f (a^{-1})] f (a)] = f (a^{-1} a) = f (e).

Also a' [f (a^{-1}) = [f (a)] [f (a^{-1})] = f (aa^{-1}) = f (e).

∴ f (a^{-1}) is the inverse of a' in G'. Thus every element of G' possesses inverse.

∴ G' is a group.

FIELD

Definition : *Suppose F is a non-empty set equipped with two binary operations called addition and multiplication and denoted by '+' and '.' respectively i.e. , for all a, b ∈ F we have a + b ∈ F and a.b ∈ F. Then this algebraic structure (F, +, .) is called a field, if the following postulates are satisfied :*

F_1. Addition is commutative, *i.e.*,

$$a + b = b + a \ \forall \ a, b \in F.$$

F_2. Addition is associative, *i.e.*,

$$(a + b) + c = a + (b + c) \ \forall \ a, b, c \in F.$$

F_3. $\exists$ an element denoted by 0 (called zero) in F such that

$$a + 0 = a \ \forall \ a \in F.$$

F_4. To each element a in F there exists an element –a in F such that $a + (-a) = 0$.

F_5. Multiplication is commutative, *i.e.*,

$$a.b = b.a \ \forall \ a, b \in F.$$

E_6. Multiplication is associative, *i.e.*,

$$a.(b.c) = (a.b).c \ \forall \ a, b, c \in F.$$

F_7. $\exists$ a non-zero element denoted by 1, (called one) in F such that $a.1 = a \ \forall \ a \in F$.

F_8. To every non-zero element a in F there corresponds an element a^{-1} (or 1/a) in F, such that $a.a^{-1} = 1$.

F_9. Multiplication is distributive with respect to addition *i.e.*, for all a, b, c in F, we have $a.(b + c) = a.b + a.c$.

The clement 0 is identity element for addition composition in F. It is called the zero element of the field. Obviously (F, +) is an abelian group. If $0 \neq a \in F$, then a is called a non-zero element of F. The element 1 is identity element for multiplication composition in F. It is called the unity of the field. In a field each non-zero element is invertible *i.e.*, possesses inverse for multiplication composition.

Note : In future we shall denote the multiplication composition in a field F not by the symbol '.' but by multiplication notation. Thus we shall omit '.' and we shall write ab in place of a.b.

Subfield

Definition : Let F be a field. A non-empty subset of the set F is said to be a subfield of F if K is closed with respect to the operations of addition and multiplication in F and K itself is a field for these operations.

CHARACTERISTIC OF FIELD

Every field is an integral domain. Therefore the characteristic of a field F is 0 or n > 0 according as any non-zero element (in particular the unit element 1) of F is of order 0 or n.

Thus in order to find the characteristic of a field F, we should find the order of the unit element 1 of F when regarded as a member of the additive group of F. If the order of 1 is zero then f is of characteristic 0. If the order of 1 is finite, say, n then the characteristic of F is n.

Then characteristic of the field of real numbers is 0.

The characteristic of the finite field $(I_7, +_7, \times_7)$ is 7 where $I_7 = \{0, 1, 2, 3, 4, 5, 6\}$.

Conditions for a Subfield

Theorem:

The necessary and sufficient conditions for a non-empty subset K of a field F to be a subfield of F are

1. $a \in K, b \in K \Rightarrow a - b \in K$,
2. $a \in K, 0 \neq b \in K \Rightarrow ab^{-1} \in K$.

Proof:

The conditions are necessary: Suppose K is a subfield of the field F. Now K is a group with respect to addition. Therefore $b \in K \Rightarrow -b \in K$. Also K is closed with a respect to addition.

$$\therefore \quad a \in K, b \in K \Rightarrow a + (-b) \in K \Rightarrow a - b \in K.$$

Now each non-zero element of K possesses multiplicative inverse. Therefore $0 \neq b \in K \Rightarrow b^{-1} \in K$.

But K is closed with respect to multiplication.

$$\therefore \quad a \in K, 0 \neq b \in K \Rightarrow ab^{-1} \in K.$$

Hence the conditions are necessary.

The conditions are sufficient: Suppose K is a non-empty subset of F and the conditions (1) and (2) are satisfied.

As we have proved in subrings, we can prove that with the help of condition (1), (K, +) is an abelian group. (Give the same proof here].

Now let a be any non-zero element of K. Then from (2) we have $a \in K$ $0 \neq a \in K \Rightarrow aa^{-1} \in K \Rightarrow 1 \in K$

Now $1 \in K$, therefore again from (2), we have

$$1 \in K \quad 0 \neq a \in K \Rightarrow 1a^{-1} \in K \Rightarrow a^{-1} \in K.$$

$\therefore$ Each non-zero element of K possesses multiplicative inverse.

Now let $a \in K$ and $0 \neq a \in K$. Then $b^{-1} \in K$.

From (2), we have

$$a \in K, 0 \neq b^{1} \in K \Rightarrow a (b^{-1})^{-1} \in K \Rightarrow ab \in K$$

Also if $b = 0$, then $ab = 0$ and $0 \in K$.

$$\therefore \quad ab \in K \ \forall \ a, b, \in K.$$

Associativity of multiplication and distributivity of multiplication over addition must hold in K since they hold in F.

ORDERED INTEGRAL DOMAINS

Definition: *An integral domain (D, +, ·) is said to be ordered if D contains a subset D_+ such that*

1. *D_+ is closed with respect to addition and multiplication as defined on D.*
2. *$\forall a \in D$, one and only one of $a = 0$, $a \in D_+$, $-a \in D_+$ holds (principle of Trichotomy).*

The elements of D_+ are called the positive elements of D, all other non-zero elements of D are called negative elements of D.

Ordered Field

A field (F, +, ·) is to be ordered if it is ordered as an integral domain.

The integral domain (I_+, +, ·) of all integers is ordered.

The set I_+ of all positive integers is the set of the positive elements of this integral domain. We know that the sum and product of two positive integers is again a positive integer *i.e.*, I_+ is closed with respect to addition and multiplication. If $a \in I$, then either is a zero or positive or negative *i.e.*, either $a = 0$ or $a \in I_+$ or $-a \in I_+$.

The field of rational numbers is an ordered field. The field of real numbers is also an ordered field. But the field of complex numbers is not an ordered field.

Theorem 1:

The field (C, +, ·) of complex numbers is not ordered.

Proof:

Suppose C is an ordered field and C_+ is the set of positive elements of this field. The additive identity *i.e.*, the zero element is $0 + i0$.

Now $i \neq 0$.

By the principle of trichotomy either $i \in D_+$ or $-i \in C_+$.

Now C_+ is closed with respect to multiplication.

$\therefore i \in C_+, -1 \in C_+ \Rightarrow i(-1) = -i \in C_+$.

Thus if $i \in C_+$, its additive inverse $-i$ also belongs to C_+. This contradicts the principle of trichotomy *i.e.*, the condition (2) of the definition of an ordered integral domain.

Similarly if we assume that $-1 \in C_+$, we can show that its additive inverse i also belongs to C_+. This again contradicts the principle of trichotomy.

Hence the field of complex numbers is not an ordered field.

Theorem 2:

The field $(I_p, +_p, \times_p)$ where p is a prime and $I_p = \{0, 1, ..., p-1\}$ is not ordered.

Proof:

Suppose I_p is an ordered field and P is the set of positive elements of the field. The zero element of this field is 0.

Now $1 \neq 0$. The additive inverse of 1 is $p - 1$. According to the definition of ordered field either $1 \in P$ or its additive inverse $p - 1 \in P$.

But P is closed with respect to $+_p$.

Therefore $1 \in P \Rightarrow 1 +_p 1 +_p 1 +_p$ ...upto $p - 1$ times $\in P$

$$\Rightarrow p - 1 \in P.$$

This contradicts the principle of trichotomy. Similarly assuming that $p - 1 \in P$ we can show that its additive inverse 1 also belongs to P.

This again contradicts the principle of trichotomy.

Hence the given field is not an ordered field.

Theorem 3:

Let D be an integral domain with unity element 1. If D is an ordered integral domain show that 1 is a positive element of D

Proof:

Let D be an ordered integral domain with unity element 1. Let D_+ denote the set of positive elements of D.

Suppose $1 \notin D_+$.

Now $1 \neq 0$. Since $1 \notin D_+$, therefore by the definition of an ordered integral domain,

$-1 \in D_+$

$\Rightarrow (-1)(-1) \in D_+$ [$\because$ D_+ is closed with respect to multiplication]

$\Rightarrow 1 \in D_+$ which is a contradiction.

Hence $1 \in D_+$ *i.e.*, 1 is a positive element of D.

POLYNOMIALS OVER A FIELD

Theorem:

If F is a field, then the set F [x] of all polynomials over F is an integral domain.

Proof:

Every field is a integral domain. We shall call the set F [x] as the polynomial domain over the field F.

Polynomials Over a Integral Domain

Theorem:

If D is an integral domain, then the polynomials ring D [x] is also an integral domain.

Proof:

Let D be a commutative ring without zero divisors and with unity element 1. As proved in d [x] is also a ring. To prove that D [x] is an integral domain we should prove that (1) D [x] is commutative, (2) is without zero divisors and (3) possesses the unity element.

D [x] is commutative. Let $f(x) = a_0 + a_1x + a_2x^2 + \ldots$ and $g(x) = b_0\ b_1x + b_2x^2 + \ldots$ be any two elements of D [x].

If n is any non-negative integer, then the coefficient of x^n in f (x) g (x)

is $= \sum_{i+j=n} a_ib_j = \sum_{i+j=n} b_ja_i,$ since D is commutative

= Coefficient of x^n in g (x) f (x).

$\therefore$ f (x) g (x) f (x). Hence D [x] is commutative ring.

If is the unity element of D, then the constant polynomial $1 + 0x + 0x^2 + 0x^3 + \ldots$ is the unity element of D [x]. We have

$$[a_0 + a_1x + a_2x^2 + \ldots]\ [1 + 0x + 0x^2 + 0x^3 + \ldots]$$

$$= (a_01) + (a_11)\ x + (a_21)\ x^2 + \ldots = a_0 + a_1x + a_2x^2 + \ldots$$

$\therefore$ The polynomial $1 + 0x + 0x^2 + \ldots$ or simply 1 is the unity element of [x].

D [x] *is without zero divisors.* Let

$$f(x) = a_0 + a_1x + a_2x^2 + \ldots + a_mx^m,\ a_m \neq 0$$

$$g(x) = b_0 + b_1x + b_2x^2 + \ldots + b_nx^n,\ b_m \neq 0$$

be two non-zero elements of D [x].

Then f (x) cannot be a zero polynomial *i.e.*, the zero element of D [x]. The reason is that at least one coefficient of f (x) g (x) namely $a_m b_n$ of x^{m+n} is $\neq 0$ because a_m, b_n are non-zero elements of D and D is without zero divisors.

Hence D [x] is an integral domain.

VECTOR SPACE PROPERTIES

Theorem 1:

Let V(F) be a vector space and 0 be the zero vector of V. Then

(i) $a0 = 0 \; \forall \; \alpha \in F.$

(ii) $0\alpha = 0 \; \forall \; \alpha \in V.$

(iii) $a(-\alpha) = -(a\alpha) \; \forall \; \alpha \in F, \; \forall \; \alpha \in V.$

(iv) $(-a)\, a = -(a\alpha) \; \forall \; \alpha \in F, \; \forall \; \alpha \in V.$

(v) $a(\alpha - \beta) = a\alpha - a\beta \; \forall \; \alpha \in F$ and $\forall \; \alpha, \beta \in V.$

(vi) $a\alpha = 0 \Rightarrow a = 0$ or $\alpha = 0.$

Proof:

The proofs is very easy and follows the lines of the analogues result proved for rings for this reason we give it briefly and with few explanations. Here

(i) We have $a0 = a(0 + 0)$ $\quad [\because 0 = 0 + 0]$

$= a0 + a0.$

$\therefore 0 + a0 + a0$ $\quad [\because a0 \in V \text{ and } 0 + a0 = a0]$

Now V is an abelian group with respect to addition.

Hence by right cancellation law in V, we get $0 = a0$.

(ii) We have $0\alpha = (0 + 0)\alpha$ $\quad [\because 0 = 0 + 0]$

$= 0\alpha + 0\alpha.$

$\therefore 0 + 0\alpha = 0\alpha + 0\alpha.$ $\quad [\because 0\alpha \in V \text{ and } 0 + 0\alpha = 0\alpha]$

Now V is an abelian group with respect to addition of vectors. Therefore by right cancellation law in V, we get $0 = 0\alpha$.

(iii) We have $a[\alpha + (-\alpha)] = a\alpha + a(-\alpha)$

$\Rightarrow a0 = a\alpha + a(-\alpha)$

$\Rightarrow 0 = a\alpha + a(-\alpha)$ $\quad [\because a0 = 0]$

$\Rightarrow$ (–a α is the additive inverse of aα)

$\Rightarrow$ a (– α) = – (aα).

(iv) We have [a + (– a)] α = aα + (– a) α

$\Rightarrow$ 0α = aα + (– a) α

$\Rightarrow$ 0 = aα + (– a) α

$\Rightarrow$ (– a) α is the additive inverse of aα

$\Rightarrow$ (– a) α = – (aα).

(v) We have (α – β) = a [α + (– β)] = aα + a (– β)

= aα + [– (aβ) [$\because$ a (– β) = – (aβ)]

= aα – aβ.

(vi) Let aα = 0 and a ≠ 0. Then a^{-1} exists because a is a non-zero element of the field F.

$\therefore$ aα = 0 $\Rightarrow$ a^{-1} (aα) = a^{-1} 0

$\Rightarrow$ (a^{-1} a) α = 0

$\Rightarrow$ 1α = 0 $\Rightarrow$ α = 0.

Again let aα = 0 and α ≠ 0. Then to prove that a = 0. Suppose α ≠ 0. Then a^{-1} exists.

$\therefore$ aα = 0 $\Rightarrow$ a^{-1} (aα) = a^{-1}0

$\Rightarrow$ (a^{-1} a) α = 0

$\Rightarrow$ 1α = 0 $\Rightarrow$ α = 0.

Thus, we get a contradiction that α must be a zero vector. Therefore a must be equal to 0. Hence α ≠ 0 and aα = 0 $\Rightarrow$ a = 0.

Theorem 2:

Let V (F) be a vector space. Then

(i) If a, b ∈ F and α is anon-zero element of V, we have

$$a\alpha = b\alpha \Rightarrow a = \beta.$$

(ii) If α, β ∈ V and α is a non-zero element of F, we have

$$a\alpha = a\beta \Rightarrow a = b.$$

Proof:

(i) We have aα = bα $\Rightarrow$ aα – bα = 0 $\Rightarrow$ (a – b) α = 0.

But α ≠ 0. Therefore (a – b) α = 0 $\Rightarrow$ a – b = 0 $\Rightarrow$ a – b.

(ii) We have aα = aβ $\Rightarrow$ aα – aβ = 0 $\Rightarrow$ a (α – β) = 0.

But a ≠ 0.

Therefore $a(\alpha - \beta) = 0$

$\Rightarrow \alpha - \beta = 0 \Rightarrow \alpha = \beta.$

VECTOR SUBSPACES

Definition: *Let V a vector space over the field F and $W \subseteq V$. Then W is called a subspace of V if W itself is a vector space over F with respect to the operations of vector addition and scalar multiplication in V.*

Theorem 1:

The necessary and sufficient condition for a non empty subset W of a vector space V (F) to be a subspace of V is

$$a, b \in F \text{ and } \alpha, \beta \in W \Rightarrow a\alpha + b\beta \in W.$$

Proof:

The condition necessary. If W is a subspace of V, then W must be closed under scalar multiplication and vector addition.

Therefore $\quad a \in F, \alpha \in W \Rightarrow a\alpha \in W$

and $\quad b \in F, \beta \in W \Rightarrow b\beta \in W.$

Now $a\alpha \in W, b\beta \in W \Rightarrow a\alpha + b\beta \in W.$

Hence the condition is necessary.

The condition is sufficient. Now suppose W is a non-empty subset of V satisfying the given condition *i.e.*, a, b, $\in$ F and

$$\alpha, \beta \in W \Rightarrow a\alpha + b\beta \in W.$$

Taking $a = 1, b = 1$, we see that if $\alpha, \beta \in W$ then

$1\alpha + 1\beta \in W$

$\Rightarrow \alpha + \beta \in W. \qquad [\because \alpha \in W \Rightarrow \alpha \in V \text{ and } 1\alpha = \alpha \text{ in } V]$

Thus W is closed under vector addition.

Now taking $a = -1, b = 0$, we see that if $\alpha \in W$, then

$(-1)\alpha + a\alpha \in W$ [in place of β we have taken α]

$\Rightarrow -(1\alpha) + 0 \in W \Rightarrow -\alpha \in W.$

Thus the additive inverse of each element of W is also in W.

Taking $a = 0, b = 0$, we see that if $a \in W$, then

$0\alpha + 0\alpha \in W \Rightarrow 0 + 0 \in W \Rightarrow 0 \in W.$

Thus the zero vector of V belongs to W. It will also be the zero vector of W.

Since the elements of W are also the elements of V, therefore vector addition will be associative as well as commutative in W. Thus W is an abelian group with respect to vector addition.

Now taking $\beta = 0$, we see that if a, b, $\in$ F and $\alpha \in$ W, then

$$a\alpha + b0 \in W \text{ i.e. } a\alpha + 0 \in W \text{ i.e., } a\alpha \in W.$$

Thus W is closed under scalar multiplication.

The remaining postulates of a vector space will hold in W since they hold in V of which W is a subset. Hence W (F) is a subspace of V (F).

Note: If we are to prove that a subset W of a vector space V is a subspace of V, then it is sufficient to prove that

$$a, b \in F \text{ and } \alpha, \beta, \in W \Rightarrow a\alpha + b\beta \in W.$$

Theorem 2:

The necessary and sufficient conditions for a non-empty subset W of a vector space V (F) to be a subspace of V are

(i) $\alpha \in W, \beta \in W \Rightarrow \alpha - \beta \in W.$

(ii) $a \in F, \alpha \in W \Rightarrow a\alpha \in W.$

Proof:

The conditions are necessary. Let if W is a subspace of V, then W is an abelian group with respect to vector addition. Therefore $\alpha \in W, \beta \in W \Rightarrow \alpha - \beta \in W$. Also W must be closed under scalar multiplication. Therefore the condition (ii) is also necessary.

The conditions are sufficient. Again let W is a non-empty subset of V satisfying the two given conditions. From condition (i), we have

$$\alpha \in W, \alpha \in W \Rightarrow \alpha - \alpha \in W \Rightarrow 0 \in W.$$

Hence, the zero vector of V belongs to W and it will also be the zero vector of W.

Now $0 \in W, \alpha \in W \Rightarrow 0 - \alpha \in W \Rightarrow - \alpha \in W.$

Thus the additive inverse of each element of W is also in W.

Again $\alpha \in W, \beta \in W \Rightarrow \alpha \in W, - \beta \in W$

$\Rightarrow \quad \alpha - (- \beta) \in W \Rightarrow \alpha + \beta \in W.$

Thus, W is closed with respect to vector addition.

Since the elements of W are also the elements of V, therefore vector addition will be commutative as well as associative in W. Hence W is an abelian group under vector addition. Also from condition (ii), W is closed under scalar multiplication. The remaining postulates of a vector space will hold in W since they hold in V which is a superset of W. Hence W is a subspace of V.

Theorem 3:

The necessary and sufficient condition for a non-empty subset W of a vector space V (F) to be a subspace of V is that W is closed under addition and scalar multiplication in V.

Proof:

If W itself is a vector space over F with respect to vector addition and scalar multiplication in V, then W must be closed with respect to these two compositions.

The condition is sufficient. Now let that W is a non-empty subset of V and W is closed under vector addition and scalar multiplication in V.

Let $\alpha \in W$, if 1 is the unity element of F, then $-1 \in F$. Now W is closed under scalar multiplication. Therefore, we have

$$-1 \in F, \alpha \in W \Rightarrow (-1)\alpha \in W \Rightarrow -(1\alpha) \in W$$

$$\Rightarrow -\alpha \in W \qquad [\because \alpha \in W \Rightarrow \alpha \in V \text{ and } 1\alpha = \alpha \text{ in } V].$$

Thus the additive inverse of each element of W is also in W.

Hence W is closed under vector addition.

Therefore $\alpha \in W, -\alpha \in W \Rightarrow \alpha + (-\alpha) \in W$

$\Rightarrow 0 \in W$ where 0 is the zero vector of V.

Hence the zero vector of V is also the zero vector of W. Since the elements of W are also the elements of V, hence, vector W is an abelian group with respect to vector addition. Also, we know that W is closed under scalar multiplication. The remaining postulates of a vector space will hold in W since they hold in V of which W is a subset.

Hence W itself is a vector space for the two compositions.

$\therefore$ W is a subspace of V.

ELEMENTARY BASIC CONCEPT

Definition: A non-empty set V is said to be a vector space over a field F. It $\forall$ is an abelian group under an operation which we denoted by +, and if for every $a \in F$, $v \in V$ there is defined an element, written av, in V subject to:

(i) $\alpha (V + w) = dv + dw$

(ii) $(\alpha + \beta)v = \alpha v + \beta v$

(iii) $\alpha (\beta v) = (\alpha\beta) v$

(iv) $1.v = v.$

for all $\alpha, \beta \in F$, $v\ w \in V$ (where the 1 represents the unit element of F under multiplication).

Note: In axiom (i) above the + is that of V, whereas, in axiom (ii) on the left hand side it is that of F and on the right hand side it is that of V.

We shall consistly use the following notations:

(a) F will be a field.

(b) Lower Case Greek Letters will be elements of F; we shall often refer to element of F an scalars.

(c) Capital Latin Letters will denote vector space over F.

(d) Lower Case Latin Letters will denote elements of veçtor space we shall often call elements of a vector space vectors.

Definition: *Let (F, + .) be a field. The elements of F will be called* **scalars.** *Let V be a non-empty set whose elements will be called* **vectors.** *Then V is a vector space over the field F, if*

1. *There is defined an internal composition in V called* **addition of vectors** *and denoted by '+'. Also for this composition V is an abelian group.*
2. *There is an external composition in V over F called* **scalar multiplication** *and denoted multiplicatively i.e.,* $a\alpha \in V$ *for all* $a \in F$ *and for all* $\alpha \in V$. *In other words V is closed with respect to scalar multiplication.*
3. *The two compositions i.e., scalar multiplication and addition of vectors satisfy the following postulates:*

(i) $a(\alpha + \beta) = a\alpha + a\beta\ \forall a \in F$ *and* $\forall \alpha, \beta, \in V.$

(ii) $(a + b)\ \alpha = a\alpha + b\alpha\ \forall a, b, \in F$ *and* $\forall \alpha \in V.$

(iii) $(ab)\ \alpha = a\ (b\alpha)\ \forall a, b \in F$ *and* $\forall \alpha \in V.$

(iv) $1\alpha = \alpha\ \forall \in V$ *and 1 is the unity element of the field F.*

When V is a vector space over the field F, we shall say that V (F) is a vector space. If the field F is understood we can simply say that V is a vector space.

(i) $\alpha + \beta \in V$ for all $\alpha, \beta, \in V$.

(ii) $\alpha + \beta = \beta + \alpha$ for all $\alpha, \beta, \in V$

(iii) $\alpha + (\beta + \gamma) = (\alpha + \beta) + \gamma$ for all $\alpha, \beta, \gamma \in V$.

(iv) There exists an element $0 \in V$ such that

$$0 + \alpha = \alpha \ \forall \ \alpha \in V.$$

The element $0 \in V$ will be called the *zero vector.* It is the additive identity in V.

(v) To every vector $\alpha \in V$ there exists a vector $-\alpha \in V$ such that $-\alpha + \alpha = 0$. Thus each vector should possess additive inverse. The vector $-\alpha$ is called the negative of the vector α.

DIRECT SUM OF SPACES

Vector Space as Direct Sum of Subspaces

Definition: *Let V(F) be a vector space and let $W_1, W_2, \ldots, W_m$ be subspaces of V. Then V is said to be the direct sum of $W_1, W_2, \ldots, W_m$ if every element $a \in V$ can be written in one and only one way as $a = a_1 + a_2 + \ldots + a_m$ where*

$$a_1 \in W_1, a_2 \in W_2, \ldots, a_m \in W_m.$$

If a vector space V(F) is a direct sum of its two subspaces W_1 and W_2 then we should have not only $V = W_1 + W_2$ but also that each vector V can be uniquely expressed as sum of an element of W_1 and an element of W_2. Symbolically the direct sum is represented by the notation $V = W_1 \oplus W_2$.

Disjoint Subspaces

Definition: *Two subspaces W_1 and W_2 of the vector space V(F) are said to be disjoint if their intersection is the zero subspace i.e., if $W_1 \cap W_2 = \{0\}$.*

Theorem:

The necessary and sufficient condition for a vector space V(F) to be direct sum of its two subspaces W_1 and W_2 are that

(i) $V = W_1 + W_2$

and *(ii)* $W_1 \cap W_2 = \{0\}$

i.e., *W_1 and W_2 are disjoint.*

Proof:

Let V be a direct sum of its two subspaces W_1 and W_2. Then each element of V is expressible as sum of an element of W_1 and an element of W_2. Therefore we have $V = W_1 + W_2$.

Let, if possible $0 \neq \alpha \in W_1 \cap W_2$. Then $\alpha \in W_1, \alpha \in W_2$. Also $\alpha \in V$ and we can write

$$\alpha = 0 + \alpha \text{ where } 0 \in W_1, \alpha \in W_2$$

and $\alpha = \alpha + 0$ where $\alpha \in W_1$, $0 \in W_2$.

Thus, $\alpha \in V$ can be expressed in at least two different ways as sum of an element of W_1 and an element of W_2. This contradicts the fact that V is a direct sum of W_1 and W_2.

Hence 0 is the only vector common to both W_1 and W_2 *i.e.*, $W_1 \cap W_2 = \{0\}$. Thus, the conditions are necessary.

The conditions are sufficient. Let $V = W_1 + W_2$ and $W_1 \cap W_2 = \{0\}$. Then to show that V is direct sum of W_1 and W_2.

$V = W_1 + W_2 \Rightarrow$ that each element V can be expressed as sum of an element of W_1 and an element of W_2. Now to show that this expression is unique.

Let, if possible,

$$\alpha = \alpha_1 + \alpha_2, \alpha \in V, \alpha_1 \in W_1, \alpha_2 \in W_2$$

and $\alpha = \beta_1 + \beta_2$, $\beta_1 \in W_1$, $\beta_2 \in W_2$.

Then to show that $\alpha_1 = \beta_1$ and $\alpha_2 - \beta_2$.

We have $\alpha_1 + \alpha_2 = \beta_1 + \beta_2$

$\Rightarrow$ $\alpha_1 - \beta_1 = \beta_2 - \alpha_2$.

Since W_1 is a subspace, therefore

$$\alpha_1 \in W_1, \beta_1 \in W_1$$

$\Rightarrow$ $\alpha_1 - \beta_1 \in W_1$.

Similarly $\beta_2 - \alpha_2 \in W_2$.

$\therefore \alpha_1 - \beta_1 = \beta_2 \quad \alpha_2 \subset W_1 \cap W_2$.

But 0 is the only vector which belongs to $W_1 \cap W_2$. Therefore

$$\alpha_1 - \beta_1 = 0 \Rightarrow \alpha_1 = \beta_1.$$

Also $\beta_2 - \alpha_2 = 0 \Rightarrow \alpha_2 = \beta_2$

Thus each vector $\alpha \in V$ is uniquely expressible as sum of an element W_1 and an element of W_2.

Hence $V = W_1 \oplus W_2$.

Dimension of a Direct Sum

Theorem:

If a finite dimensional vector space V (F) is a direct sum of its two subspaces W_1 and W_2, then

$$dim\ V = dim\ W_1 + dim\ W_2.$$

Proof:

Let dim W_1 = m and dim W_2 = l. Also let the sets of vectors $S_1 = \{\alpha_1, \alpha_2, ..., \alpha_m\}$ and $S_2 = \{\beta_1, \beta_2, ..., \beta_l\}$ be some bases of W_1 and W_2 respectively.

We have dim W_1 + dim W_2 = m + l.

In order to prove that dim V = dim W_1 + dim W_2, we should therefore prove that dim V = m + l. We claim that the set

$$S = S_1 \cup S_2 = \{\alpha_1, \alpha_2, ..., \alpha_m, \beta_1, \beta_2, ..., \beta_l\}$$

is a basis of V.

First we show that the set S is linearly independent. Let

$$a_1\alpha_1 + a_2\alpha_2 + ... + a_m\alpha_m + b_1\beta_2 + b_2\beta_2 + ... + b_l\beta_l = 0$$

$\Rightarrow$ $a_1\alpha_1 + a_2\alpha_2 + ... + a_m\alpha_m = -(b_1\beta_1 + b_2\beta_2 + ... + b_l\beta_l)$.

Now $a_1\alpha_1 + a_2\alpha_2 + ... + a_m\alpha_m \in W_1$

and $-(b_1\beta_1 + b_2\beta_2 + ... + b_l\beta_l) \in W_2$. Therefore

$a_1\alpha_1 + a_2\alpha_2 + ... + a_m\alpha_m \in W_1 \cap W_2$

and $-(b_1\beta_1 + b_2\beta_2 + ... + b_l\beta_l) \in W_1 \cap W_2$.

But V is the direct sum of W_1 and W_2. Therefore 0 is the only vector belonging to $W_1 \cap W_2$. Then we have

$$a_1\alpha_1 + a_2\alpha_2 + ... + a_m\alpha_m = 0,$$

$$b_1\beta_1 + b_2\beta_2 + ... + b_l\beta_l = 0.$$

Since both the sets $\{\alpha_1, \alpha_2, ..., \alpha_m\}$ and $\{\beta_1, \beta_2, ..., \beta_l\}$ are linearly independent, therefore we have

$$a_1 = 0, a_2 = 0, ..., a_m = 0,$$

$$b_1 = 0, b_2 = 0, ..., b_l = 0.$$

Therefore S is linearly independent.

Now we shall show that L(S) = V.

Let α be any element of V. Then

α = an element of W_1 + an element of W_2

= a linear combination of S_1 + a linear combination of S_2

= a linear combination of S.

$\therefore$ L(S) = V.

$\therefore$ S is a basis of V. Therefore dim V = m + l.

Hence the theorem.

Complementary Subspaces

Definition. *Let V(F) be a vector space and W_1, W_2 be two subspace of V. Then the subspace W_2 is called a complement of W_1 in V if V is the direct sum of W_1 and W_2.*

Existence of Complementary Subspaces

Theorem:

Corresponding to each subspace W_1 of a finite dimensional vector space V(F), there exists a subspace W_2 such that V is the direct sum of W_1 and W_2.

Proof:

Let dim $W_1 = m$. Let the set $S_1 = \{\alpha_1, \alpha_2, ..., \alpha_m\}$ be a basis of W_1. Since S_1 is a linearly independent subset of V, therefore S_1 can be extended to form a basis of V.

Let the set $S = \{\alpha_1, \alpha_2, ..., \alpha_m, \beta_1, \beta_2, ..., \beta_l\}$ be a basis of V. Let W_2 be the sub-space of V generated by the set $S_2 = \{\beta_1, \beta_2, ..., \beta_l\}$.

We shall prove that V is the direct sum of W_1 and W_2. For this we shall prove that $V = W_1 + W_2$ and $W_1 \cap W_2 = \{0\}$.

Let α be any element of V. Then we can express

α = a linear combination of S

= a linear combination of S_1 + a linear combination of S_2

= an element of W_1 + an element of W_2.

$\therefore V = W_1 + W_2$.

Again let $\beta \in W_1 \cap W_2$. The β can be expressed as a linear combination of S_1 and also a linear combination of S_2. So we have

$$\beta = a_1\alpha_1 + a_2\alpha_2 + ... + a_m\alpha_m = b_1\beta_1 + b_2\beta_2 + ... b_l\beta_l.$$

$$\therefore a_1\alpha_1 + a_2\alpha_2 + ... + a_m\alpha_m - b_1\beta_1 - b_2\beta_2 - ... b_l\beta_l = 0$$

$\Rightarrow a_1 = 0, a_2 = 0, ..., a_m = 0, b_1 = 0, b_2 = 0, ..., b_l = 0$ since $\alpha_1, \alpha_2, ..., \alpha_m, \beta_1, \beta_2, ..., \beta_l$ are linearly independent.

$\therefore \beta = 0$ (zero vector).

Thus $W_1 \cap W_2 = \{0)$ Hence V is the direct sum of W_1 and W_2.

COORDINATES

Let V (F) be a finite dimensional vector space. Let $B = \{\alpha_1, \alpha_2, ..., \alpha_m\}$ be an ordered basis for V. By an ordered basis we mean that the vectors of B have been enumerated in some well-defined way *i.e.*, the vectors occupying the first, second,..., n[th] places in the set B are fixed.

Let $\alpha \in V$. Then there exists a unique n-tuple $(x_1, x_2,..., x_n)$ of scalars such that $\alpha = x_1\alpha_1 + x_2\alpha_2 + ... + x_n\alpha_n = \sum_{i=1}^{n} x_i a_i$.

The n-tuple $(x_1, x_2,..., x_n)$ is called the n-tuple of coordinates of α relative to the ordered basis B. The scalar x_i is called the ith coordinate of α relative to the ordered basis B.

It should be noted that for the same basis set B, the coordinates of the vector α are unique only with respect to a particular ordering of B. The basis set B can be ordered in several ways. The coordinates of α may change with change in ordering of B.

FINITE DIMENSIONAL VECTOR SPACE

Definition: *The vector space V(F) is said to be **finite dimensional** or **finitely generated** if there exists a **finite** subset of S of V such that V = L (S).*

The vector space V_n (F) of n-tuples is a finite dimensional vector space.

The vector space F[x] of all polynomials over a field F is not finite dimensional. There exists no finite subset S of F [x] which spans F[x]. *A vector space which is no finitely generated may be referred to as an infinite dimensional space.* Thus, the vector space F[x] of all polynomials over a field F is infinite dimensional.

Existence of Basis of a Finite Dimensional Vector Space

Theorem:

There exists a basis for each finite dimensional vector space.

Proof:

Let V(F) be a finitely generated vector space. Let S $\{\alpha_1, \alpha_2,..., \alpha_m\}$ be a finite subset of V such that L(S) = V. Without loss of generality we may suppose that no member of S is 0.

If S is linearly independent, then S itself is a basis of V. If S is linearly dependent, then there exists, a vector $\alpha_t \in S$ which can be expressed as a linear combination of the preceding vectors $\alpha_1, \alpha_2,..., \alpha_{i-1}$.

If we omit this vector α_i from S, then the remaining set S' of m–1 vectors

$$\alpha_1, \alpha_2,..., \alpha_{i-1}, \alpha_{i+1},..., \alpha_m$$

also generates V *i.e.* V = L(S').

For if α is any element of V, then L(S) = V implies that α can be written as a linear combination of $\alpha_1, \alpha_2, ..., \alpha_m$. In this linear combination we can replace α_1 by a linear combination of $\alpha_1, \alpha_2, ..., \alpha_{i-1}$. Then α will be a linear combination of $\alpha_1, \alpha_2, ..., \alpha_{i-1}, \alpha_{i+1} ..., \alpha_m$. Thus V will be equal to L(S').

If S' is linearly independent, then S' will be a basis of V. If S' is linearly dependent, then proceeding as above are shall get a now set of m–2 vectors which generates V. Continuing this process we shall, after a finite number of steps, obtain a linearly independent subset of S which generates V and which is therefore a basis of V.

At the most it may happen that we shall be left with a subset of S which contains only one non-zero vector and which spans V. We know that a set containing a single non-zero vector is definitely linearly independent and so it will form a basis of V.

If a finite set S of vectors spans a finite dimensional vector space V(F), there exists a subset of S which forms a basis of V.

Invariance of Number of Elements in the Basis of a Finite Dimensional Vector Space

Theorem:

If V(F) is a finite dimensional vector space, then any two bases of V have the same number of elements.

Proof:

Suppose V(F) is a finite dimensional vector space. Then V definitely possesses a basis. Let $S_1 = \{\alpha_1, \alpha_2, ..., \alpha_m\}$ and $S_2 = \{\beta_1, \beta_2, ..., \beta_n\}$ be two bases of V. We shall prove that m = n.

Since V = $L(S_1)$ and $\beta \in$ V, therefore β_1 can be expressed as a linear combination of $\alpha_1, \alpha_2, ..., \alpha_m$. Consequently the set $S_3 = \{\beta_1, \alpha_1, \alpha_2, ..., \alpha_m\}$ which also obviously generates V(F) is linearly dependent.

Therefore there exists a member $\alpha_i \neq \beta_1$ of this set S_3 such that α_i is a linear combination of the preceding vectors $\beta_1, \alpha_1, \alpha_2, ..., \alpha_{i-1}$. If we omit the vector α_i from S_3 then V is also generated by the remaining set

$$S_4 = \{\beta_1, \alpha_1, \alpha_2, ..., \alpha_{i-1}, \alpha_{i+1}, ..., \alpha_m\}.$$

Since V = $L(S_4)$ and $\beta_2 \in$ V, therefore β_2 can be expressed as a linear combination of the vectors belonging to S_4. Consequently the set

$$S_5 = \{\beta_2, \beta_1, \alpha_1, \alpha_2, ..., \alpha_{i-1}, \alpha_{i+1}, ..., \alpha_m\}$$

is linearly dependent. Therefore there exists a member α_j of this set S_5 such that α_j is a linear combination of the preceding vectors.

Obviously α_j will be different from β_1 and β_2 since $\{\beta_1, \beta_2\}$ is a linearly independent set. If we exclude the vector α_j from S_5, then the remaining set will generate V(F).

We may continue to proceed in this manner. Here each step consists in the exclusion of an α and the inclusion of a β in the set S_1.

Obviously the set S_1 of α's cannot be exhausted before the set S_2 of β's, otherwise V(F) will be a linear span of a proper subset of S_2 and thus S_2 will become linearly dependent.

Therefore we must have $m \not\in n$

Interchanging the roles of S_1 and S_2 we shall get that

$$n < m. \text{ Hence } m = n.$$

Example:

For the vector space $V_3(F)$, both the sets

$$S_1 = \{(1, 0, 0), (0, 1, 0), (0, 0, 1)\}$$

and $$S_2 = \{(1, 0, 0), (1, 1, 0), (1, 1, 1)\}$$

are bases as can be easily seen. Both these contain the same number of elements i.e., 3.

Dimension of a Finitely Generated Vector Space

Definition. *The number of elements in any basis of a finite dimensional vector space V(F) is calt d the dimension of the vector space V(F) and will be denoted by dim V.*

The vector space $V_n(F)$ is of dimension n. The vector space $V_3(F)$ is of dimension 3.

If a field F is regarded as a vector space over F, then F will be of dimension 1 and the see S = (1) consisting of unity element of F alone is a basis of F.

In fact every non-zero element of F will form a basis of F.

SOME PROPERTIES OF FINITE DIMENSIONAL VECTOR SPACES

Theorem 1:

(Extension Theorem) : *Every linearly independent subset of a finitely generated vector space V(F) forms a part of a basis of V.*

Or

Every linearly independent of a finitely generated vector space V(F) is either a basis of V or can be extended to form a basis of V.

Proof:

Let $S = \{\alpha_1, \alpha_2, ..., \alpha_m\}$ be a linearly independent subset of a finite dimensional vector space V(F). If dim V = n, then V has a finite basis say, $\{\beta_1, \beta_2, ..., \beta_n\}$. Consider the set

$$S_1 = \{\alpha_1, \alpha_2, ..., \alpha_m, \beta_1, \beta_2, ..., \beta_n\}.$$

Obviously $L(S_1) = V$. Since the α's can be expressed as linear combination of the β's therefore the set S_1 is linearly dependent.

Therefore there is some vector of S_1 which is a linear combination of its preceding vectors. This vector cannot be any of the α's since the α's are linearly independent. Therefore this vector must be some β, say, β_i. Now omit the vector β_i from S_i and consider the set

$$S_2 = \{\alpha_1, \alpha_2, ..., \alpha_m, \beta_1, \beta_2, ..., \beta_{i-1}, \beta_{i+1}, ..., \beta_n\}.$$

Obviously $L(S_2) = V$. If S_2 is linearly independent then S_2 will be a basis of V and it is the required extended set which is a basis of V. If S_2 is not linearly independent, then repeating the above process a finite number of times, we shall get a linearly independent set containing $\alpha_1, \alpha_2, ..., \alpha_m$ and spanning V. This set will be basis of V and it will contain S. Since each basis of V contains the same number of elements, therefore exactly n – m elements of the set of β's will be adjoined to S so as to form a basis of V.

Theorem 2:

Let $S = \{\alpha_1, \alpha_2, ..., \alpha_n\}$ be a basis of a finite dimensional vector space V(F) of dimension n. Then every element α of V can be uniquely expressed as $\alpha = a_1\alpha_1 + a_2\alpha_2 + ... + a_n\alpha_n$ where $a_1, a_2, ..., a_n \in F$.

Proof:

Since S is a basis of V, therefore L(S) = V. Therefore any vector $\alpha \in V$ can be expressed as $\alpha = a_1\alpha_1 + a_2\alpha_2 + ...1 + a_n\alpha_n$.

The show uniqueness, let us suppose that

$$\alpha = b_1\alpha_1 + b_2\alpha_2 + ... + b_n\alpha_n.$$

Then we must show that $a_1 = b_1, a_2 = b_2, ..., a_n = b_n$.

We have $a_1\alpha_1 + a_2\alpha_2 + ... + a_n\alpha_n = b_1\alpha_1 + b_2\alpha_2 + ... + b_n\alpha_n$

$\Rightarrow \quad (a_1 - b_1)\alpha_1 + (a_2 - b_2)\alpha_2 + ... + (a_n - b_n)\alpha_n = 0$

$\Rightarrow \quad a_1 - b_1 = 0, a_2 - b_2 = 0, ..., a_n - b_n = 0$ since $\alpha_1, \alpha_2, ..., \alpha_n$ are linearly independent

$\Rightarrow \quad a_1 = b_1, a_2 = b_2, ..., a_n = b_n$.

Hence the theorem.

Theorem 3:

If W_1, W_2 are two subspaces of a finite dimensional vector space V(F), then

$$dim\ (W_1 + W_2) = dim\ W_1 + dim\ W_2 + dim\ (W_1 \cap W_2).$$

Proof:

Let dim $(W_1 \cap W_2) = k$ and let the set

$$S = \{\gamma_1, \gamma_2, \gamma_3, ..., \gamma k\}$$

be a basis of $W_1 \cap W_2$. The $S \subseteq W_1$ and $S \subseteq W_2$.

Since S is linearly independent and $S \subseteq W_1$, therefore S can be extended to form a basis of W_1. Let

$$\{\gamma_1, \gamma_2, ..., \gamma_k, \alpha_1, \alpha_2, ..., \alpha_m\}$$

be a basis of W_1. Then dim $W_1 = k + m$. Similarly let

$$\{\gamma_1, \gamma_2, ..., \gamma_k, \beta_1, \beta_2, ..., \beta_t\}$$

be a basis of W_2. Then dim $W_2 = k + t$.

$$\therefore \dim W_1 + \dim W_2 - \dim (W_1 \cap W_2) = (m + k) + (k + t) - k$$
$$= k + m + t.$$

$\therefore$ to prove the theorem we must show that

$$\dim (W_1 + W_2) = k + m + t.$$

We claim that the set

$$S_1 = \{\gamma_1, \gamma_2, ..., \gamma k, \alpha_1, \alpha_2, ..., \alpha_m, \beta_1, \beta_2, ..., \beta_1\}$$

is a basis of $W_1 + W_2$.

First we show that S_1 is linearly independent. Let

$$c_1\gamma_1 + c_2\gamma_2 + ... + c_k\gamma_k + a_1\alpha_1 + a_2\alpha_2 + ... + a_m\alpha_m + b_1\beta_1 + b_2\beta_2 + ... + b_t\beta_t = 0 \quad ...(1)$$

$$\Rightarrow b_1\beta_1 + b_2\beta_2 + ... + b_t\beta_t = -(c_1\gamma_1 + ... + c_k\gamma_k + a_1\alpha_1 + ... + a_m\alpha_m) \quad ...(2)$$

Now $-(c_1\gamma_1 + ... + c_k\gamma_k + a_1\alpha_1 + ... + a_m\alpha_m) \in W_1$ since it is a linear combination of a basis of W_1. Again

$$b_1\beta_1 + b_2\beta_2 + ... + b_t\beta_t \in W_2$$

since it is a linear combination of elements belonging to a basis of W_2.

Also by virtue of the equality (2), $b_1\beta_1 + ... + b_t\beta_t \in W_1$. Therefore $b_1\beta_1 + b_2\beta_2 + ... + b_t\beta_t \in W_1 \cap W_2$. Therefore it can be expressed as a linear combination of the basis of $W_1 \cap W_2$. Thus we have a relation of the form

$$b_1\beta_1 + b_2\beta_2 + ... + b_t\beta_t = d_1\gamma_1 + d_2\gamma_2 + ... + d_k\gamma_k$$

$$\Rightarrow \quad b_1\beta_1 + b_2\beta_2 + ... + b_t\beta_t - d_1\gamma_1 - d_2\gamma_2 - ... - d_k\gamma_k = 0.$$

But $\beta_1, \beta_2,..., \beta_t, \gamma_1, ..., \gamma_k$ are linearly independent vectors.

Therefore we must have $b_1 = 0, b_2 = 0,..., b_t = 0$.

Putting these values of β's in (1), it reduces to

$$c_1\gamma_1 + c_2\gamma_2 + ... + c_k\gamma_k + a_1\alpha_1 + a_2\alpha_2 + ... + a_m\alpha_m = 0$$

$$\Rightarrow \quad c_1 = 0, c_2 = 0,..., c_k = 0, a_1 = 0, a_2 = 0,..., a_m = 0$$

since the vectors $\gamma_1, \gamma_2,..., \gamma_k, \alpha_1, \alpha_2,..., \alpha_m$ are linearly independent.

Thus the relation (1) implies that

$$c_1 = 0, c_2 = 0,..., c_k = 0, a_1 = 0,..., a_m = 0, b_1 = 0,..., b_t = 0.$$

Therefore the set S_1 of vectors $\gamma_1,..., \gamma_k, \alpha_1,..., \alpha_m, \beta_1,...., \beta_t$ is linearly independent.

Now to show that $L(S_1) = W_1 + W_2$.

Since $W_1 + W_2$ is a subspace of V and each element of S_1 belongs to $W_1 + W_2$, therefore $L(S_1) \subseteq W_1 + W_2$.

Again let α be any element of $W_1 + W_{-2}$. Then

α = some element of W_1 + some element of W_2

= a linear combination of a basis of W_1 + a linear combination of a basis of W_2

= a linear combination of S_1.

$\therefore \alpha \in L(S_1)$. Hence $W_1 + W_2 \subseteq L(S_1)$.

$\therefore L(S_1) = W_1 + W_2$.

$\therefore S_1$ is a basis of $W_1 + W_2$ and consequently

$\dim (W_1 + W_2) = k + m + t$.

Hence the theorem.

Dimension of a Subspace

Theorem 1:

Each subspace W of a finite dimensional vector space V(F) of dimension n is a finite dimensional space with dim $m \leq n$.

Also V = W iff dim V = dim W.

Proof:

Let V(F) be a finite dimensional vector space of dim n. Let W be a subspace of V. Any subset of W containing (n + 1) or more vectors is also a subset of V and any (n + 1) vectors in V are linearly dependent. Therefore any linearly independent set of vectors in W can contain, at the most n vectors. Let

$$S = \{\alpha_1, \alpha_2,...., \alpha_m\}$$

be a linearly independent subset of W with a maximal number of elements. We claim that S is a basis of W. The proof is as follows:

(i) By assumption S is a linearly independent subset of W.

(ii) L (S) = W as we shall just show.

Let α be any element of W.

Then the (m + 1) vectors $\alpha, \alpha_1,..., \alpha_m$ belonging to W are linearly dependent because we have supposed that the largest independent subset of W contains m vectors.

Now $\{\alpha_1, \alpha_2,...., \alpha_m, \alpha\}$ is a linearly dependent set. Therefore there exists a vector belonging to it which can be expressed as a linear combination of the preceding vectors. Since $\alpha_1, \alpha_2,..., \alpha_m$ are linearly independent, therefore this vector cannot be any of these m vectors. So it must be α itself. Thus, α can be expressed as a linear combination of $\alpha_1, \alpha_2,..., \alpha_m$.

Hence L(S) = W.

$\therefore$ S is a basis of W.

$\therefore$ dim W = m and $m \leq n$.

Now if V = W, then every basis of V is also a basis of W.

Hence dim V = dim W = n.

Conversely let dim W = dim V = n. Then to prove that W = V.

Let S be a basis of W. Then L(S) = W and S contains n vectors. Since S is also a subset of V and S contains n linearly independent vectors, therefore S will also be a basis of V. Therefore L(S) = V. Hence W = V.

Theorem 2:

Each set of (n + 1) or more vectors of a finite dimensional vector space V(F) of dimension n is linearly dependent.

Proof:

Let V(F) be a finite dimensional vector space of dimension n. Let S be a linearly independent subset of V containing (n + 1) or more vectors. Then S can be extended to form a basis of V. Thus we shall bet a basis of V containing more than n vectors. But every basis of V will contain exactly n vectors. Hence, our assumption is wrong. Therefore if S contains (n + 1) or more vectors, then S must be linearly dependent. From this theorem we conclude that if S contains m vectors and S is linearly independent then

$$m \leq n.$$

Theorem 3:

If V(F) is a finite dimensional vector space of dimension n, then any set of a n linearly independent vectors in V forms a basis of V.

Proof:

Let $S = \{\alpha_1, \alpha_2, ..., \alpha_n\}$ be a linearly independent subset of finite dimensional vector space V(F) of dimension n. If S is not a basis of V, then it can be extended to form a basis of V. Thus we shall get a basis of V containing more than n vectors. But every basis of V must contain exactly n vectors. Therefore our assumption is wrong and S must be a basis of V.

Theorem 4:

If a set S of n vectors of a finite dimensional vector space V(F) of dimension n generators V(F), then S is a basis of V.

Proof:

Let V(F) be a finite dimensional vector space of dimension n. Let $S = (\alpha_1, \alpha_2, ..., \alpha_n\}$ be a subset of V such that L(S) = V.

If S is linearly independent, then S will form a basis of V. If S is not linearly independent, then there will exist a proper subset of S which will form a basis of V.

Thus we shall get a basis of V containing less than n elements. But every basis of V must contain exactly n elements. Hence S cannot be linearly dependent and so S must be a basis of V.

Note: *If is a finite dimensional vector space of dimension n, then V cannot be generated by fewer than n vectors.*

ISOMORPHISM OF VECTOR SPACE

Definition: *Let U(F) and V(F) be two vector spaces. Then a mapping $f: U \to V$ is called an isomorphism of U onto V if*

(i) f is one-one,

(ii) f is onto,

(iii) $f(a\alpha + b\beta) = af(\alpha) + bf(\beta)\ \forall\, a, b \in F,\ \forall\, \alpha, \beta \in U.$

Also then the two vector space U and V are said to be isomorphic and symbolically we write $U(F) \cong V(F)$.

The vector space V(F) is also called the isomorphic image of the vector space U(F).

If f is a homomorphism of U(F) into V(F), then f will become an isomorphism of U into V if f is one-one. Also in addition if f is onto V, then f will become a isomorphism of U onto V.

Isomorphism of Finite Dimensional Vector Spaces

Theorem 1:

Two finite dimensional vector spaces over the same field are isomorphic if and only if they are of the same dimension.

Proof:

First suppose that U(F) and V(F) are two finite dimensional vector spaces each of dimension n.

Then to prove that U (F) $\cong$ V(F).

Let the sets of vectors $\{\alpha_1, \alpha_2, \ldots, \alpha_n\}$ and $\{\beta_1, \beta_2, \ldots, \beta_n\}$ be the bases of U and V respectively.

Any vector $\alpha \in U$ can be uniquely expressed as

$$\alpha = a_1\alpha_1 + a_2\alpha_2 + \ldots + a_n\alpha_n.$$

Let f: U $\rightarrow$ V be defined by

$$f(\alpha) = a_1\beta_1 + a_2\beta_2 + \ldots + a_n\beta_n.$$

Since in the expression for α as a linear combination of $\alpha_1, \alpha_2, \ldots, \alpha_n$ the scalars $\alpha_1, \alpha_2, \ldots, \alpha_n$ are unique, therefore the mapping f is well-defined *i.e.*, $f(\alpha)$ is a unique element of V.

f is one-one. We have

$$f(a_1\alpha_1 + a_2\alpha_2 + \ldots + a_n\alpha_n) = f(b_1\alpha_1 + b_2\alpha_2 + \ldots + b_n\alpha_n)$$

$$\Rightarrow a_1\beta_1 + a_2\beta_2 + \ldots + a_n\beta_n = b_1\beta_1 + b_2\beta_2 + \ldots + b_n\beta_n$$

$$\Rightarrow (a_1 - b_1)\ \beta_1 + \ldots + (a_n - b_n)\ b_n = 0' \text{ (zero vector of V)}$$

$$\Rightarrow a_1 - b_1 = 0,\ a_2 - b_2 = 0, \ldots, a_n - b_n = 0 \text{ because}$$

$\beta_1, \beta_2, \ldots, \beta_n$ are linearly independent

$$\Rightarrow a_1 = b_1,\ a_2 = b_2, \ldots, a_n = b_n$$

$$\Rightarrow a_1\alpha_1 + a_2\alpha_2 + \ldots + a_n\alpha_n + b_1\alpha_1 + b_2\alpha_2 + \ldots + b_n\alpha_n$$

$\therefore$ f is one-one.

f is onto V. If $a_1\beta_1 + a_2\beta_2 + \ldots + a_n\beta_n$ is any element of V, then $\exists$ an element $a_1\alpha_1 + a_2\alpha_2 + \ldots + a_n\alpha_n \in U$ such that

$f(a_1\alpha_1 + a_2\alpha_2 + ... + a_n\alpha_n) = a_1\beta_1 + a_2\beta_2 + ... + a_n\beta_n$.

$\therefore$ f is onto V.

f is a linear transformation. We have

$f[a(a_1\alpha_1 + a_2\alpha_2 + ... a_n\alpha_n) + b(b_1\alpha_1 + b_2\alpha_2 + ... + b_n\alpha_n)$

$= f[(aa_1 + bb_1)\alpha_1 + (aa_2 + bb_2)\alpha_2 + ... + (aa_n + bb_n)\alpha_n)]$

$= (aa_1 + bb_1)\beta_1 + (aa_2 + bb_2)\beta_2 + ... + (aa_n + bb_n)\beta_n$

$= a(a_1\beta_1 + a_2\beta_2 + ... + a_n\beta_n) + b(b_1\beta_1 + b_2\beta_2 + ... + b_n\beta_n)$

$= af(a_1\alpha_1 + a_2\alpha_2 + ... + a_n\alpha_n) + bf(b_1\alpha_1 + b_2\alpha_2 + ... + b_n\alpha_n)$.

$\therefore$ f is a linear transformation.

Hence f is an isomorphism of U onto V. $\therefore$ $U \cong V$.

Conversely, let U (F) and V(F) be two isomorphic finite dimensional vector spaces. Then to prove that dim U = dim V. Let dim U = n. Let $S = \{\alpha_1, \alpha_2, ..., \alpha_n)$ be a basis of U. If f is an isomorphism of U onto V, we shall show that $S' = \{f(\alpha_1), f(\alpha_2), ..., f(\alpha_n)\}$ is a basis of V. Then V will also be of dimension n.

First we show that S' is linearly independent.

Let $a_1 f(\alpha_1) + a_2 f(\alpha_2) + ... + a_n (f(\alpha_n) = 0$, (zero vector of V)

$\Rightarrow f(a_1\alpha_1 + a_2\alpha_2 + ... + a_n\alpha_n) = 0'$ [$\because$ f is linear transformation]

$\Rightarrow a_1\alpha_1 + a_2\alpha_2 + ... + a_n\alpha_n = 0$ [$\because$ f is one-one and f(0) = 0' where 0 is zero vector of U]

$\Rightarrow a_1 = 0, a_2 = 0, ..., a_n = 0$ since $\alpha_1, \alpha_2, ..., \alpha_n$ are linearly independent.

$\therefore$ S' is linearly independent.

Now to prove that L(S') = V. For this we shall prove that any vector $\beta \in V$ can be expressed as a linear combination of the vectors of the set S'. Since f is onto V, therefore $\beta \in V \Rightarrow$ there exists $\alpha \in U$ such that $f(\alpha) = \beta$.

Let $\alpha = c_1\alpha_1 + c_2\alpha_2 + ... + c_n\alpha_n$.

Then $\beta = f(\alpha) = f(c_1\alpha_1 + c_2\alpha_2 + ... + c_n\alpha_n)$

$= c_1 f(\alpha_1) + c_2 f(\alpha_2) + ... + c_n f(\alpha_n)$.

Thus β is a linear combination of the vectors of S'.

Hence V = L(S'). Therefore S' is a basis of V. Since S' contains n vectors, therefore dim V = n.

Note: While proving the converse, we have proved that *if f is an isomorphism of U onto V then f maps a basis of U onto a basis of V.*

Theorem 2:

Every n-dimensional vector space V(F) is isomorphic to $V_n(F)$.

Proof:

Let $\{\alpha_1, \alpha_2, ..., \alpha_n\}$ be any basis of V(F). Then every vector $\alpha \in V$ can be uniquely expressed as

$$\alpha = a_1\alpha_1 + a_2\alpha_2 + ... + a_n\alpha_n,\ a_i \in F.$$

The ordered n-tuple $(a_1, a_2, ..., a_n) \in V_n\ (F)$.

Let $f =: V(F) \rightarrow V_n\ (F)$ be defined by $f(\alpha) = (a_1, a_2, ..., a_n)$.

Since in the expression of α as a linear combination of $\alpha_1, \alpha_2, ..., \alpha_n$ the scalars $a_1, a_2, ..., a_n$ are unique, therefore $f(\alpha)$ is a unique element of $V_n(F)$ and thus the mapping f is well-defined.

f is one-one. Let $\alpha = a_1\alpha_1 + a_2\alpha_2 + ... + a_n\alpha_n$ and

$\beta = b_1\alpha_1 + b_2\alpha_2 + ... + b_n\alpha_n$ be any two elements of V. We have

$$f(\alpha) = f(\beta)$$

$$\Rightarrow \quad f(a_1\alpha_1 + a_2\alpha_2 + ... + a_n\alpha_n) = f(b_1\alpha_1 + b_2\alpha_2 + ... + b_n\alpha_n)$$

$$\Rightarrow \quad (a_1, a_2, ..., a_n) = (b_1, b_2, ..., b_n)$$

$$\Rightarrow \quad a_1 = b_1,\ a_2 = b_2, ..., a_n = b_n \Rightarrow \alpha = \beta. \qquad \therefore \text{ f is one-one.}$$

f is onto $V_n\ (F)$. Let $(a_1, a_2, ..., a_n)$ be any element of $V_n\ (F)$. Then there exists an element $a_1\alpha_1 + a_2\alpha_2 + ... + a_n\alpha_n \in V\ (F)$ such that $f(a_1\alpha_1 + a_2\alpha_2 + ... + a_n\alpha_n) = (a_1, a_2, ..., a_n)$.

$\therefore$ f is onto $V_n\ (F)$.

f is a linear transformation. If $a, b \in F$ and $\alpha, \beta \in V(F)$, we have $f(a\alpha + b\beta)$

$$= f[a(a_1\alpha_1 + a_2\alpha_2 + ... + a_n\alpha_n) + b(b_1\alpha_1 + b_2\alpha_2 + .. + b_n\alpha_n)]$$

$$= f[(aa_1 + bb_1)\alpha_1 + (aa_2 + bb_2)\alpha_2 + ... + (aa_n + bb_n)\alpha_n]$$

$$= (aa_1 + bb_1, aa_2 + bb_2, ..., aa_n + bb_n)$$

$$= (aa_1, aa_2, ..., aa_n) + (bb_1, bb_2, ..., bb_n)$$

$$= a(a_1, a_2, ..., a_n) + b(b_1, b_2, ..., bb_n)$$

$$= af(a_1\alpha_1 + a_2\alpha_2 + ... + a_n\alpha_n) + bf(b_1\alpha_1 + b_2\alpha_2 + ... + b_n\alpha_n)$$

$$= af(\alpha) + bf(\beta).$$

$\therefore$ f is an linear transformation.

$\therefore$ f is an isomorphism.

of V(F) onto $V_n(F)$. Hence $V(F) \cong V_n(F)$.

HOMOMORPHISM OF VECTOR SPACES OR LINEAR TRANSFORMATIONS

Definition: *Let U(F) and V(F) be two vector spaces. Then a mapping $f: U \to V$ is called a homomorphism or a linear transformation of 'U into V if:*

(i) $f(\alpha + \beta) = f(\alpha) + f(\beta), \forall \alpha, \beta, \in U$

and (ii) $f(a\alpha) = af(\alpha) \ \forall a \in F, \ \forall \alpha \in U.$

The conditions (i) and (ii) can be combined into a single condition $f(a\alpha + b\beta) = af(\alpha) + bf(\beta) \ \forall a, b \in F$ and $\forall \alpha, \beta, \in U$.

If F is a homomorphism of U onto V, then V is called a homomorphic image of U.

Theorem 1:

If f is a homomorphism of U(F) into V(F), then

(i) $f(0) = 0'$ where 0 and 0' are the zero vectors of U and V respectively.

(ii) $f(-\alpha) = -f(\alpha) \ \forall \alpha \in U.$

Proof:

(i) Let $\alpha \in U$. Then $f(\alpha) \in V$. Since 0' is the zero vector of V, therefore $f(\alpha) + 0' = f(\alpha) + f(\alpha + 0) = f(\alpha) + f(\alpha) + f(0)$

Now V is an abelian group with respect to addition of vectors. Therefore $f(\alpha) + 0' = f(\alpha) + f(0) \Rightarrow 0' = f(0)$ by left cancellation law.

(ii) If $\alpha \in U$, then $-\alpha \in U$. Also we have

$$0' = f(0) = f[\alpha + (-\alpha)] = f(\alpha) + f(-\alpha).$$

Now $f(\alpha) + f(-\alpha) = 0' \Rightarrow f(-\alpha) =$ additive inverse of $f(\alpha)$

$$\Rightarrow f(-\alpha) = -f(\alpha).$$

Remark: The students may use without any confusion the same symbol 0 to denote both the zero vectors of U and V. In the relation $f(0) = 0$, the zero in the left hand side is the zero vector of U and its f-image *i.e.*, the zero on the right hand side is the zero vector of V.

Theorem 2:

The kernel of a homomorphism is a subspace.

Proof:

Let U(F) and V(F) be two vector spaces and let f be a homomorphism of U into V.

Let 0' be the zero vector of V and 0 be the zero vector of U.

Let W be the kernal of f *i.e.* $W = \{\alpha \in U: f(\alpha) = 0'\}$.

To prove that W is a sub space of U.

Let α, β be any two elements of W. Then $f(\alpha) = 0'$ and $f(\beta) = 0'$.

If α, β are any two elements of F, then we have

$$f(a\alpha + b\beta) = af(\alpha) + bf(\beta) \quad [\because \text{ f is a homomorphism of U into V}]$$

$$= a0' + b0' = 0' + 0' = 0'.$$

$\therefore a\alpha + b\beta \in W$.

Thus $\quad a, b \in F$

and $\quad \alpha, \beta, \in W$

$\Rightarrow \quad a\alpha + b\beta \in W$.

$\therefore$ W is a subspace of U.

Kernal of a Homomorphism

Definition: *Let f be a homomorphism of a vector space U(F) into a vector space V(F). The kernal W of f is defined as*

$$W = \{\alpha \in U: f(\alpha) = 0' \text{ where } 0' \text{ is the zero vector of } V\}$$

Thus, the kernal W of f is subset of U consisting of those elements of U which are mapped under f onto the zero vector of V. Since f(0), 0', therefore at least $0 \in W$. Thus W is not empty.

SOME THEOREMS OF LINEAR DEPENDENCE AND LINEAR INDEPENDENCE

Theorem 1:

If in a vector space V(F), a vector β is a linear combination of the set of vectors $\alpha_1, \alpha_2, \alpha_3, \ldots, \alpha_n$, then the set of vectors $\beta, \alpha_1 \alpha_1, \ldots, \alpha_n$ is linearly dependent.

Proof:

Since b is a linear combination of $\alpha_1, \alpha_2, .., \alpha_n$, therefore there exist scalars $a_1, a_2, \ldots, a_n$ such that

$$\beta = a_1\alpha_1 + a_2\alpha_2 + \ldots a_n\alpha_n$$

$$\Rightarrow \quad \beta_1 - a_1\alpha_1 - a_2\alpha_3 - \ldots - a_n\alpha_n = 0. \qquad \ldots(1)$$

In the relation (1) the scalar coefficient of β is 1 which is $\neq 0$. Hence in the relation (1) not all the scalar coefficients are 0. Therefore the set of vectors $\beta, \alpha_1, \alpha_2, \alpha_2, \ldots, \alpha_n$ is linearly dependent.

Theorem 2:

Let V(F) be a vector space. If $\alpha_1, \alpha_2, ..., \alpha_n$ are non-zero vector $\in V$ then either they are linearly independent or some α_k, $2 \leq k \leq n$, is a linear combination of the preceding ones, $\alpha_1, \alpha_2, ..., \alpha_{k-1}$.

Proof:

If $\alpha_1, \alpha_2, \alpha_3, ..., \alpha_n$ are linearly independent, we are nothing to prove. So let $\alpha_1, \alpha_2, ..., \alpha_n$ be linearly dependent. Then there exists a relation of the form

$$a_1\alpha_1 + a_2\alpha_2 + ... + a_n\alpha_n = 0 \qquad ...(1)$$

where not all the scalar coefficients $a_1, a_2, ..., a_n$ are 0.

Let k be the largest integer for which $a_k \neq 0$ *i.e.*, $a_{k+1} = 0, a_{k+2} = 0, ..., a_n = 0$ and $a_k \neq 0$.

There is no harm in this assumption because at the most if $a_n \neq 0$ then $k = n$.

Also $2 \leq k$. Because if $a_2 = 0, a_3 = 0, ..., a_n = 0$, then $a_1\alpha_1 = 0$ and $\alpha_1 \neq 0 \Rightarrow a_1 = 0$. This contradicts the fact that not all the a's are 0.

Now the relation (1) reduces to

$$a_1\alpha_1 + a_2\alpha_2 + ... + a_k\alpha_k = 0, \text{ where } a_k \neq 0$$

$$\Rightarrow \quad a_k\alpha_k = - a_1\alpha_1 - a_2\alpha_2 - ... - a_{k-1}\alpha_{k-1}$$

$$\Rightarrow \quad a_k^{-1}(a_k\alpha_k) = a_k^{-1}(- a_1\alpha_1 - a_2\alpha_2 - ... - a_{k-1}\alpha_{k-1})$$

$$\Rightarrow \quad \alpha_k = (- a_k^{-1} a_1)\alpha_1 + (- a_k^{-1} a_1)\alpha_2 + ... + (- a_k^{-1} a_{k-1})\alpha_{k-1}.$$

Thus a_k is a linear combination of its preceding vectors.

Theorem 3:

The set of non-zero vectors $\alpha_1, \alpha_2, ..., \alpha_n$ of V(F) is linearly dependent if some a_k, $2 \leq k \leq n$, is a linear combination of the proceeding ones.

Proof:

If some α_k, $2 \leq k \leq n$ is a linear combination of the proceeding ones, $\alpha_1\ \alpha_2, ..., \alpha_{k-1}$, then $\exists$ scalars $a_1, a_2, ..., a_{k-1}$ such that $\alpha_k = a_1\alpha_1 + a_2\alpha_2 + ... + a_{k-1}\alpha_{k-1}$

$$\Rightarrow \quad 1\alpha_k - a_1\alpha_1 - a_2\alpha_2 - ... - ak_{k-1}\ \alpha_{k-1} = 0 \qquad ...(1)$$

$\Rightarrow$ the set $\{\alpha_1, \alpha_2, ..., \alpha_k\}$ is linearly dependent because in the linear combination (1) the scalar coefficient $1 \neq 0$.

Hence, the set $\{\alpha_1, \alpha_2, .., \alpha_n\}$ of which $\{\alpha_1, \alpha_2, ..., \alpha_k\}$ is a subset must be linearly dependent.

BASIS OF A VECTOR SPACE

Definition: *A subset S of a vector space V(F) is said to be a basis of V(F), if*

(i) S consists of linearly independent vectors,

(ii) S generates V(F) i.e., L(S) = V i.e., each vector in V is a linear combination of a finite number of elements of S.

LINEAR COMBINATION OF VECTORS: LINEAR SPAN OF A SET

Linear Combination

Definition: *Let V (F) be a vector space. If $\alpha_1, \alpha_2, \ldots, \alpha_n \in V$, then any vector*

$$\alpha = a_1\alpha_1 + a_2\alpha_2 + \ldots + a_n\alpha_n \text{ where } \alpha_1, \alpha_2, \ldots, \alpha_n.$$

is called a linear combination of the vector $\alpha_1, \alpha_2, \ldots, \alpha_n$.

Linear Span

Definition: *Let V (F) be a vector space and S be any non-empty subset of V.*

Then the linear span of S is the set of all linear combinations of finite sets of elements of S and is denoted by L(S). Thus we have

$$L(S) = \{a_1\alpha_1 + a_2\alpha_2 + \ldots + a_n\alpha_n : \alpha_1, \alpha_2, \ldots, \alpha_n$$

is any arbitrary finite subset of S and $a_1, a_2, \ldots, a_n$ is any arbitrary finite subset of F}.

Theorem 1:

The linear span L(S) of any subset of S of a vector space V (F) is a subspace of V generated by S i.e. L (S) = {S}.

Proof:

Let α, β be any two elements of L(S).

Then $\alpha = a_1\alpha_1 + a_2\alpha_2 + \ldots + a_m\alpha_m$

and $\beta = b_1\beta_1 + b_2\beta_2 + \ldots + b_n\alpha_n$

where the a's and b's are elements of F and the α's and β's are element of S.

If a, b, be any two elements of F, then

$$a\alpha + b\beta = a\,(a_1\alpha_1 + a_2\alpha_2 + \ldots + a_m\alpha_m) + (b_1\beta_1 + b_2\beta_2 + \ldots + b_n\beta_n)$$

$= a(a_1\alpha_1) + a(a_2\alpha_2) + \ldots + a(a_m\alpha_m) + b(b_1\beta_1) + b(b_2\beta_2) + \ldots + b(b_n\beta_n)$

$= (aa_1)\alpha_1 + (aa_2)\alpha_2 + \ldots + (aa_m)\alpha_m + (bb_1)\beta_1 + (bb_2)\beta_2 + \ldots + (bb_n)\beta_n.$

Thus, $a\alpha + b\beta$ has been expressed as a linear combination of a finite set $\alpha_1, \alpha_2, \ldots \alpha_m, \beta_1, \beta_2, \ldots, \beta_n$ of the elements of S, Consequently $a\alpha + b\beta \in L(S)$.

Thus $a, b, \in F$ and $\alpha, \beta \in L(S) \Rightarrow a\alpha + b\beta \in L(S)$.

Hence L(S) is a subspace of V(F).

Also each element of S belongs to L(S) because if $\alpha_r \in S$, then $\alpha_r = 1\alpha_r$ and this implies that $\alpha_r \in L(S)$. Thus L(S) is a subspace of V and S is contained in L(S).

Now if W is any subspace of V containing S, then each element of L(S) must be in W because W is to be closed under vector addition and scalar multiplication. Therefore L(S) will be contained in W.

Hence L(S) = {S) *i.e.* L(S) is the smallest subspace of V containing S.

Note: If in any case we are to prove that L(S) = V, then we should prove that $V \subseteq L(S)$ because $L(S) \subseteq V$ since I (S) is a subspace of V. In order to prove that $V \subseteq L(S)$, we should prove that each element of V can be expressed as a linear combination of a finite number of elements of S. Then each element of V will also be an element of L(S) and we shall have $V \subseteq L(S)$.

Finally $V \subseteq L(S)$ and $L(S) \subseteq V \Rightarrow L(S) = V$.

Example:

The subset S = {(1, 0, 0), (0, 1, 0) (0, 0, 1)} of $V_3(F)$ generates or spans the entire vector space $V_3(F)$ i.e., L(S) = V.

Solution:

Hint: If (a, b, c) be any element of V, then

(a, b, c) = (1, 0, 0) + b (0, 1, 0) + c (0, 0, 1).

Thus $(a, b, c) \in L(S)$. Hence $V \subseteq L(S)$. Also $L(S) \subseteq V$.

Hence L(S) = V.

LINEAR SUM OF TWO SUBSPACES

Definition: *Let W_1 and W_2 be two subspaces of the vector space V (F). Then the linear sum of the subspaces W_1 and W_2 denoted by $W_1 + W_2$ is the set of sums $\alpha_1 + \alpha_2$ such that $\alpha_1 \in W_1$, $\alpha_2 \in W_2$.*

Thus $W_1 + W_2 = \{\alpha_1 + \alpha_2 : \alpha_1 \in W_1, \alpha_2 \in W_2\}$.

Theorem 1:

If W_1 and W_2 are subspaces of the vector space V(F), then

(i) $W_1 + W_2$ is a subspace of V (F).

(ii) $W_1 + W_2 = \{W_1 \cup W_2\}$

i.e., $L(W_1 \cup W_2) = W_1 + W_2$

Proof:

(i) Let α, β, be any two elements of $W_1 + W_2$.

Then $\alpha = \alpha_1 + \alpha_2$ and $\beta = \beta_1 + \beta_2$ where $\alpha_1, \beta_1 \in W_1$ and $\alpha_1, \beta_2 \in W_2$. If a, b, ∈ F, we have

$$a\alpha + b\beta = a(\alpha_1 + \alpha_2) + b(\beta_1 + \beta_2) = (a\alpha_1 + b\beta_1) + (a\alpha_2 + \beta b_2)$$

Since W_1 is a subspace of V, therefore a, b, ∈ F

and $\alpha_1, \beta_1 \in W_1 \Rightarrow a\alpha_1 + b\beta_1 \in W_1$

Similarly $a\alpha_2 + b\beta_2 \in W_2$.

Consequently $a\alpha + b\beta = (a\alpha_1 + b\beta_1) + (a\alpha_2 + b\beta_2) \in W_1 + W_2$.

Thus a, b ∈ F and $\alpha, \beta \in W_1 + W_2 \Rightarrow a\alpha + b\beta \in W_1 + W_2$.

Hence $W_1 + W_2$ is a subspace of V(F).

(ii) Since W_2 contains the zero vector, therefore if $\alpha_1 \in W_1$, then we can write $\alpha_1 = \alpha_1 + 0 \in W_1 + W_2$. Thus $W_1 \subseteq W_1 + W_2$. Similarly $W_2 \subseteq W_1 + W_2$ hence $W_1 \cup W_2 \subseteq W_1 + W_2$. Therefore $W_1 + W_2$ is a subspace of V(F) containing $W_1 \cup W_2$.

Now to prove that $W_1 + W_2 = \{W_1 \cup W_2\}$ we should prove that $W_1 + W_2 \subseteq L(W_1 \cup W_2)$ and $L(W_1 \cup W_2) \subseteq W_1 + W_2$.

Let $\alpha = \alpha_1 + \alpha_2$ be any element of $W_1 + W_2$. Then $\alpha \in W_1$ and $\alpha_2 \in W_2$. Therefore $\alpha_1, \alpha_2 \in W_1 \cup W_2$. We can write

$$\alpha_1 + \alpha_2 = 1\alpha_1 + 1\alpha_2.$$

Thus, $\alpha_1 + \alpha_2$ is a linear combination of a finite number of elements α_1, $a_2 \in W_1 \cup W_2$.

Therefore $\alpha_1 + \alpha_2 \in L(W_1 \cup W_2)$.

∴ $W_1 + W_2 \subseteq L(W_1 \cup W_2)$.

Also $L(W_1 \cup W_2)$ is the smallest subspace containing $W_1 \cup W_2$ and $W_1 + W_2$ is a subspace containing $W_1 \cup W_2$. Therefore $L(W_1 \cup W_2)$ must be contained in $W_1 + W_2$. Consequently

$$L(W_1 \cup W_2) \subseteq W_1 + W_2.$$

Hence $W_1 + W_2 = L(W_1 \cup W_2) = \{W_1 \cup W_2\}$.

Theorem 2:

If S, T are subsets of V (F), then

(i) $S \subseteq T \Rightarrow L(S) \subseteq L(T)$.

(ii) $L(S \cup T) = L(S) + L(T)$.

(iii) $L(L(S)) = L(S)$.

Proof:

(i) Let $\alpha = a_1\alpha_1 + a_2\alpha_2 + \ldots a_n\alpha_n \in L(S)$, where $\{\alpha_1, \alpha_2, \ldots, \alpha_n\}$ is a finite subset of S. Since $S \subseteq T$, therefore $\{\alpha_1, \alpha_2, \ldots, \alpha_n)$ is a finite subset of T. So $\alpha \in L(T)$.

Thus $\alpha \in L(S) \Rightarrow \alpha \in L(T)$.

$\therefore L(S) \subseteq L(T)$.

(ii) Let α be any element of $L(S \cup T)$. Then

$$\alpha = a_1\alpha_1 + a_2\alpha_2 + \ldots + a_m\alpha_m + b_1\beta_1 + b_2\beta_2 + \ldots + b_p\beta_p$$

where $(\alpha_1, \alpha_2, \ldots, \alpha_m, \beta_1, \beta_2, \ldots, \beta_p)$ is a finite subset of $S \cup T$ such that $\{\alpha_1, \alpha_2, \ldots, \alpha_m)$ Í S and $\{\beta_1, \beta_2, \ldots, \beta_p\} \subseteq T$.

Now $a_1\alpha_1 + a_2\alpha_2 + \ldots + a_m\alpha_m \in L(S)$ and

$b_1\beta_1 + b_2\beta_2 + \ldots + b_p\beta_p \in L(T)$.

Therefore $\alpha \in L(S) + L(T)$. Consequently $L(S \cup T) \subseteq L(S) + L(T)$.

Now let γ be any element of $L(S) + L(T)$. Then $\gamma = \beta + \delta$ where $\beta \in L(S)$ and $\delta \in L(T)$. Now β will be a linear combination of a finite number of elements of S and δ will be a linear combination of a finite number of elements of T. Therefore $\beta + \delta$ will be a linear combination of a finite number of elements of $S \subseteq T$. Thus $\beta + \delta \in L(S \cup T)$. Consequently

$$L(S) + L(T) \subseteq L(S) \cup T).$$

Hence $L(S \cup T) = L(S) + L(T)$.

(iii) L(L(S)) is the smallest subspace of V containing L(S). But L(S) is a subspace of V. Therefore the smallest subspace of V containing L(S) is L(S) itself.

Hence L(L(S) = L(S).

COMMUTATIVE BANACH ALGEBRAS STRUCTURE

The function spaces we encountered earlier led us to begin a study of Banach spaces for their own sake, and as we proceeded, we found our attention focusing more and more closely on the properties of their operators. Except for a few elementary notions about metric spaces, we used very little genuine topology. As a matter of fact, our treatment of spectral theory was completely independent of topology, for in the finite-dimensional case, all linear transformations are continuous.

The primarily concerned with topological spaces and the continuous functions carried by them. In the following three chapters, these two apparently diverse trains of thought-the topological and the algebraic—are united by a single elegant concept : that of a Banach algebra. Our remarks in the last section of the previous chapter suggested that there may be important links between algebras of operators on Hilbert spaces and algebras of the type $\mathcal{C}(X)$, where X is a compact Hausdorff space. Banach algebras are the systems which enable us to establish these connections on a firm footing. They are also interesting in that they constitute a field of study in which a wide variety of mathematical ideas meet in significant contact.

The Definition and Some Examples

A *Banach algebra* is a complex Banach space which is also an algebra with identity 1, and in which the multiplicative structure is related to the norm by the following requirements:

(1) $\|xy\| \leq \|x\| \, \|y\|$;

(2) $\|1\| = 1$.

It follows from (1) that multiplication is jointly continuous in any Banach algebra, that is, that if $x_n \to x$ and $y_n \to y$, then $x_n y_n \to xy$ (proof:

$$\|x_n y_n - xy\| = \|x_n (y_n - y) + (x_n - x)y\|$$

$$\leq \|x_n\| \, \|y_n - y\| + \|x_n - x\| \, \|y\|)$$

A *Banach subalgebra* of a Banach algebra A is a closed subalgebra of A which contains 1. The Banach subalgebras of A are precisely those subsets of A which are themselves Banach algebras with respect to the same algebraic operations, the same identity, and the same norm.

The definition of a Banach algebra is sometimes given without the restriction that the scalars are the complex numbers. The complex case, however, is the only one that concerns us, and by framing the definition as we do, we avoid the necessity of treating the additional complications which arise in the real case. We have further assumed, for the sake of simplicity, that every Banach algebra has an identity. It is possible, at a considerable sacrifice of clarity, to develop ;most of the important ideas without this assumption, and this is done whenever the primary purpose of the theory is the study of group algebras of locally compact but not discrete groups. Since our attention will be directed chiefly to the structure of operator algebras, there is no need for us to strain for the added generality obtained by not requiring the presence of an identity.

The Banach algebras of principal interest to us ;are described in the following examples. The reader will notice that they all consist of functions or operators and that the linear operations in all of them are defined pointwise. They can be classified in a general way into *function algebras, operator algebras*, or *group algebras*, according as multiplication is defined pointwise, by composition, or by convolution.

REGULAR AND SINGULAR ELEMENTS

Let A be a Banach algebra. We denote the set of regular elements in A by G, and its complement, the set of singular elements, by S. It is clear that G contains 1 and is a group, and that S contains 0. Several important issues depend on the character of G and S. Our first result along these lines is

Theorem 1:

Every element x for which $\|x-1\| < 1$ *is regular, and the inverse of such an element is given by the formula* $x^{-1} = 1 + \sum_{n=1}^{\infty} (1-x)^n$.

Proof:

If we put $r = \|x-1\|$, so that $r < 1$, then

$$\left\|(1-x)^n\right\| \le \|1-x\|^n = r^n$$

shows that the partial sums of the series $\sum_{n=1}^{\infty} (1-x)^n$ form a Cauchy sequence in A. Since A is complete, these partial sums converge to an element of A, which we denote by $\sum_{n=1}^{\infty} (1-x)^n$. If we define y by $y = 1 + \sum_{n=1}^{\infty} (1-x)^n$, then the joint continuity of multiplication in A implies that

$$y-xy=(1-x)y=(1-x)+\sum_{n=2}^{\infty}(1-x)^n=\sum_{n=2}^{\infty}(1-x)^n=y-1,$$

so $xy = 1$. Similarly, $yx = 1$.

We now use this as a tool to prove

Theorem 2:

G is an open set, and therefore S is a closed set.

Proof:

Let x_0 be an element in G, and let x by any element in A such that It is clear that $\|x-x_0\| < 1/\|x_0^{-1}\|$. It is clear that

$$\left\|x_0^{-1}x - 1\right\| = \left\|x_0^{-1}(x - x_0)\right\| \leq \left\|x_0^{-1}\right\| \left\|x - x_0\right\| < 1,$$

so we see by Theorem 1 that $x_0^{-1}x$ is in G. Since $x = x_0 (x_0^{-1}x)$, it follows that x is also in G, so G is open.

It was the every Banach space is locally connected, so A is also locally connected. A direct application of theorem yields the fact that the components of G are themselves open sets.

As our final result, we have

Theorem 3:

The mapping $x \to x^{-1}$ of G into G is continuous and is therefore a homeomorphism of G onto itself.

Proof:

Let x_0 be an element of G, and x another element of G such that $\left\|x - x_0\right\| < 1 \Big/ \left(2\left\|x_0^{-1}\right\|\right)$. Since

$$\left\|x_0^{-1}x - 1\right\| = \left\|x_0^{-1}(x - x_0)\right\| \leq \left\|x_0^{-1}\right\| \left\|x - x_0\right\| < \frac{1}{2},$$

we see by Theorem A that $x_0^{-1}x$ is in G and

$$x^{-1}x_0 = \left(x_0^{-1}x\right)^{-1} = 1 + \sum_{n=1}^{\infty} \left(1 - x_0^{-1}x\right)^n.$$

Our conclusion now follows from

$$\left\|x^{-1} - x_0^{-1}\right\| = \left\|\left(x^{-1}x_0 - 1\right)x_0^{-1}\right\| \leq \left\|x_0^{-1}\right\| \left\|x^{-1}x_0 - 1\right\| = \left\|x_0^{-1}\right\|$$

$$\left\|\sum_{n=1}^{\infty}\left(1 - x_0^{-1}x\right)^n\right\| \leq \left\|x_0^{-1}\right\| \sum_{n=1}^{\infty} \left\|1 - x_0^{-1}x\right\|^n$$

$$= \left\|x_0^{-1}\right\| \left\|1 - x_0^{-1}x\right\| \sum_{n=0}^{\infty} \left\|1 - x_0^{-1}x\right\|^n$$

$$= \frac{\left\|x_0^{-1}\right\| \left\|1 - x_0^{-1}x\right\|}{1 - \left\|1 - x_0^{-1}x\right\|}$$

$$< 2\left\|x_0^{-1}\right\| \left\|1 - x_0^{-1}x\right\| \leq 2\left\|x_0^{-1}\right\|^2 \left\|x - x_0\right\|.$$

If x is an element in A, it should always be kept in mind that the regularity or singularity of x depends on A as well as on x itself. If x is regular in A, and if we pass to a Banach subalgebra A' of a which also contains x, then x may lose its inverse and become singular in A'. By the same token, if x is singular in A, and if A is regarded as a Banach subalgebra of a larger Banach algebra A", then x may acquire an inverse and become regular in A". In the next section, we study certain element in A which are singular and remain singular with respect to all possible enlargements of A.

Topological Divisors of Zero

An element z in our Banach algebra A is called topological divisor of zero if there exists a sequence $\{z_n\}$ in A such that $\|z_n\| = 1$ and either $zz_n \to 0$ or $z_n z \to 0$. It is clear that every divisor or zero is also a topological divisor of zero. We denote the set of all topological divisors of zero by z.

Theorem 1:

Z is a subset of S.

Proof:

Let z be an element of Z and $\{z_n\}$ a sequence such that $\|z_n\| = 1$ and (say) $zz_n \to 0$. If z were in G, then by the joint continuity of multiplication we would have $z^{-1}(zz_n) = z_n \to 0$, contrary to $\|z_n\| = 1$

Our next theorem relates to the manner in which Z is distributed within S.

Theorem 2:

The boundary of S is a subset of Z.

Proof:

Since S is closed, its boundary consists of all points in S which are limits of convergent sequences in G. We show that if z is such a point, that is, if z is in S and there exists a sequence $\{r_n\}$ in G such that $r_n \to z$, then z is in Z. First, we see from $r_n^{-1} z - 1 = r_n^{-1}(z - r_n)$ that the sequence $\{r_n^{-1}\}$ is unbounded; for otherwise, we would have

$$\left\| r_n^{-1} z - 1 \right\| < 1$$

for some n, so that $r_n^{-1}z$, and therefore $z = r_n(r_n^{-1}z)$, would be regular. Since $\{r_n^{-1}\}$ is unbounded, we may assume that $\left\| r_n^{-1} \right\| \to \infty$. If z_n is now defined by $z_n = r_n^{-1} / \left\| r_n^{-1} \right\|$, then our conclusion follows from the observations that $\| z_n \| = 1$ and

$$zz_n = \frac{zr_n^{-1}}{\left\| r_n^{-1} \right\|} = \frac{1 + (z - r_n) r_n^{-1}}{\left\| r_n^{-1} \right\|} = \frac{1}{\left\| r_n^{-1} \right\|} + (z - r_n) z_n \to 0.$$

In order to understand the significance of these facts, let us suppose that A is imbedded as Banach subalgebra in a larger Banach algebra A'. As we remarked in the previous section, an element which is singular in A may cease to be so in A'. However, if it is a topological divisor of zero in A, then it is in A' as well, so it is singular in A'. The topological divisors of zero in A are thus "permanently singular," in the sense that they are singular and remain so with respect to every possible enlargement of the containing Banach algebra. Theorem B tells us that no matter what happens to S as a whole in such a process, its boundary is "permanent" in this sense.

THE SPECTRUM

Let T be an operator on a non-trivial Hilbert space. In the previous chapter, we defined the spectrum of T to be the set

$$\sigma (T) = \{\lambda : T - \lambda l \text{ is singular}\},$$

and we devoted a good deal of attention to the geometric ideas leading to this concept. We found— at least in the finite-dimensional case-that a number in σ (T) is a *value assumed* by T, in the sense that t acts on some non-zero vector as if it were scalar multiplication by that number. We shall see later that this formulation of the meaning of the spectrum has a much wider significancy than we might at first suspect.

Let us now consider an element x in our general Banach algebra A. By analogy with the above, we define the spectrum of x to be the following subset of the complex plane:

$$\sigma (x) = \{\lambda : x - \lambda l \text{ is singular}\}.$$

Wherever it is desirable to express the fact that the spectrum of x depends on A as well as x, we use the notation $\sigma_A(x)$. It is easy to see that $x - \lambda l$ is a continuous function of λ with values in A; and since the set of singular elements in A is closed, it follows at one that σ (x) is closed. We further observe that σ (x) is a subset of the closed disc $\{z:|z| \leq \|x\|\}$, for it λ is a complex number such that $|\lambda| > \|x\|$, then $\|x/\lambda\| < 1$, $\|1-(1-x/\lambda)\| < 1$, $1 - x/\lambda$ is regular, and therefore $x - \lambda l$ is regular.

Our first task is to establish the fact that σ (x) is always non-empty, and for this we need a few preliminary notions.

The resolvent set of x, denoted by ρ (x), is the complement of σ (x); it is clearly an open subset of the complex plane which contains $\{z:|z| > \|x\|\}$. The resolvent of x is the function with values in A defined on ρ (x); by

$$x(\lambda) = (x - \lambda l)^{-1}.$$

Tells us that $x(\lambda)$ is a continuous function of λ; and the fact that $x(\lambda) = \lambda^{-1}(x/\lambda - 1)^{-1}$ for $\lambda \neq 0$ implies that $x(\lambda) \to 0$ as $\lambda \to \infty$. If λ and μ are both in ρ (x) , then

$$\begin{aligned} x(\lambda) &= x(\lambda)\,[x - \mu 1]\,x(\mu) \\ &= x(\lambda)\,[x - \lambda 1 + (\lambda - \mu)1]\,x(\mu) \\ &= [1 + (\lambda - \mu)\,x(\lambda)]\,x(\mu) \\ &= x(\mu) + (\lambda - \mu)x(\lambda)\;x(\mu), \end{aligned}$$

so
$$x(\lambda) - x(\mu) = (\lambda - \mu)x(\lambda)\;x(\mu)$$

This relation is called the resolvent equation.

Theorem 1:

σ (x) is non-empty.

Proof:

Let f be a functional on A—that is, an element of the conjugate space A*–and define $f(\lambda)$ by $f(\lambda) = f(x(\lambda)$. It is clear that $f(\lambda)$ is a complex function which is defined and continuous on the resolvent set ρ (x). The resolvent equation shows that

$$\frac{f(\lambda) - f(\mu)}{\lambda - \mu} = f(x(\lambda)x(\mu),$$

and it follows from this that

$$\lim_{\lambda - \mu} \frac{f(\lambda) - f(\mu)}{\lambda - \mu} = f\left(x(\mu)^2\right),$$

so $f(\lambda)$ has a derivative at each point of ρ (x). Further,

$$|f(\lambda)| \leq \|f\|\,\|x(\lambda)\|,$$

so $f(\lambda) \to 0$ as $\lambda \to \infty$. We now assume that σ (x) is empty, so that ρ (x) is the entire complex plane. Liouville's theorem from complex analysis allows us to conclude that $f(\lambda) = 0$ for all λ. Since f is an arbitrary functional on A, $x(\lambda) = 0$ for all λ. This is impossible, for no inverse can equal 0, and therefore it cannot be true that σ (x) is empty.

If the reader is surprised by the appearance of Liouville's theorem in such a context, he should recall two facts. A which is a special case of the above result, required the use of the fundamental theorem of algebra. And second, the fundamental theorem of algebra is most commonly proved as a simple consequence of Liouville's theorem. It is therefore only to be expected that some tool from analysis comparable in depth with Liouville's theorem should be necessary for the proof of Theorem A. Now that we know that σ (x) is non-empty, we also know that it is a compact subspace of the complex plane. The number r(x) defined by

$$r(x) = \sup \{|\lambda| : \lambda \varepsilon \sigma (x)\}$$

is called the spectral radius of x. It is clear that $0 \leq r(x) \leq \|x\|$. The concept of the spectral radius will be useful in certain parts of our later work.

We recall that a division algebra is an algebra with identity in which each non-zero element is regular. The most important single consequence of Theorem A is

Theorem 2:

If A is a division algebra, then it equals the set of all scalar multiples of the identity.

Proof:

We must show that if x is an element of A, then x equals $\lambda 1$ for some scalar λ. Suppose, on the contrary, that $x \neq \lambda 1$ for every λ. Then $x - \lambda 1 \neq 0$ for every λ, $x - \lambda 1$ is regular for every λ, and therefore $\sigma(x)$ is empty. This contradicts Theorem 1 and completes the proof.

The mapping $\lambda 1 - \lambda$ is clearly an isometric isomorphism of the set of all scalar multiples of the identity onto the Banach algebra C of all complex numbers.

We may therefore identify this set with C; and in terms of this identification, Says that *any Banach algebra which is a division algebra equals C.* This fact is the foundation on which we build the structure theory presented in the next.

It is obvious that C itself which is the simplest of all Banach algebras, is a division algebra, so Theorem B characterizes C as the only Banach algebra with this property. In the next two theorems, we give some other interesting characterizations of C among all possible Banach algebras.

Since 0 is a divisor of zero, it is a topological divisor of zero in every Banach algebra. In the Banach algebra C, 0 is plainly the only topological divisor of zero. Conversely, we have

Theorem 3:

If 0 is the only topological divisor of zero in A, then A = C.

Proof:

Let x be an element of A. Its spectrum $\sigma(x)$ is non-empty, so it has a boundary point λ; and $x-\lambda 1$ is easily seen to be a boundary point of the set S of all singular elements. By Theorem 2, $x - \lambda 1$ is a topological divisor of zero, so it follows from our hypothesis that $x - \lambda 1 = 0$ or $x = \lambda 1$.

The basic link between multiplication in A and the norm is given by the inequality $\|xy\| \leq \|x\|\|y\|$, and when A = C, this inequality; can be reversed. The following result shows to what extent this reversibility is true in general.

Theorem 4:

If the norm in A satisfies the inequality $\|xy\| \leq K\,\|x\|\|y\|$ *for some positive constant K, then A = C.*

Proof:

In the light of Theorem 3, it suffices to observe that the hypothesis here implies that 0 is the only topological divisor of zero.

We next look into the question of what happens to the spectrum of an element x in A when A is enlarged.

Theorem 5:

If A is a Banach subalgebra of a Banach algebra A', then the spectra of an element x in A with respect of A and A' are related as follows : (1) $\sigma_{A'}(x) \subseteq \sigma_A(x)$*; (2) each boundary point of* $\sigma_A(x)$ *is also a boundary point of* $\sigma_{A'}(x)$.

Proof:

If $x-\lambda 1$ is singular in A', then it is certainly singular in A, so (1) is clear. To prove (2), we let λ be a boundary point $\sigma_A(x)$. It is easy to see that $x-\lambda 1$ is a boundary point of the set of singular elements in A, so by Theorem 66 - B, it is a topological divisor of zero in A.

It is therefore a topological divisor of zero in A' as well, so it is singular in A' and λ is in $\sigma_{A'}(x)$. The fact that λ is actually a boundary point of $\sigma_{A'}(x)$ is immediate from (1), so the proof of (2) is complete.

This result shows that in general the spectrum of an element shrinks when its containing Banach algebra is enlarged, and further, that since its boundary points cannot be lost in this process, it must shrink by "hollowing out."

An illuminating example of this phenomenon is provided by the disc algebra A of all complex functions which the defined and continuous on D = $\{z : |z| \leq 1\}$ and analytic in the interior of this set. If f is a function in A, then the maximum modulus theorem from complex analysis implies that

$$\|f\| = \sup\{|f(z)| : |z| \leq 1\}$$

$$= \sup\{|f(z)| : |z| = 1\}$$

This allows us to identify A with the Banach algebra of all the restrictions of its functions to the boundary of D, which is a Banach subalgebra of $A' = \mathcal{C}(\{z:|z| = 1\})$. If we now consider the element f in A defined by f(z) = z, then it is easy to see that $\sigma_A(f)$ equals D and that $\sigma_{A'}(f)$ equals the boundary of D.

The Formula for the Spectral Radius

Let x be an element in our general Banach algebra A, and consider its spectral radius r(x), which is defined by

$$r(x) = \sup\{|\lambda| : \lambda \varepsilon \sigma_A(x)\}.$$

Now let A' be the Banach subalgebra of A generated by x, that is, the closure of the set of all polynomials in x. E shows that r(x) has the same value if it is computed with respect to A':

$$r(x) = \sup\{|\lambda| : \lambda \varepsilon \sigma_{A'}(x)\}$$

This suggests quite strongly that r(x) depends only on the sequence of powers of x. The formula for r(x) is given in Theorem A below, and our purpose in this section to prove it. It is convenient to begin with the following preliminary result.

Lemma:

$$\sigma(x^n) = \sigma(x)^n.$$

Proof:

Let λ be a non-zero complex number and $\lambda_1, \lambda_2, \ldots, \lambda_n$ its distinct nth roots, so that

$$x^n - \lambda 1 = (x - \lambda_1 1)(x - \lambda_2 1) \ldots (x - \lambda_n 1).$$

The statement of the lemma follows easily from the fact that $x^n - \lambda 1$ is singular $\Leftrightarrow x - \lambda_i 1$ is singular for at least one i.

Theorem:

$$r(x) = \lim \|x^n\|^{1/n}.$$

Proof:

Our lemma shows that $r(x^n) = r(x)^n$, and since $r(x^n) \leq \|x^n\|$, we have $r(x)^n \leq \|x^n\|$ or $r(x) \leq \|x^n\|^{1/n}$ for every n. To conclude the proof, it suffices to show that if a is any real number such that $r(x) < \alpha$, then $\|x^n\|^{1/n} \leq \alpha$ for all but a finite number of n's, and this we now do.

A and our work in Sec. 67 that if $|\lambda| > \|x\|$, then

$$x(\lambda) = (x-\lambda 1)^{-1} = \lambda^{-1}\left(\frac{x}{\lambda}-1\right)^{-1}$$

$$= -\lambda^{-1}\left(1-\frac{x}{\lambda}\right)^{-1}$$

$$= -\lambda^{-1}\left[1+\sum_{n=1}^{\infty}\frac{x^n}{\lambda^n}\right]. \quad (1)$$

If f is any functional of A, then (1) yields

$$f(x(\lambda)) = -\lambda^{-1}\left[f(1)+\sum_{n=1}^{\infty} f\left(\frac{x^n}{\lambda^n}\right)\right]$$

$$= -\lambda^{-1}\left[f(1)+\sum_{n=1}^{\infty} f(x^n)\lambda^{-n}\right] \quad (2)$$

for all $|\lambda| > \|x\|$. We saw in the proof of Theorem 67 - A that f(x(λ) is an analytic function in the region $|\lambda| > r(x)$; and since (2) is its Laurent expansion for $|\lambda| > \|x\|$. We know from complex analysis that this expansion is valid for $|\lambda| > r(x)$. If we now let ∝ be any real number such that r(x) ∝ < ∝, then it follows from the preceding remark that the series $\sum_{n=1}^{\infty} f(x^n/\alpha^n)$ converges, so its terms form a bounded sequence. Since this is true for every f in A*, an application of Theorem 51 - B shows that the elements x^n/α^n form a bounded sequence in A. Thus

$$\left\|\frac{x^n}{\sigma^n}\right\| \le K$$

or $\|x^n\|^{1/n} \le K^{1/n\alpha}$ for some positive constant K and every n. Since $K^{1/n} \alpha \le \alpha$ for every sufficiently large n, we have $\|x^n\|^{1/n} \le \alpha$ for all but a finite number of n's, and the proof is complete.

The applications we make of this formula will appear in the next

THE STRUCTURE OF BANACH ALGEBRAS

The set $\mathcal{C}(X)$ of all bounded continuous complex functions defined on a topological space X is the simplest of the really interesting Banach algebras. Our purpose in this chapter is to prove the famous *Gelfand-Neumark theorem,* which says that every commutative Banach algebra A of a certain type is essentially identical with $\mathscr{C}(X)$ for a suitable compact Hausdroff space X. More precisely, we shall prove that a compact Hausdorff space X can be built out of the inner structure of A, that X is accompanied by a natural mapping of A into $\mathscr{C}(X)$, and that this mapping is one-to-one onto and preserves all the structure assumed to be present in A.

THE GELFAND MAPPING

Let A be an arbitrary commutative Banach algebra. Our first theorem below is the principal source of the source of the structure theory of A, and the remainder of the chapter will be devoted entirely to shaping its consequences into the elegant form of the Gelfand-Neumark theorem.

Theorem 1:

If M is a maximal ideal in A, then the Banach algebra A/M is a division algebra, and therefore equals the Banach algebra C of complex numbers. The natural homeomorphism x →x + M of A onto A/M = C assigns to each element x in A ∝ complex number x(M) defined by

$$x(M) = x + M,$$

and the mapping x → x (M) has the following properties :

(1) $(x + y)(M) = x(M) + y(M);$

(2) $(\propto x)(M) = \propto x(M);$

(3) $(xy)(M) = x(M)y(M);$

(4) $x(M) = 0 \Leftrightarrow x \in M;$

(5) $1(M) = 1;$

(6) $|x(M)| \leq \|x\|.$

Proof:

That A/M is indeed a Banach algebra. Since A contains an identity, M is maximal as a ring ideal and therefore, A/M is a division algebra. We now appeal to conclude that A/M equals C. (Actually, of course, A/M equals the set of all scalar multiples of its own identity, but we identify this set with C in accordance with the remarks following. Finally, properties (1) to (5) are obvious consequences of the nature of the homeomorphism under discussion, and (6) follows from

$$|x(M)| = |x + M| = \|x + M\| = \inf\{\|x + m\| : m \varepsilon M\} \leq \|x\|.$$

If is interesting to observe that proof depends, either directly or indirectly, on virtually every major theorem in the previous chapter. We also note that the ultimate reason for assuming that A is commutative lies in which is definitely not true in the non-commutative case.

The language of Theorem 1 is oriented toward the idea that x(M) is a function of x for each fixed M. The notation, however, suggests that we reverse this point of view and that for each fixed x we regard x(M) as a complex function defined on the set $\mathcal{M}$ of all maximal ideals in A. This is the direction in which we now proceed.

If x is a given element of A, we denote by $\hat{x}$ the function defined on $\mathcal{M}$ by $\hat{x}(M) = x(M)$, and we put $\hat{A} = \{\hat{x} : x \varepsilon A\}$. Our next step is to defined a topology for $\mathcal{M}$ in such a manner that every function in $\hat{A}$ is continuous. The most natural way of doing this is to introduce the weak topology generated by $\hat{A}$. It will be recalled that this is the weakest topology on $\mathcal{M}$ relative to with every function $\hat{x}$ is continuous and that a typical subbasic open set has the form

$$S(\hat{x}, M_0, \in) = \{M : M \varepsilon \mathcal{M} \text{ and } |\hat{x}(M) - \hat{x}(M_0)| < \in\}.$$

We call the topological space $\mathcal{M}$ the space of maximal ideals, or the maximal ideal space, and the mapping $x \to \hat{x}$ of A onto $\hat{A}$ will be referred to as the Gelfand mapping.

We are now in a position to reformulate to extend it, in such a way that the Galfand mapping is displayed as the object of central importance.

Theorem 2:

The Gelfand mapping $x \to \hat{x}$ is a norm-decreasing (and therefore continuous) homeomorphism of A into $\mathcal{C}(\mathcal{M})$ with the following properties:

(1) the image $\hat{A}$ of A is a subalgebra of $\mathcal{C}(\mathcal{M})$ which separates the points of @ and contains the identity of $\mathcal{C}(\mathcal{M})$;

(2) the radical R of A equals the sat of all elements x for which $\hat{x} = 0$, so $x \to \hat{x}$ is an isomorphism $\Leftrightarrow$ A is semi-simple;

(3) an element x in A is regular $\Leftrightarrow$ it does not belong to any maximal ideal $\Leftrightarrow$ $\hat{x}(M) \neq 0$ for every M;

(4) if x is an element of A, then its spectrum equals the range of the function $\hat{x}$ and its spectral radius equals the norm of $\hat{x}$, that is, $\sigma(x) = \hat{x}(\mathcal{M})$ and $r(x) = \sup |\hat{x}(M)| = \|\hat{x}\|$.

Proof:

The definition of the topology on $\mathcal{M}$ guarantees that each function $\hat{x}$ is continuous, and part theorem 1 shows that $\hat{x}$ is bounded and that

$\|\hat{x}\| = \sup|\hat{x}(M)| \leq \|x\|$, so $x \rightarrow \hat{x}$ is a norm-decreasing mapping of A into $\mathcal{C}(\mathcal{M})$. The fact that this mapping is a homeomorphism is immediate from parts (1), (2), and (3) of Theorem 1.

Since $x \rightarrow \hat{x}$ is a homeomorphism, $\hat{A}$ is obviously a subalgebra of $\mathcal{C}(\mathcal{M})$. The stated properties of $\hat{A}$ follow readily from parts (4) and (5) of Theorem 1: if $M_1 \neq M_2$, and if (say) x is in M_1 but not in M_2, then $\hat{x}(M_1) = 0$ and $\hat{x}(M_2) \neq 0$; and $\hat{1}(M) = 1$ for every M.

If we recall that R is the intersection of all the M's, then the proof of (2) is easy : we have only to notice that part (4) of Theorem A tells us that $\hat{x}(M) = 0$ for every $M \Leftrightarrow x$ is in every M.

To prove (3), it suffices - in view of part (4) of Theorem A - to show that x is regular $\Leftrightarrow$ it does not belong to any M. It is elementary that a regular element cannot lie in any proper ideal, so we confine our attention to showing that if x is singular, then it does belong to some M. We prove this by observing that the singularity of x implies that $Ax = \{yx : y \varepsilon A\}$ is a proper ideal which contains x and can therefore be imbedded be in a maximal ideal M which also contains x.

Finally, we use (3) to prove (4). By the definition of the spectrum of x, we have $\lambda \, \varepsilon \sigma(x) \Leftrightarrow x - \lambda 1$ is singular $\Leftrightarrow (x - \lambda 1)(M) = 0$ for at least one $M \Leftrightarrow (\hat{x} - \lambda 1(M) = 0$ for at least one $M \Leftrightarrow \hat{x}(M) = \lambda$ for at least one M, so $\sigma(x)$ equals the range of $\hat{x}$. The rest of (4) follows from this statement and the definition of the spectral radius.

We add the final touch to this portion of the theory by showing that @@ is a compact Hausdorff space. The reader will recall that if A* is the conjugate space of A, then its closed unit sphere

$$S^* = \{f : f \, \varepsilon \, A^* \text{ and } \|f\| \leq 1\}$$

is a compact Hausdorff space in the weak* topology.

Our strategy is to identify $\mathcal{M}$, both as a set and as a topological space, with a closed subspace of S*.

A *multiplicative functional* on A is a functional f in the ordinary sense-that is, an element of the conjugate space A*-which is non-zero and satisfies the additional condition $f(xy) = f(x)f(y)$. Theorem A shows that to each M in $\mathcal{M}$ there corresponds a multiplicative functional f_M defined by $f_M(x) = x(M)$. It is important for us to know that $M \rightarrow f_M$ is a one-to-one mapping of $\mathcal{M}$ onto the set of all multiplicative functional. It will facilitate our work if we begin by proving the

Lemma:

If f_1 and f_2 are multiplicative functionals on A with the same null space M, then $f_1 = f_2$.

Proof:

We first show that $f_1 = \alpha f_2$ for some scalar α. Let x_0 be an element of A which is not in M. If x is an arbitrary element of A, it is easy to see that x can be expressed uniquely in form $x = m + \beta x_0$ with m in M (set $\beta = f_2(x)/f_2(x_0)$, put $m = x - \beta x_0$, and observe that $f_2(m) = 0$). It now follows that

$$f_1(x) = f_1(m) + \beta f_1(x_0) = \beta f_1(x_0) = [f_1(x_0)/f_2(x_0)]\ f_2(x),$$

So $f_1 = \alpha f_2$ with $\alpha = f_1(x_0)/f_2(x_0)$. We complete the proof by showing that α equals 1. Let x be an element not in M, so that $f_2(x) \neq 0$. Then $\alpha f_1(x)^2 = \alpha f_2(x^2) = f_1(x^2) = f_1(x^2) = [\alpha^2 f_2(x^2) = \alpha^2 f_2(x^2)$ implies that

$$\alpha^2 = \alpha$$

so $\alpha = 0$ or $\alpha = 1$. Since $f_1 \neq 0$, we conclude that $\alpha = 1$.

We now use this to prove

Theorem 3:

$M \to f_M$ is a one-to-one mapping of the set $\mathcal{M}$ of all maximal ideals in A onto the set of all its multiplicative functionals.

Proof:

The mapping is easily seen to be one-to-one, for if $M_1 \neq M_2$, and if (say) x is in M_1 and not in M_2, then $f_{M1}(x) = 0$ and $f_{M2}(x) \neq 0$. To prove that it is onto, let f be an arbitrary multiplicative functional, and consider its null space $M = \{x : f(x) = 0\}$. It is clear by the assumed properties of that M is a proper closed ideal in A. Furthermore, M is maximal, for if it were properly contained in a proper ideal I, then f(I) would be a non-trivial ideal in C, contrary. Since f and f_M are multiplicative functionals with the same null space, the lemma just proved implies that $f = f_M$, and our proof is complete.

In some of its more concrete applications, this theorem is used to replace the algebraic problem of determining the maximal ideals in A by the analytic problem of finding its multiplicative functionals. Its importance for our current task of showing that $\mathcal{M}$ is a compact Hausdorff space is that it enables us to regard $\mathcal{M}$ as a subset of A*. We can say even more than this, for parts (5) and (6) of Theorem 1 tell us that every multiplicative functional f_M has norm 1, so $\mathcal{M}$ is a subset of the closed unit sphere S*. We recalled earlier that S* is a compact Hausdorff space with respect to the weak* topology, which is the weak topology generated by all the functions f_x defined on S* by $F_x(f) = f(x)$. We now observe that when F_x is restricted to $\mathcal{M}$, it is precisely $\hat{x}$, for

$$F_x(f_M) = f_M(x) = x(M) = \hat{x}(M).$$

Therefore, the topology which $\mathcal{M}$ has as a subspace of S* is exactly its topology as the space of maximal ideals. These considerations permit us to regard $\mathcal{M}$ as a subspace of S*.

Theorem 4:

The maximal ideal space $\mathcal{M}$ is a compact Hausdorff space.

Proof:

In view of the above discussion, it suffices to show that $\mathcal{M}$ is a closed subspace of S*. We accomplish this by forming the subspace X of S* defined by

$$X = \cap_{x,y \varepsilon A} \{f : f \varepsilon S^* \text{ and } f(xy) = f(x)\, f(y)\}.$$

It is evident that X is simply Q@ together with the zero functional; and since we have

$$\begin{aligned} X &= \cap_{x,y \varepsilon A} \{f : f \varepsilon S^* \text{ and } f(xy) - f(x)\, f(y) = 0\} \\ &= \cap_{x,y \varepsilon A} \{f : f \varepsilon S^* \text{ and } F_{xy}(f) - F_x(f)\, F_y(f) = 0\} \\ &= \cap_{x,y \varepsilon A} \{f : f \varepsilon S^* \text{ and } (F_{xy} - F_x F_y)\,(f) = 0\}, \end{aligned}$$

it is easy to see that X is closed in S* (note that each of the sets last written has this property). We next remark that F_1 is continuous on X and equals 1 on $\mathcal{M}$ and 0 at the zero functional. It follows from this that $\mathcal{M}$ is closed in X and is therefore closed in S*.

It is worthy of notice that the topology we imposed on $\mathcal{M}$ is the only one which makes it into a compact Hausdorff space on which all the functions $\hat{x}$ are continuous, for any stronger compact Hausdorff must equal the given one.

In essence, this tells us that every commutative semi-simple Banach algebra is isomorphic to an algebra of continuous complex functions on a suitable compact Hausdorff space.

APPLICATIONS OF THE FORMULA $R(X) = LIM \left\|x^n\right\|^{1/n}$

We continue our study of an arbitrary commutative Banach algebra A and of the Gelfand mapping $X \to \hat{x}$ of A onto the subalgebra $\hat{A}$ of $\mathcal{C}(\mathcal{M})$. Our first theorem provides a simple way of guaranteeing that this mapping preserves norms.

Theorem 1:

The following conditions on A are all equivalent to one another:

(1) $\|x^2\| = \|x\|^2$ for every x ;

(2) $r(x) = \|x\|$ for every x ;

(3) $\|\hat{x}\|\ \|x\|$ for every x .

Proof:

It follows from condition (1) that

$$\|x^4\| = \left\|(x^2)^2\right\| = \|x^2\|^2 = \|x\|^4$$

and, in general, that $\|x^{2k}\| = \|x\|^{2k}$ for every positive integer k. The formula for the spectral radius now yields

$$r(x) = \lim \|x^n\|^{1/n} = \lim \|x^{2k}\|^{1/2k} = \lim \|x\| = \|x\|,$$

so (1) implies (2). The fact that (2) implies (1) is immediate from

$\|x^2\| = r(x^2) = r(x)^2 = \|x\|^2$ In view of the equation $r(x) = \|\hat{x}\|$

Our next problem is to devise a way of making sure that the representing algebra $\hat{A}$ comes as close as it can to exhausting $\mathcal{C}(\mathcal{M})$, and we accomplish this by introducing the following property. A is said to be self-adjoint if for each x in A there exists an element y in A such that $\hat{y}(M) = \overline{\hat{x}(M)}$ for every M.

Theorem 2:

If A is self-adjoint, then $\hat{A}$ is dense in $\mathcal{C}(\mathcal{M})$.

Proof:

By part (1) of T know that $\hat{A}$ is a subalgebra of $\mathcal{C}(\mathcal{M})$ which separates the points of $\mathcal{C}$ and contains the identity function. Our hypothesis now tell us that the closure of $\hat{A}$ is a closed subalgebra of $\mathcal{C}(\mathcal{M})$ which separates points, contains the identity function, and contains the conjugate of each of its functions. Theorem 2 (the complex Stone-Weierstrass theorem) shows that this closure equals $\mathcal{C}(\mathcal{M})$, so $\hat{A}$ itself is dense in $\mathcal{C}(\mathcal{M})$.

If we put together the results obtained in the above two theorems, we have,

Theorem 3:

If A is self-adjoint, and if $\|x^2\| = \|x\|^2$ *for every x, then the Gelfand mapping* $x \to \hat{x}$ *is an isometric isomorphism of A onto* $\mathcal{C}(\mathcal{M})$.

Proof:

By Theorem A, the mapping $x \to \hat{x}$ preserves norms. It is therefore an isometric isomorphism of A onto $\hat{A}$, and we see from this that $\hat{A}$ is closed in $\mathcal{C}(\mathcal{M})$. Since $\hat{A}$ is dense in $\mathcal{C}(\mathcal{M})$ by Theorem B, it follows that $\hat{A}$ equals $\mathcal{C}(\mathcal{M})$, and the proof is complete.

This theorem lacks a certain simplicity which it ought to have, for the condition of self-adjointness is rather far removed from the intrinsic structure of A. Our work in the next two sections will remedy this defect and at the same time will establish closer connections with the operator algebras to which we apply our final result.

INVOLUTIONS IN BANACH ALGEBARS

A Banach algebra A is called a Banach *-algebra if it has an involution, that is, if there exists a mapping $x \to x^*$ of A into itself with the following properties :

(1) $(x + y)^* = x^* + y^*$;

(2) $(\alpha x)^* = \overline{\alpha} x^*$;

(3) $(xy)^* = y^*x^*$;

(4) $x^{**} = x$.

It is an easy consequence of (4) that the involution $x \to x^*$ is actually a one-to-one mapping of A onto itself. We also note that $0^* = 0$ and $1^* = 1$, as we see from $0 + x^* = x^*$ $(0 + x)^* = 0^* + x^*$ and $1^* = 11^* = 1^{**}1^* = (11^*)^* = (1^*)^* = 1^{**} = 1$. The element x^* is called the adjoint of x, and a subalgebra of A is said to be self-adjoint if it contains the adjoint of each of its elements. If A' is also a Banach *-algebra, and if f is an isomorphism of A onto A', then f is called a *-isomorphism if it preserves the involution in the sense that $f(x^*) = f(x)^*$

We naturally want the involution in a Banach *-algebra to be linked in some useful way to the norm. The property $\|x^*\| = \|x\|$ clearly implies that the involution is continuous; for if $x_n \to x$, then

$$\|x_n^* - x^*\| = \|(x_n - x)^*\| = \|x_n - x\|$$

Shows that $x_n^* \to x^*$. A much stronger relation between the involution and the norm is given by the condition

$$\|x^*x\| = \|x\|^2$$

and any Banach *-algebra which satisfies it is called a B*-algebra. It is easy to see that we have $\|x^*\| = \|x\|$ in every B*-algebra; for

$$\|x\|^2 = \|x^*x\| \leq \|x^*\|\,\|x\|$$

shows that $\|x\| \leq \|x^*\|$ for every x, so $\|x^*\| \leq \|x^{**}\| = \|x\|$, and therefore $\|x^*\| = \|x\|$. It follows from this that relations $\|x^*x\| = \|x^*\|\,\|x\|$ is also true.

Several of the Banach algebras described in ask Banach *-algebras with respect to natural involutions. If X is any topological space, then $\mathcal{C}(X)$ is clearly a commutative B*-algebra relative to the involution defined by f*(x) $= \overline{f(x)}$. The disc algebra, however, is not, for it f is the function defined by f(z) = z, then f* (z) = $\bar{z}$ is not analytic at any point . If H is a non-trivial Hilbert space, then $\mathcal{B}(H)$ is a B*-algebra with ;the adjoint operation T → T* taken as the involution. Since C*-algebras are the self-adjoint Banach subalgebras of $\mathcal{B}(H)$'s, they too are B*-algebras. Finally, the group algebra $L_1(G)$ of a finite group G is a Banach*-algebra with respect to the involution defined by $f^*(g_i) = \overline{f(g_i^{-1})}$, and it is easy to see that $\|f^*\| = \|f\|$.

It should be reasonably clear that Banach*-algebras (and especially Banach*-algebras) are modeled along lines suggested by $\mathcal{B}(H)$. We have already called the element x* in such an algebra the adjoint of x. By analogy, we say that x is self-adjoint if x = x*, normal if xx* = x*x, and a projection if x = x* and $x^2 = x$.

Theorem:

If x is a normal element in a Banach-algebra, then* $\|x^2\| = \|x\|^2$.

Proof:

It is obvious that $\|x^2\| \leq \|x\|^2$. The inequality in the other direction is a consequence of the following computation :

$$\|x^*\|^2\|x\|^2 = (\|x^*\|\,\|x\|)^2 = \|x^*x\|^2 = \|(x^*x)^*x^*x\| = \|x^*xx^*x\|$$

$$= \|x^*x^*xx\| = \|(x^*)^2x^2\| = \|(x^2)^*x^2\| = \|(x^2)^*\|\,\|x^2\|$$

$$= \|(x^2)^*\|\;\|x^2\| \leq \|x^*\|^2\,\|x^2\|.$$

This result suggests more strongly than ever that there are close connections between B*-algebras and algebras of operators on Hilbert spaces. We describe the true state of affairs in this matter at the end of the next section.

THE GEFLAND-NEUMARK THEOREM

We are now in a position to give Theorem 3 its final form.

Theorem 1:

If A is a B-algebra, then the Gelfand mapping $x \to \hat{x}$ is an isometric *-isomorphism of A onto the commutative B*-algebra $\mathcal{C}(\mathcal{M})$.*

Proof:

Since A is commutative, each of its elements is normal, and it follows that $\|x^2\| = \|x\|^2$ for every x. By now suffices to show that $\hat{x}^*(M) = \overline{\hat{x}(M)}$ for each x in A and M in $\mathcal{M}$.

Our first step is to prove that if x is self-adjoint, then $\hat{x}(M)$ is real for every M. We assume the contrary, namely, that there exists an M such that $\hat{x}(M) = \sigma + i\beta$ with $\beta \neq 0$. Since x ;is self-adjoint,

$$y = (x - \sigma l)/\beta$$

is also self-adjoint. We further note that $\hat{y}(M) = i$, so $y - il$ is in M. It is obvious from the properties of the involution in A that

$$M^* = \{m^* : m \varepsilon M\}$$

is a maximal ideal; and since it contains $(y - il)^* = y + il$, we see that $\hat{y}(M^*) = - i$. If K is any positive number, then

$$(y - iKl)\ (M^*) = - i(1 + K)$$

and $(y + iKl)\ (M) = - i(l + K)$. It follows from this that $1 + K \leq \|y - iKl\| \leq \|y - iKl\|$ and, similarly, that $1 + K \leq \|y + iKl\|$. On multiplying these two inequalities, we obtain

$$(1+K)^2 \leq \|y - iKl\|\ \|y + iKl\| = \|(y + iKl)^*\| \|y + iKl\|$$

$$= \|(y + iKl)^*\ (y + iKl)\| = \|(y - iKl)(y + iKl)\| = \|\ \| \|y + iKl\|$$

$$= \|y^2 + k^2 l\| \leq \|y^2\| + K^2$$

so $1 + 2K \leq \|y^2\|$. Since K is arbitrary, this is impossible, and this portion of our proof is complete.

We now conclude the proof by showing that if x is any element of A, then $\hat{x}^*(M) = \overline{\hat{x}(M)}$ for every M. It is clear that $y = (x + x^*)/2$ and $z =$

$(x - x^*)/(2i)$ are self-adjoint, and that $x = y + iz$; and therefore, by the result of the above paragraph, we have

$$\hat{x}^*(M) = (y - iz)(M) = \hat{y}(M) - i\hat{z}(M) = \overline{\hat{y}(M)} - \overline{i\hat{z}(M)}$$
$$= \overline{\hat{y}(M) + i\hat{z}(M)} = \overline{\hat{x}(M)}.$$

We already know that if X is any compact Hausdorff space, then $\mathscr{C}(X)$ is a commutative B*-algebra. The theorem just proved—it is called the *Gelfand-Neumark representation theorem*—tells us that commutative B*-algebras are simply abstract $\mathscr{C}(X)$'s, in the sense that every such algebra is abstractly identical with $\mathscr{C}(X)$ for a suitable compact Hausdorff space X.

There is another Gelfand-Neumark theorem of great interest, which applies to arbitrary B*-algebras. We observed in the previous section that every C*-algebra. The converse of this is also true, for if A is a B*-algebra, then exists a Hilbert space H with the property that A is isometrically *-isomorphic to a C*-algebra of operators on H. General B*-algebras are therefore abstract C*-algebras. The proof of this theorem evidently requires that a suitable Hilbert space be constructed out of the given structure of A. The details of this construction are beyond the scope of this book, and we content ourselves with merely stating the facts.

Some Special Commutative Banach Algebras

Our discussion foreshadowed a generalized form of the spectral theorem, and the principal purpose of the present chapter is to formulate and prove this result. We begin with some additional material relating to Banach algebras of continuous functions. In particular, we keep the promise made in showing that the Stone-Cech compactifcation of a completely regular space is essentially unique.

IDEALS IN $\mathscr{C}(X)$ AND THE BANACH-STONE THEOREM

Let X be a compact Hausdorff space, and consider the commutative B*-algebra $\mathscr{C}(X)$. If $\mathscr{M}$ is the space of maximal ideals in $\mathscr{C}(X)$, then the developments of the previous chapter lead us to expect that $\mathscr{M}$ can be identified with X and that the Gelfand mapping is the identity mapping of $\mathscr{C}(X)$ onto itself.

In order to substantiate this conjecture, we begin by observing that to each point x in X there corresponds a proper ideal M_x in $\mathscr{C}(X)$, defined by

$$M_x = \{f : f \varepsilon \mathscr{C}(X) \text{ and } f(x) = 0\}.$$

M_x is easily seen to be maximal and is thus an element of $\mathscr{M}$, for it is the null space of the multiplicative functional f_{Mx} defined by $f_{Mx}(f) = f(x)$, which assigns to each function in $\mathscr{C}(X)$ its value at x. Since X is compact

Hausdorff, and therefore normal, Urysohn's lemma tells us that for each point $y \neq x$ there exists a function f in $\mathscr{C}(X)$ such that $f(x) = 0$ and $f(y) \neq 0$. This shows that $x \rightarrow M_x$ is a one-to-one mapping of X into $\mathscr{M}$. Our next step is to prove that this mapping is onto, and for this it clearly suffices to show that if M is any maximal ideal in $\mathscr{C}(X)$, then there exists a point in X at which every function in M vanishes. We assume the contrary, namely, that for each point x in X there exists a function f in M such that $f(x) \neq 0$. Since f is continuous, x has a neighbourhood at on point of which f vanishes. We now vary x to obtain an open cover for X, and we used compactness to infer that this open cover has a finite subcover. Let $f_1, f_2, ..., f_n$ be the corresponding functions in M. M is an ideal, so the function $g = \sum_{i=1}^{n} f_i\bar{f}_i = \sum_{i=1}^{n} |f_i|^2$ is also in M; and by the manner of its construction, it clearly has the property that $g(x) > 0$ for every x. It follows that g is a regular elemer.. of $\mathscr{C}(X)$, and this contradicts the fact that it lies in ;the proper ideal M. We therefore conclude that $x \rightarrow M_x$ is a one-to-one mapping of X onto $\mathscr{M}$.

These considerations enable us to identify the set $\mathscr{M}$ with the set X, and in terms of this identification, we regard X and $\mathscr{M}$ as two possibly different compact Hausdorff spaces built on the same underlying set of points. By our work in the previous chapter, we know that the Gelfand mapping $f \rightarrow \hat{f}$ is a one-to-one mapping of $\mathscr{C}(\mathscr{M})$. If we use the notation established there, then we find that

$$\hat{f}(M_x) = f(M_x) = fM_x(f) = f(x),$$

so $\hat{f} = f$ and $\mathscr{C}(\mathscr{M}) = \mathscr{C}(X)$. We now recall that any compacted Hausdorff topology on a non-empty set is uniquely determined as the weak topology generated by the set of all its continuous complex functions. If follows from this that X and $\mathscr{M}$ are equal as topological spaces.

We summarize the results of this discussion in

Theorem 1:

Let X be a compact Hausdorff space and $\mathscr{M}$ the space of maximal ideals in the commutative B-algebra $\mathscr{C}(X)$. Then to each point x in X there corresponds a maximal ideal M_x defined by*

$$M_x = \{f : f \varepsilon \mathscr{C}(X) \text{ and } f(x) = 0\},$$

and $x \rightarrow M_x$ is a one-to-one mapping of X onto $\mathscr{M}$ If this mapping is used to identify $\mathscr{M}$ with X, then $\mathscr{M}$ and X are equal as topological spaces, $\mathscr{C}(X)$ equals $\mathscr{C}(X)$, and the Gelfand mapping $f \rightarrow \hat{f}$ is the identify mapping of $\mathscr{C}(X)$ onto itself.

The main idea of this theorem is that the maximal ideals in @(X) correspond in a natural way to the points of X. Our next step is to extend this idea and to obtain a similar characterization of the proper closed ideals in $\mathcal{C}(X)$.

We again consider a compact Hausdorff space X, and we begin our discussion with the observation that to each non-empty closed subset F of X there corresponds a proper closed ideal I(F) in $\mathcal{C}(X)$, defined by

$$I(F) = \{f : f \in \mathcal{C}(X) \text{ and } f(F) = 0\}.$$

If x is any point not in F, then it follows from the complete regularity of X that there exists a function f in $\mathcal{C}(X)$ such that $f(x) \neq 0$ and $f(F) = 0$. This shows that $F \rightarrow I(F)$ is a one-to-one mapping of the class of all non-empty closed subsets of X into the set of all proper closed ideals in $\mathcal{C}(X)$. We shall prove that this mapping is onto, that is, that every proper closed ideal in $\mathcal{C}(X)$ arises in this way from some F.

Let I be a proper closed ideal in $\mathcal{C}(X)$. WE may assume that I is not the zero ideal, for this ideal clearly arises from the full space X. We define F by

$$F = \{x : f(x) = 0 \text{ for every } f \in I\}.$$

It is easy to see that F is a proper closed subset of X; and since I is contained in some maximal ideal, it follows that F is non-empty. Our task is to prove that I(F) = I, and since it is obvious that $I \subseteq I(F)$, the real problem is to prove that $I(F) \subseteq I$. If f is any function which vanishes on F, we must show that f lies in I. We may evidently assume that $f \neq 0$, so that $\{x : f(x) = 0\}$ is a proper subset of X.

In the first part of our proof, we assume that f vanishes on some open set G which contains F. Since $f \neq 0$, G' is non-empty and is thus a compact subspace of X. For each point x in G', there exists a function g in I such that $g(x) \neq 0$. The technique used in the proof of Theorem A can now be applied again, to yield a finite number of functions $g_1, g_2, \ldots, g_n$ in I with the property that at least one is non-zero at every point of G'. We next define a function g_0 by $g_0 = \sum_{i=1}^{n} g_i\bar{g}_i = \sum_{i=1}^{n} |g_i|^2$, and we observe that g_0 is in I and that $g_0(x) > 0$ for every x in G'. By the Tietze extension theorem, the function whose values on G' are given by $1/g_0(x)$ can be extended to a function h in $\mathcal{C}(X)$. It is easily seen that g_0h is in I, that it equals 1 on G', and that $f = fg_0h$, so f is in I.

We now turn to the general case. For each $\epsilon > 0$, the sets K and L defined by $K = \{x : |f(x)| \leq \epsilon/2\}$ and $L = \{x : |f(x)| \geq \epsilon\}$ are disjoint closed subsets of X. K is clearly non-empty, and since $f \neq 0$, L is also non-empty for every sufficiently small ϵ. We assume that ϵ has been chosen at least this small, so that K and L constitute a disjoint pair of closed subspaces of X.

By Urysohn's lemma, there exists a function h in $\mathcal{C}(X)$ such that g(K) = 0, g(L) = 1, $0 \le g(x) \le 1$ for every x. We now define a function h in $\mathcal{C}(X)$ by h = fg, and we note that

$$\|f-h\| = \|f(1-g)\| \le \epsilon$$

It is evident that h vanishes on the set $G = \{x : |f(x)| < \epsilon/3\}$; and since G is an open set which contains F, it follows from the preceding paragraph that h is in I.

This shows that for every sufficiently small positive number ϵ there exists a function h in I such that $\|f-h\| \le \epsilon$, and since I is closed, we conclude that f is in I.

We give the following formal statement of our result.

Theorem 2:

Let X be a compact Hausdorff space. Then to each non-empty closed set F in X there corresponds a proper closed ideal I(F) in $\mathcal{C}(X)$, defined by $I(F) = \{f : f \in \mathcal{C}(X) \text{ and } f(F) = 0\}$; and further, $F \to I(F)$ is a one-to-one mapping of the class of all non-empty closed subsets of X onto the set of all proper closed ideals in $\mathcal{C}(X)$.

As an easy consequence of this, we have

Theorem 3:

If X is a compact Hausdorff space, then every closed ideal in $\mathcal{C}(X)$ is the intersection of the maximal ideals which contain it.

Proof:

Since the intersection of the empty set of maximal ideals is $\mathcal{C}(X)$ itself, we may confine our attention to a proper closed ideal I. By Theorem B, I = I(F) for some non-empty closed set F.

It is clear that the maximal ideals which contain I are precisely those associated with the points of F. It therefore suffices to observe that a function in $\mathcal{C}(X)$ vanishes on F $\Leftrightarrow$ it vanishes at each point of F.

We have seen in Theorem A that the points and the topology of a compact Hausdorff space X can be recovered from the maximal ideals in $\mathcal{C}(X)$.

Since the maximal ideals in $\mathcal{C}(X)$ are objects of a purely algebraic nature, it follows that the compact Hausdorff space X is fully determined, both as a set and as a topological space, by the algebraic structure of $\mathcal{C}(X)$. These observations lead us directly to

Theorem 4:

(the Banach-stone Theorem). Two compact Hausdorff spaces X and Y are homeomorphic ⇔ their corresponding function algebras $\mathcal{C}(X)$ and $\mathcal{C}(Y)$ are isomorphic.

THE RADICAL AND SEMI-SIMPLICITY

Our final preliminary task is to reach a clear understanding of what is meant by the statement that our Banach algebra A is semi-simple. For this, it is necessary to give an adequate definition of the radical of A, and this in turn depends on a detailed analysis of its ideals.

We recall that an ideal in A was defined in Sec. 45 to be a subset I with the following three properties :]

(1) I is a linear subspace of A;

(2) $i \in I \Rightarrow xi \in I$ for every element $x \in A$;

(3) $i \in I \Rightarrow ix \in I$ for every element $x \in A$.

If I is assumed only to satisfy conditions it is called a left ideal (or a right ideal). For the sake of clarity and emphasis, an ideal in our previous sense—one which satisfies all three of these conditions—is often called a two-sided ideal. In the commutative case, of course, these three concepts coincide with one another.

The properties of the ideals in A are closely related to the properties of its regular and singular elements. In our work so far, the statement that an element x in A is regular has meant that there exists an element y such that $xy = yx = 1$. For our present purposes, it is useful to refine this notion slightly, as follows. We say that x is left regular if there exists an element y such that $yx = 1$; and if x is not left regular, it is called left singular. The terms right regular and right singular are defined similarly. If x is both left regular and right regular, so that there exist elements y and z such that $yx = 1$ and $xz = 1$, then the relation

$$y = y1 = y(xz) = (yx)z = 1z = z$$

shows that x is regular in the ordinary sense and that $x^{-1} = y = z$.

The concept of maximality for two-sided ideals was introduced by analogy, we define a maximal left ideal in A to be a proper left ideal which is not properly contained in any other proper left ideal. A straightforward application of Zorn's lemma shows a that any proper left ideal can be imbedded in a maximal left ideal; and since the zero ideal {0} is a proper left ideal, maximal left ideals certainly exist. We now define the radical R of A to be the intersection of all its maximal left ideals. It will be convenient to abbreviate this definition by writing $R = \cap$ MLI. R is clearly a proper left ideal.

These ideas can be formulated just as easily for right ideals as for left ideals, and there is no reason for giving preference to either side over the other. The purpose of the following chain of lemmas is to show that R is also the intersection of all the maximal right ideals in A, that is, that $R = \cap$ MRI.

Lemma 1:

If r is an element of R, then 1 – r is left regular.

Proof:

We assume that 1 – r is left singular, so that

$$L = A\,(1 - r) = \{x - xr : x \varepsilon A\}$$

is a proper left ideal which contains 1 – r. We next imbed L in a maximal left ideal M, which of course also contains 1 – r. Since r is in R, it is also in M, and therefore 1 = (1 – r) + r is in M. This implies that M = A, which is a contradiction.

Lemma 2:

If r is element of R, then 1 – r is regular.

Proof:

By the lemma just proved, there exists an element s such that s(1 – r)=1, so s is right regular and s = 1 – (–s)r. The fact that R is a left ideal implies that (–s) r is in R along with r, and another application of the preceding lemma shows that 1 – (–s)r = s is left regular. Since s is both left regular and right regular, it is regular with inverse 1 – r, so 1 – r is also regular.

Lemma 3:

If r is an element of R, then 1 – xr is regular for every x.

Proof:

R is a left ideal, so xr is in R and the statement follows from the lemma just proved.

Lemma 4:

If r is an element of A with the property that 1 – xr is regular for every x, then r is in R.

Proof:

We assume that r is not in R, so that r is not in some maximal left ideal M. It is easy to see that the set

$$M + Ar = \{m + xr : m \varepsilon M \text{ and } x \varepsilon A\}$$

is a left ideal which contains both M and r, so M + Ar = a and

$$m + xr = 1$$

for some m and x. It now follows that 1 – xr = m is a regular element in M, and this is impossible, for no proper ideal can contain any regular element.

The effect of these lemmas is to establish the equality of two sets :

$$\cap \text{ MLI} = \{ r : 1 - xr \text{ is regular for every } x \}. \qquad \text{...(1)}$$

Precisely the same arguments, when applied to maximal right ideals, show that

$$\cap \text{ MRI} = \{ r : 1 - xr \text{ is regular for every } x \}. \qquad \text{...(2)}$$

We now prove that all four of these sets are the same by showing that the two sets on the right of (1) and (2) are equal to one another. By symmetry, it evidently suffices to prove the

Lemma 5:

If 1 – xr is regular, then 1 – rx is also regular.

Proof:

We assume that 1 – xr is regular with inverse

$$s = (1 - xr)^{-1}.$$

This means, of course, that (1 – xr)s = s(1 – xr) = 1. We leave it to the reader to show, by a simple computation, that

$$(1 - rx) + (1 - rsx)\,(1 - rx) = 1,$$

so that 1 – rx is regular with inverse 1 + rsx. (The formula for $(1 - rx)^{-1}$ is less mysterious than it looks, as the reader can see by inspecting the meaningless but suggestive expressions.

$$s = (1 - xr)^{-1} = 1 + xr + (xr)^2 \ ...$$

and $\quad (1 - rx)^{-1} = 1 + rx + (rx)^2 + (rx)^3 + \ldots = 1 + rx + rxrx$

$$+ rxrxrx + \ldots = 1 + r(1 + xr + xrxr + \ldots)\,x = 1 + rsx.)$$

We summarize our results in

Theorem 1:

The radical R of A equals each of the four sets in (1) and (2) and is therefore a proper two-sided ideal.

A is said to be semi-simple if its radical equals the zero ideal {0}, that is, if each non-zero element of A is outside of some maximal left ideal.

It will be observed that the ideas discussed above are purely algebraic in nature. They can be applied not only to our Banach algebra A, but also to any algebra or ring with identity. Our interest, however, is in A, and we now bring to bear upon these notions the results, notably, the fact that the set S of all singular elements in A is closed.

We begin by noting that if l is any ideal in A (left, right, or two-sided), then by the joint continuity of the algebraic operations, its closure $\bar{I}$ is an ideal of the same king. Next, since any proper ideal is contained in the proper closed set S, the closure of any proper ideal is a proper ideal of the same kind. It is an easy step from these facts to

Theorem 2:

Every maximal left ideal in A is closed.

Proof:

If any maximal left ideal L is not closed, then L is a proper subset of the proper left ideal $\bar{L}$; and this cannot happen, for it contradicts the maximality of L.

Taken together, the above two theorems yield

Theorem 3:

The radical R of A is a proper closed two-sided ideal. We shall also need

Theorem 4:

If l is a proper closed two-sided ideal in A, then the quotient algebra A/I is a Banach algebra.

Proof:

A tells us that A/l is a non-trivial complex Banach space with respect to the norm defined by

$$\|x+I\| = \inf\{\|x+i\| : i \varepsilon I\}.$$

Further, A/I is clearly an algebra with identity 1 + I, and

$$\|1+I\| = \inf\{\|1+i\| : i \varepsilon I\} \le \|1\| = 1.$$

The multiplicative inequality for the norm is easily proved as follows:

$$\|(x+I)(y+I)\| = \|xy+I\| = \inf\{\|xy+i\| : i \varepsilon I\}$$

$$\le \inf\{\|(x+i_1)(y+i_2)\| : i_1, i_2 \varepsilon I\}^-$$

$$\le \inf\{\|(x+i_1)\| \, \|(y+i_2)\| : i_1, i_2 \varepsilon I\}$$

$$= \left[\inf \left\{\|x + i_1\| : i_1 \varepsilon I\right\}\right] \left[\inf \left\{\|(y+i_2)\| : i_2 \varepsilon I\right\}\right]$$
$$= \|x + I\| \, \|y + I\|.$$

All that remains is to show that $\|1 + I\| = 1$; and since we already have $\|1 + I\| \leq 1$, this is an immediate consequence of the fact that

$$\|1 + I\| = \left\|(1 + I)^2\right\| \leq \|1 + I\|^2 \text{ implies } 1 \leq \|1 + I\|.$$

As a final result, we state

Theorem 5:

A/R is a semi-simple Banach algebra.

Proof:

It suffices to observe that the natural homeomorphism $x \to x + R$ of A onto A/R induces a one-to-one correspondence between the maximal left ideals in A and those in A/R.

We shall be concerned almost exclusively with commutative Banach algebras. An algebra of this kind is of course much easier to handle than one which is not commutative, for all its ideals are two-sided and its radical is simply the intersection of its maximal ideals. Our reason for studying the general case here is that when it becomes necessary to assume commutativity, as it will in the next section, we want the force of this assumption, and the issues that depend on it, to be quite clear.

THE STONE-CECH COMPACTIFICATION

It is natural to wonder what can be said along the lines in the case of a topological space X which is not necessarily compact Hausdorff. Regardless of the properties of X, we know from our previous work that $\mathcal{C}(X)$ is a commutative B*-algebra, that its maximal ideal space $\mathcal{M}$ is a compact Hausdorff space, and that $x \to M_x$ is a mapping of X into $\mathcal{M}$. Our difficulty is that without restrictions of some kind on X, we know practically nothing about the properties of the mapping $x \to M_x$. If it happens that this mapping is one-to-one and is also a homeomorphism of X onto a subspace of $\mathcal{M}$, then we observe that X is necessarily completely regular. It is therefore reasonable to assume at the outset that X is completely regular, and we shall see that several interesting conclusions follow this hypothesis.

Theorem 1:

Let X be a completely regular space and $\mathcal{M}$ the space of maximal ideals in the commutative B-algebra $\mathcal{C}(X)$. Then the mapping $x \to M_x$ is a* *homeomorphism of X onto a subspace of $\mathcal{M}$. Furthermore, if this mapping is used to identify X with its image in $\mathcal{M}$, then (1) X is a dense subspace of $\mathcal{M}$; (2) each function in $\mathcal{C}(X)$ has a unique extension to a function in $\mathcal{C}(\mathcal{M})$; and (3) if Y is a compact Hausdorff space with the properties of $\mathcal{M}$ stated in (1) and (2), then there exists a homeomorphism of $\mathcal{M}$ onto Y which leaves the points of X fixed.*

Proof:

The fact that $x \rightarrow M_x$ is one-to-one is immediate from the complete regularity of X, so we may identify X as a set with its image in $\mathcal{M}$. The subset X of $\mathcal{M}$ has two topologies : its own, and its relative topology as a subspace of $\mathcal{M}$. The following arguments show that these topologies are equal. We know that the Gelfand mapping $f \rightarrow \hat{f}$ is an isomorphism of $\mathcal{M}(X)$ onto $\mathcal{C}(\mathcal{M})$. Also, just as in the proof we have $\hat{f}(x) = f(x)$ for each f in $\mathcal{C}(X)$ and each x in X. These observations imply that $\mathcal{C}(X)$ is precisely the set of all restrictions to X of functions in $\mathcal{C}(\mathcal{M})$; and since both topologies are completely regular, it follows the weak topology generated by $\mathcal{C}(X)$, so they are equal and X can be regarded as a subspace of $\mathcal{M}$. These observations also show that X is dense in $\mathcal{M}$—for if $\hat{f}$ vanishes on X, then $f = 0$, $\hat{f} = 0$, and $\hat{f}$ also vanishes on $\mathcal{M}$—and that each function f in $\mathcal{C}(X)$ has a unique extension $\hat{f}$ in $\mathcal{C}(\mathcal{M})$. All that remains is to prove (3). We know that $\hat{f} \rightarrow f$ is an isomorphism of $\mathcal{C}(\mathcal{M})$ onto $\mathcal{C}(X)$; and by the assumptions about Ym, the mapping $f \rightarrow f'$, which assigns to each f in $\mathcal{C}(X)$ its extension f' in $\mathcal{C}(Y)$, is an isomorphism of $\mathcal{C}(X)$ onto $\mathcal{C}(Y)$. Thus $\hat{f} \rightarrow f \rightarrow f'$ is an isomorphism of $\mathcal{C}(\mathcal{M})$ onto $\mathcal{C}(Y)$. If x is a point of X, then this isomorphism clearly carries the maximal ideal in $\mathcal{C}(\mathcal{M})$ corresponding to x over to the maximal ideal in $\mathcal{C}(Y)$ corresponding to x ; so by the Banach-Stone theorem, it induces a homeomorphism of $\mathcal{M}$ onto Y which leaves the points of X fixed.

On comparing this result with $\mathcal{M}$ is homeomorphic, in the manner described, to the Stone-Cech compactification $\beta(X)$. In this sense, therefore, $\mathcal{M}$ and $\beta(X)$ can be considered equal to one another, and also to any other compact Hausdodrff space which contains X axs a dense subspace and has the required extension property. In effect, we have shown that the Stone-Cech compactification of a completely regular space X is unique and can equally well be regarded as the maximal ideal of $\mathcal{C}(X)$.

COMMUTATIVE C*-ALGEBRAS

In this final section, we apply the result of the preceding two chapters to the theory of operators on a non-trivial Hilbert space H. We know that $\mathcal{B}(H)$

and all its self-adjoint Banach subalgebras (that is, all C*-algebras of operators on H) are B*-algebras. As a special case of the Gelfand-Neumark theorem we therefore have

Theorem 1:

Let A be a commutative C-algebra of operators on H, and @@ it space of maximal ideals. Then the Gelfand mapping $T \to \hat{T}$ is an isometric *-isomorphism of A onto $\mathcal{C}(\mathcal{M})$.*

Proof:

If $\{T_i\}$ is a non-empty set of operators on H, then the smallest Banach subalgebra of $\mathcal{B}$(H) which contains every T_i is called the Banach subalgebra of $\mathcal{B}$(H) generated by the T_i's.

It is easy to see that this Banach subalgebra of $\mathcal{B}$(H) is the closure of the set of all polynomials in the T_i's. If N is a normal operator on H, then the Banach subalgebra of $\mathcal{B}$(H) generated by N and N* is clearly a commutative C*-algebra, and is called the *commutative C*-algebra generated by N.* We now specialize.

Theorem 2:

Let N be a normal operator on H, and A the commutative C-algebra generated by N. If $\mathcal{M}$ is the space of maximal ideals in A, then the Gelfand mapping $T \to \hat{T}$ is an isometric *-isomorphism of A onto $\mathcal{C}(\mathcal{M})$.*

Proof:

As it stands, this result is only a beginning. In order to exploit it effectively, our first task is to show that the spectrum of an operator in A–which is understood to be its spectrum as an element of $\mathcal{B}$(H)—equals its spectrum as an element of A. In proving this, we shall need the following preliminary fact.

Lemma:

Let X be a compact Hausdorff space and A a Banach subalgebra of $\mathcal{C}(X)$. If f is a real function in A which is regular in $\mathcal{C}(X)$, then it is also regular in A.

Proof:

The range of f is clearly a compact subspace of the real line which does not contain 0. If $\epsilon > 0$ is given, then by the Weierstrass approximation theorem there exists a polynomial p such that $|p(t) - 1/t| < \epsilon$ for every t in f(X). It follows from this that $|p(f(x)) - 1/f(x)| < \in$ for every x, so $\|p(f) - 1/f\| < \in$. Since p(f) is in A and A is closed, we conclude that 1/f is in A.

Theorem 3:

Let A be a commutative C-algebra of operators on H. If an operator T in A is regular in $\mathscr{B}(H)$, then it is also regular in A, and therefore the spectrum of T as an operator on H equals its spectrum as an element of A.*

Proof:

We begin by considering the special case in which T is assumed to be self-adjoint. Let B be the Banach subalgebra of $\mathscr{B}(H)$ generated by T and T^{-1}. Since T and T^{-1} are self-adjoint and commute with one another, it is evident that B is commutative C*-algebra; and if $\mathscr{M}$ is its space of maximal ideals then B is isometrically *-isomorphic to $\mathscr{C}(\mathscr{M})$ and T is represented by a real function in $\mathscr{C}(\mathscr{M})$. $C = A \cap B$ is a Banach subalgebra of B; and is therefore isomorphic to a Banach subalgebra of $\mathscr{C}(\mathscr{M})$. Sine T is in C and is regular in B, our lemma shows that T^{-1} is also in C and therefore lies in A.

We now turn to the general case, in which T is not assumed to be self-adjoint. It is clear that $U = TT^*$ is a self-adjoint operator in A, and since it has an inverse $U^{-1} = (TT^*)^{-1} = (T^*)^{-1} T^{-1} = (T^{-1})^* T^{-1}$ in $\mathscr{B}(H)$, we know from the preceding paragraph that U^{-1} is in A. We now make use of the commutativity of A to write the relation $UU^{-1} = I$ in the form $T(T^*U^{-1})T = I$. This shows that $T^{-1} = T^*U^{-1}$, so T^{-1} lies in A and the proof is complete.

This result tells us, in particular, that if N is a normal operator on H, then its spectrum $\sigma(N)$ equals its spectrum as an element of the commutative C*-algebra generated by N. Our next step is to provide a concrete representation for the space of maximal ideals in this algebra.

Theorem 4:

Let N be a normal operator on H, A the commutative C-algebra generated by N, and $\mathscr{M}$ the space of maximal ideals in A. Then the function $\hat{N}$ in $\mathscr{C}(\mathscr{M})$ which corresponds to N under the Gelfand mapping is a homeomorphism of $\mathscr{M}$ onto $\sigma(N)$.*

Proof:

It follows from Theorem C and part (4) that $\sigma(N)$ is precisely the range of the continuous function $\hat{N}$ defined on $\mathscr{M}$. Since both $\mathscr{M}$ and $\sigma(N)$ are compact Hausdorff spaces, it suffices by Theorem 5 to show that $\hat{N}$ is one-to-one. Let M_1 and M_2 be points of $\mathscr{M}$ such that $\hat{N}(M_1) = \hat{N}(M_2)$. Then we also have

$$\hat{N}^*(M_1) = \overline{\hat{N}(M_1)} = \overline{\hat{N}(M_2)} = \hat{N}^*(M_2),$$

so each of the function $\hat{N}$ and $\hat{N}^*$ takes equal values at M_1 and M_2. Since A is the closure of the set of all polynomials in N and N*, every function in $\mathcal{C}(\mathcal{M})$ is a uniform limit of polynomials in $\hat{N}$ and $\hat{N}^*$, and therefore every function in $\mathcal{C}(\mathcal{M})$ takes equal values at M_1 and M_2. We conclude the proof by observing that since $\mathcal{C}(\mathcal{M})$ separates the points of $\mathcal{M}$, it follows that $M_1 = M_2$.

In accordance with this result, we may identify $\mathcal{M}$ with the compact subspace σ(N) of the complex plane; and when this identification is carried out, it is easy to see that $\hat{N}(z) = z$ for every z in $\mathcal{M}$. We summarize our conclusions in

Theorem 5:

Let N be a normal operator on H with spectrum σ(N), and let A be the commutative C*-algebra generated by N. Then the space $\mathcal{M}$ of maximal ideals in A equals σ(N), and the Gelfand mapping $T \to \hat{T}$ of A onto $\mathcal{C}(\mathcal{M})$ is an isometric *-isomorphism which carries N into the function whose values are given by $\hat{N}(z) = z$ for every z in $\mathcal{M}$.

This theorem has a number of simple consequences, of which the following are only a few:

(1) $N = 0 \Leftrightarrow \sigma(N) = \{0\}$;

(2) N is singular $\Leftrightarrow$ σ(N) contains 0;

(3) $\sigma(N^*) = \overline{\sigma(N)}$

(4) N is self-adjoint $\Leftrightarrow$ σ(N) lies on the real line;

(5) N is unitary $\Leftrightarrow \sigma(N) \subseteq \{z : |z| = 1\}$;

(6) N is a projection $\Leftrightarrow \sigma(N) \subseteq \{0, 1\}$. If it happens that H is finite-dimensional, so that σ(N) consists of a finite number of distinct complex numbers $\lambda_1, \lambda_2, \ldots, \lambda_m$, then we can write

$$\hat{N} = \sum_{i=1}^{m} \lambda_i \hat{P}_i,$$

where $\hat{p}_i$ is the function in $\mathcal{C}(\mathcal{M})$ defined by $\hat{p}_i(\lambda_j) = \delta_{ij}$. It is evident from this that

$$N = \sum_{i=1}^{m} \lambda_i P_i,$$

where the P_i's are non-zero pairwise orthogonal projections in A such that

$\sum_{i=1}^{m} P_i = I$. This is precisely the spectral resolution of N treated in actually contains the finite-dimensional spectral theorem. We therefore have solid grounds for regarding Theorem E as the generalized form of the spectral theorem discussed in the last paragraph of our promises are fulfilled.

Fixed Point Theorems and Some Applications to Analysis

Let t be a continuous mapping of the closed interval [– 1, 1] into itself. Fig. 39 suggests that the graph of f must touch or cross the indicated diagonal, or more precisely, that there must exist a point x_0 in [– 1, 1] with the property that $f(x_0) = x_0$. The proof is easy.

We consider the continuous function F defined on [–1,1] by F (x) = f(x) – x, and we observe that F (–1) ≥ 0 and that F (1) ≤ 0. If now follows from the Weierstrass intermediate value theorem that exists a point x_0 in [– 1, 1] such that $F(x_0) = 0$ or $f(x_0) = x_0$. It is convenient to describe this phenomenon by means of the following terminology. A topology space X is called a *fixed point* space if every continuous mapping f of X into itself has a fixed point, in the sense that $f(x_0) = x_0$ for some x_0 in X. The remarks in the above paragraph show that [– 1, 1] is a fixed point space.

Furthermore, the closed disc $\{(xy) : x_2 + y_2 \leq 1\}$ in the Euclidean plane R^2 is also a fixed point space (for a lucid elementary proof of this. Both of these facts are special cases of

Brouwer's Fixed Point Theorem: The closed unit sphere $S = \{x : \|x\| \leq 1\}$ in R^n is a fixed point space.

There are several proofs of this classic result, but since they all depend on the methods of algebraic topology, we refer the reader to Bers. Brouwer's theorem itself is a special case of

Schauder's Fixed Point Theorem: *Every convex compact subspace of a Banach space is a fixed point space.*

For a proof, together with a discussion of other related results. Schauder's theorem was foreshadowed by the work of Birkh off and Kellogg on existence theorems in analysis. We illustrate the relevance of these ideas to such problems by given a full treatment of Picard's theorem on the existence and uniqueness of solutions of first differential equations.

We begin by considering an arbitrary metric space X with metric d. A mapping T of X into itself is called a contraction if there exists a positive real number r < 1 with the property that d(Tx, Ty) ≤ r d(x,y) for all points x and y in X. It is obvious that such a mapping is continuous. We shall need the following.

Lemma:

If T is a contraction defined on a complete metric space X, then T has a unique fixed point.

Proof:

Let x_0 be an arbitrary point in X, and write $x_1 = Tx_0$,

$$x_2 = T^2x_0 = Tx_1,$$

and, in general, $x_n = T^nx_0 = Tx_{n-1}$. if m < n, then

$$d(x_m,x_n) = d(T^mx_0, T^nx_0) = d(T^mx_0, T^mT^{n-m}x_0)$$
$$\le r_m\, d(x_0, T^{n-m}x_0) = r^m\, d(x_0, x_{n-m})$$
$$\le r^m\, [d(x_0,x_1) + d(x_1, x_2) + \ldots + d(x_{n-m-1},x_{n-m})]$$
$$\le r^m\, [d(x_0,x_1)\, [1 + r + \ldots + r^{n-m-1}]$$
$$< \frac{r^m d(x_0,x_1)}{1-r}.$$

Since r < 1, it is evident from this that $\{x_n\}$ is a Cauchy sequence, and by the completeness of X, there exists a point x in X such that $x_n \to x$.

We now use the continuity of T to infer that x is a fixed point :

$$T_x = T(\lim x_n) = \lim Tx_n = \lim x_n + 1 = x.$$

We conclude the proof by showing that x is the only fixed point. If y is also a fixed point, that is, if Ty = y, then $d(x,y) = d(Tx, Ty) \le r\, d(x,y)$; and since r < 1, this implies that d(x,y) = 0 or y = x.

This result is the key to

Picard's Theorem : If f(x,y) and $\partial f/\partial y$ are continuous in a closed rectangle $R = \{(x,y) : a_1 \le x \le a_2 \text{ and } b_1 \le y \le b_2\}$, and if (x_0,y_0) is an interior point of R, then the differential equation

$$\frac{dy}{dx} = f(x,y) \qquad ...(1)$$

has a *unique solution* y = g(x) which passes through (x_0,y_0).

Proof:

Since f(x,y) and $\partial f/\partial y$ are continuous in R, they are bounded, and consequently there exist constants K and M such that

$$|f(x,y)| \le K \qquad ...(2)$$

and
$$\left|\frac{\partial}{\partial y} f(x,y)\right| \le M \qquad ...(3)$$

for all points (x,y) in R. We next observe that if (x,y_1) and (x,y_2) are in R, then the means value theorem guarantees that

$$\left|f(x,y_1) - f(x,y_2)\right| = \left|y_1 - y_2\right|\left|\frac{\partial}{\partial y} f(x, y_1 + \theta\,(y_2 + y_1))\right| \qquad (4)]$$

for some θ such that 0 < θ 1. It now follows from (3) and (4) that

$$\left|f(x,y_1) - f(x,y_2)\right| \le M\left|y_1 - y_2\right| \qquad ...(5)$$

for all (x,y_1) and (x,y_2) in R.[1]

It is convenient at this replace our problem by an equivalent problem relating to an *integral equation.* If y = g(x) satisfies (1) and has the property that $g(x_0) = y_0$, then integrating (1) from x_0 to x yields

$$g(x) - g(x_0) = \int_{x_0}^{x} f(t, g(t))\, dt$$

or $$g(x) = y_0 + \int_{x_0}^{x} f(t, g(t))\, dt. \qquad (6)$$

Conversely, if y = g(x) satisfies (6), then it clear that $g(x_0) = y_0$, and on differentiating (6) we obtain (1). It therefore suffices to show that the integral equation (6) has a unique solution.

To accomplish this, we choose a positive number ∝ such that M∝ < 1 and the closed rectangle R' determined by $|x - x_0| \le \alpha$ and $|y - y_0| \le K\alpha$ is contained in R. We now let X be the set of all continuous real functions y = g(x) defined on the closed interval $|x - x_0| \le \alpha$ such that Ka. X is clearly a closed subspace of the complete metric space $\mathcal{C}[x_0 - \propto, x_0 + \propto]$ and is therefore itself a complete metric space. Our next step is to consider the mapping T of X into itself defined by Tg = h, where

$$h(x) = y_0 + \int_{x_0}^{x} f(t, g(t))\, dt.$$

The fact that T maps X into itself is evident from (2), for

$$\left|h(x) - y_0\right| = \left|\int_{x_0}^{x} f(t, g(t))\, dt\right| \le K\alpha.$$

Furthermore, it follows from (5) that

$$\left|h_1(x) - h_2(x)\right| = \left|\int_{x_0}^{x} \left[f(t, g_1(t)) - f(t, g_2(t))\right] dt\right|$$

$$\le M\alpha \sup \left|g_1(x) - g_2(x)\right|;$$

and since Ma < 1, this shows that T is a contraction on X. We now appeal to our lemma to conclude that the equation Tg = g has a unique solution.

Since this amounts to saying that the integral equation (6) has a unique solution, our proof is complete.

The ideas in this proof have a much wider scope than might be suspected, and can be applied to establish many other existence theorems in the theory of differential and integral equations.

Continuous Curves and the Hahn-Mazurkiewicz Theorem

A continuous curve is usually thought of as "the path of a continuously moving point," and this rather vague notion is often felt to carry with it the even vaguer attribute of "thinness," or "one-dimensionality." For the case of plane curves, Hordan (in 1887) gave precise expression to this intuitive geometric concept by means of the following definition : if f is a continuous mapping of the closed unit interval I = [0, 1] into the Euclidean plane R^2, then the subset f(I) of R^2 is called a continuous curve. The fame of Jordan's definition rests mainly on Peano's discovery (in 1890) of a continuous curve which passes through every point of a closed square. Curves of this type have come to be called *space-filling curves.*

The first three stages in the construction of a particularly simple example known as Hilbert's space-filling curve. If the square under consideration is $S = \{(x,y) : 0 \leq 1 \text{ and } 0 \leq y \leq 1\}$, then the respective curves are the images of I under continuous mappings f_1, f_2, and f_3 of I into S. The process of constructing these curves can be continued in the same way, and it yields a sequence of continuous mapping f_n of I into S. By the manner in which each curve is constructed from its predecessor, it is clear that the sequence $\{f_n\}$ converges pointwise to a mapping f of I into S; and since this convergence is evidently uniform, f is continuous a continuous curve in the sense of Jordan. Furthermore, each point of S lies in f(I), so f(I) is a space-filling curve.

Peano's discovery of space-filling curves was shock to many mathematicians of the time, for it violated all their preconceived ideas of what a continuous curve ought to be. To a few of the others, however, it presented an opportunity. It suggested the very interesting problem of determining what a continuous curve actually is, or in other words, of finding intrinsic topological properties of a subset X of R^2 which are equivalent to the existence of a continuous mapping of I onto X.

Before describing the solution of this problem, we place it in a wider context by extending Jordan's definition. A topological space X is called a *continuous curve* if X is a Hausdorff space and there exists a continuous mapping of I onto X.[1] We know that I is compact and connected any continuous curve is also compact and connected. In the lemmas below, we give two additional properties which every continuous curve must have.

It is convenient to begin by introducing the following concept. A mapping f of one topological space into another is said to be closed if it carries closed sets into closed sets, that is, if f (F) is closed whenever F is closed. We shall use the fact that a continuous mapping of a compact space into a Hausdorff space is automatically closed.

Lemma:

Every continuous curve is second countable.

Proof:

Let f be a continuous mapping of I onto a Hausdorff space X. We must show that X is second countable, that is, that it has a countable open base. I is a separable metric space, so it has a countable open base $\{B_i\}$ and it is easily seen that the class $\{G_i\}$ of all finite unions of the Bi's is also a countable open base of I. Since I is compact and X is Hausdorff, f is closed, and therefore each set $f(G_i')$ is closed,.

The class of all sets of the form $f(G_i')$ is thus a countable class of open subsets of X, so it suffices to show that these sets constitute an open base for X.

Let x be a point of X with neighbourhood G. The set $f^{-1}(\{x\})$ is closed and is therefore a compact subspace of I with neighbourhood $f^{-1}(G)$. For each point y in $f^{-1}(\{x\})$, there exists a set in $\{G_i\}$ which contains y and

A continuous curve in the sense of this definition is often called a Peano space.

is contained in $f^{-1}(G)$. By the compactness of $f^{-1}(\{x\})$ and the fact that $\{G_i\}$ is closed under the formation of finite unions, there exists a G_i such that

$$f^{-1}(\{x\}) \subseteq G_i \subseteq f^{-1}(G).$$

On taking complements, we obtain

$$f^{-1}(\{x\})' \supseteq G_i' \supseteq f^{-1}(G)' \qquad ...(1)$$

and since the complement of an inverse image equals the inverse image of the complement, we can write (1) in the form

$$f^{-1}(\{x\}') \supseteq G_i' \supseteq f^{-1}(G'). \qquad ...(2)$$

If we now apply f to all members of (2), we get

$$ff^{-1}(\{x\}') \supseteq f(G_i') \supseteq ff^{-1}(G')$$

or $\qquad \{x\}' \supseteq f(G_i') \supseteq G',$

so $\qquad \{x\} \subseteq f(G_i')' \subseteq G,$

and the proof is complete.

Lemma:

Every continuous curve is locally connected.

Proof:

Let f be a continuous mapping of I onto a Hausdorff space X. We must show that X is locally connected. Suffices to show; if C is a component of an open subspace G of X, then C is open.

Let A be a component of $f^{-1}(G)$. Then A is connected, and therefore f(A) is connected in G; and since C is a component of G, we see that f(A) is either disjoint from C or contained in C. It follows from this that $f^{-1}(C)$ is a union of components of $f^{-1}(G)$. Since $f^{-1}(G)$ is open and I is locally, tells us that the components of $f^{-1}(G)$ are open, do $f^{-1}(C)$ is open and $f^{-1}(C') = f^{-1}(C')$ is closed. We conclude the proof by observing that since the mapping f is closed, the set $ff^{-1}(C') = C'$ is closed, so C is open.

The above remarks and lemmas establish the easy half of the following famous characterization of continuous curves.

The Hahn-Mazurkiewicz Theorem 1 topological space X is a continuous curve $\Leftrightarrow$ X is a compact Hausdorff space which is second countable, connected, and locally connected.

For the remainder of the proof, we refer the reader to Wilder Additional discussions of a descriptive and historical nature can be found.

ALGEBRA OF SUBSPACE

Theorem 1:

The intersection of any two subspaces W_1 and W_2 of a vector space V(F) is also a subspace of V(F)

Proof:

Since $0 \in W_1$ and W_2 both therefore $W_1 \cap W_2$ is not empty.

Let $\alpha, \beta \in W_1 \cap W_2$ and $a, b \in F$.

Now $\alpha \in W_1 \cap W_2 \Rightarrow \alpha \in W_1$ and $\alpha \in W_2$.

and $\beta \in W_1 \cap W_2 \Rightarrow \beta \in W_1$ and $\beta \in W_2$.

Since W_1 is a subspace, therefore

$a, b \in F$ and $\alpha, \beta, \in W_1 \Rightarrow a\alpha + b\beta \in W_1$.

Similarly $a, b, \in F$ and $\alpha, \beta, \in W_2 \Rightarrow a\alpha + b\beta \in W_2$.

Now $a\alpha + b\beta \in W_1$, $a\alpha + b\beta \in W_2 \Rightarrow a\alpha + b\beta \in W_1 \cap W_2$.

Thus $a, b, \in F$ and $\alpha, \beta, \in W_1 \cap W_2 \Rightarrow a\alpha + b\beta \in W_1 \cap W_2$.

Hence $W_1 \cap W_2$ is a subspace of V(F).

Note: The union of two subspaces of V(F) may not be a subspace of V(F). For example if **R** be the field of real numbers, then W_1 = {(0, 0, z); z ∈ **R**} are two subspaces of V2(R). We have (0, 9, 3) ∈ W_1 and (0, 5, 0) ∈ W_2.

But (0, 0, 3) + (0, 5, 0) = (0, 5, 3) ∉ $W_1 \cup W_2$ since neither (0, 5, 3) ∈ W_2. Thus $W_1 \cup W_2$ is not closed under vector addition. Hence $W_1 \cup W_2$ is not a subspace of V_2(**R**).

Theorem 2:

Arbitrary intersection of subspaces i.e., the intersection of any family of subspaces of vector space is a subspace.

Proof:

Suppose V (F) be a vector space and let $\{W_t : t \in T\}$ be any family of subspaces of V. And T is an index set such that ∀ t ∈ T, Wt is a subspace of V.

Let $$U = \bigcap_{t \in T} W_t = \{x \in V : x \in W_t \; \forall \; t \in T\}$$

be the intersection of this family of subspaces of V. Then to prove that U is also a subspace of V.

We know that U ≠ ∅, since at least the zero vector 0 of V is in W_t ∀ t ∈ T.

Now let a, b, ∈ F and α, β, be any two elements of $\bigcap_{t \in T} W_t$.

Then α, β ∈ W_t ∀ t ∈ T. Since each W_t is a subspace of V, therefore $a\alpha + b\beta \in W_i$ ∀ t ∈ T. Thus $a\alpha + b\beta \in \bigcap_{t \in T} W_t$.

Thus a, b, ∈ F and α, β, $\in \bigcap_{t \in T} W_t \Rightarrow a\alpha + b\beta \in \bigcap_{t \in T} W_t$.

Hence $\bigcap_{t \in T} W_t$ is a subspace of V(F).

Theorem 3:

The union of two subspace is a subspace if and only if one is contained in the other.

Proof:

Suppose W_1 and W_2 are two subspaces of a vector space V.

Let $W_1 \subseteq W_2$ of $W_2 \subseteq W_1$. Then $W_1 \cup W_2 = W_2$ or W_1.

But W_1, W_2 are subspaces and therefore, $W_1 \cup W_2$ is also a subspace.

Conversely, suppose $W_1 \cup W_2$ is subspace.

To prove that $W_1 \subseteq W_1$ or $W_2 \subseteq W_1$.

Let us assume that W_1 is not a subset of W_2 and W_2 is also not a subset of W1.

Now W_1 is not a subset of $W_2 \Rightarrow \exists\ \alpha \in W_1$ and $\alpha \notin W_2$ ...(i)

and W_2 is not a subset of $W_1 \Rightarrow \exists\ \beta \in W_2$ and $\beta \notin W_2$ ...(ii)

From (i) and (ii), we have

$\alpha \in W_1 \cup W_2$ and $\beta \in W_1 \cup W_2$.

Since $W_1 \cup W_2$ is an subspace, therefore

$\alpha + \beta$ is also in $W_1 \cup W_2$.

But $\quad \alpha + \beta \in W_1 \cup W_2$

$\Rightarrow \quad \alpha + \beta \in W_1$ or W_2.

Suppose $\alpha + \beta \in W_1$. Since $\alpha \in W_1$ and W_2 is a subspace, therefore $(\alpha + \beta) - \alpha = \beta$ is in W_1.

But from (2), we have $\beta \notin W_1$. Thus we get a contradiction. Again suppose that $\alpha + \beta \in W_2$. Since $\beta \in W_2$ and W_2 is subspace, therefore $(\alpha + \beta) - \beta = \alpha$ is in W_2. But from (1), we have a Ï W_2. Thus here also we get a contradiction. Hence either $W_1 \subseteq W_2$ or $W_2 \subseteq W_1$.

Smallest subspace containing any subset of V(F). Let V(F) be a vector space and S be any subset of V. If U is subspace of V containing S and is itself contained in every subspace of V containing S, then U is called the *smallest subspace* of V containing S.

The smallest subspace of V containing S is also called the subspace of V generated or spanned by S and we shall denote it by the symbol {S} or by (S).

It can be easily seen that the intersection of all the subspaces of V(F) containing S is the subspace of V(F) generated by S If {S) = V, then we say that V is spanned by S.

LINEAR DEPENDENCE AND LINEAR INDEPENDENCE OF VECTORS

Linear Dependence

Definition: *Let V(F) be a vector space. A finite set $\{\alpha_1, \alpha_2 ..., \alpha_n\}$ of vectors of V is said to be linearly dependent if there exist scalars a_1, a_2, ..., $a_n \in F$ not all of them 0 (some of they may be zero) such that*

$$a_1\alpha_1 + a_2\alpha_2 + a_3\alpha_3 + ... + a_n\alpha_n = 0.$$

LINEAR INDEPENDENCE

Definition: *Let V(F) be a vector space. A finite set* $\{\alpha_1, \alpha_2 ..., \alpha_n\}$ *of vectors of V is said to be linearly independent if every relation of the form*

$$a_1\alpha_1 + a_2\alpha_2 + ... + a_n\alpha_n = 0,\ a_1 \in F,\ 1 \leq i \leq n$$

$$\Rightarrow a_1 = 0 \text{ for each } 1 \leq i \leq n.$$

Any **infinite set** *of vectors of V is said to be linearly independent if its every finite subset is linearly independent, otherwise it is linearly dependent*

SOLVED EXAMPLES

Example 1:

If $f: U \rightarrow V$ *is an isomorphism of the vector space U into the vector space V, then a set of vectors* $\{f(\alpha_1), f(\alpha_2),..., f(\alpha_r)\}$ *is linearly independent if and only if the set* $\{\alpha_1, \alpha_2,..., \alpha_r\}$ *is linearly independent.*

Solution:

U(F) and V(F) are two vector spaces over the same field F and f is isomorphism of U into V *i.e.*,

$f: U \rightarrow V$ such that

f is 1-1 and $f(a\alpha + b\beta) = af(\alpha) + bf(\beta)$

$\forall\ a, b \in F$ and $\forall\ \alpha, \beta, \in U$.

Let $\{\alpha_1, \alpha_2..., \alpha_r\}$ be a subset of U. First suppose that the vectors $\alpha_1, \alpha_2,..., \alpha_r$ are linearly independent. Then to show that the vectors $f(\alpha_1)$, $f(\alpha_2),..., f(\alpha_r)$ re also linearly independent.

We have

$a_1 f(\alpha_1) + a_2 f(\alpha_2) + ... + a_r f(\alpha_r) = 0,$ where $a_1, a_2,..., a_r \in F$

$\Rightarrow f(a_1\alpha_1 + a_2\alpha_2 + ... + a_r\alpha_r) = 0$ [$\because$ f is a linear transformation]

$\Rightarrow f(a_1\alpha_1 + a_2\alpha_2 + ... + a_r\alpha_r) = f(0)$ [$\because f(0) = 0$]

$\Rightarrow a_1\alpha_1 + a_2\alpha_2 + ... + a_r\alpha_r = 0$ [$\because$ f is 1-1]

$\Rightarrow a_1 = 0, a_2 = 0,..., \alpha_r = 0$ since the vectors $\alpha_1, \alpha_2,..., \alpha_r$ are linearly independent.

Hence the vectors $f(\alpha_1), f(\alpha_2),..., f(\alpha_r)$ are also linearly independent.

Conversely suppose that the vectors $f(\alpha_1), f(\alpha_2),..., f(\alpha_r)$ also linearly independent.

Conversely suppose that the vectors $f(\alpha_1)$ $f(\alpha_2),\ldots, f(\alpha_r)$ linearly independent. Then to show that the vectors $\alpha_1, \alpha_2,\ldots, \alpha_r$ are also linearly independent.

We have

$$a_1\alpha_1 + a_2\alpha_2 + \ldots a_r\alpha_r = 0, \text{ where } a_1, a_2,\ldots, a_r \in F$$

$$\Rightarrow \quad f(a_1\alpha_1 + a_2\alpha_2 + \ldots + a_r\alpha_r) = f(0)$$

$$\Rightarrow \quad a_1 f(\alpha_1) + a_2 f(\alpha_2) + \ldots + a_r f(\alpha_r) = 0$$

[$\because$ f is a linear transformation]

$$\Rightarrow \quad a_1 = 0, a_2 = 0,\ldots, a_r = 0$$

since the vectors $f(\alpha_1), f(\alpha_2),\ldots, f(\alpha_r)$ linearly independent.

Hence the vectors $\alpha_1, \alpha_2,\ldots, \alpha_r$ are also linearly independent.

Example 2:

If V is finite dimensional and f is a homomorphism of V onto V prove that f must be one-one, and so, an isomorphism.

Solution:

Let V (F) be a finite dimensional vector space of dimension n. Let f be a homomorphism of V onto V *i.e.*, f is a linear transformation and f is onto V. To prove that f is one-one.

Let $S = \{\alpha_1, \alpha_2,\ldots, \alpha_n\}$ be a basis of V. We shall first prove that $S' = \{f(\alpha_1), f(\alpha_2),\ldots, f(\alpha_n)$ is also a basis of V. We claim that L (S') = V. The proof is as follows:

Let α be any element of V. We shall show that α can be expressed as a linear combination of $f(\alpha_1), f(\alpha_2), \ldots f(\alpha_n)$. Since f is onto V, therefore $\alpha \in V$ implies that there exists $\beta \in V$ such that $f(\beta) = \alpha$. Now β can be expressed as a linear combination of $\alpha_1, \alpha_2,\ldots, \alpha_n$. Let

$$\beta = a_1\alpha_1 + a_2\alpha_2 + \ldots + a_n\alpha_n.$$

Then
$$\alpha = f(\beta) = f(a_1\alpha_1 + a_2\alpha_2 + \ldots + a_n\alpha_n)$$
$$= a_1 f(\alpha_1) + a_2 f(\alpha_2) + \ldots + a_n f(\alpha_n).$$

Thus α has been expressed as a linear combination of

$$f(\alpha_1), f(\alpha_2),\ldots, f(\alpha_n).$$

Therefore L(S') = V.

Since is of dimension n and S' is a subset of V containing n vectors and L(S') = V, therefore S' must be a basis of V. Therefore each vector in V can be expressed as a linear combination of the vectors belonging to S' and S'

Now, we shall that f is one-one. Let γ, δ be any two elements of V such that $\gamma = c_1\alpha_1 + c_2\alpha_2 + ... + c_n\alpha_n$, $\delta = d_1\alpha_1 + d_2\alpha_2 + ... + d_n\alpha_n$.

We have $f(\gamma) = f(\delta)$

$\Rightarrow \quad f(c_1\alpha_1 + c_2\alpha_2 + ... + c_n\alpha_n) = f(d_1\alpha_1 + d_2\alpha_2 + ... + d_n\alpha_n)$

$\Rightarrow \quad c_1 f(\alpha_1) + c_2 f(\alpha_2) + ... + c_n f(\alpha_n) = d_1 f(\alpha_1) + d_2 f(\alpha_2) + ... + d_n f(\alpha_n)$

$\Rightarrow \quad (c_1 - d_2) f(\alpha_1) + (c_2 - d_2) f(\alpha_2) + ... + (c_n - d_n) f(\alpha_n) = 0$

$\Rightarrow \quad c_1 - d_1 = 0, c_2 - d_2 = 0, ..., c_n - d_n = 0$ since $f(\alpha_1), f(\alpha_2),..., f(\alpha_n)$ are linearly independent

$\Rightarrow \quad c_1 = d_1, c_2 = d_2,..., c_n = d_n \Rightarrow \gamma = \delta.$

$\therefore$ f is one-one.

Hence f is an isomorphism of V onto V.

Example 3(a):

Let f be a linear transformation from a vector space U into a vector space V. If S is a subspace of U, prove that f (S) will be a subspace of V.

Solution:

U (F) and V (F) are two vector spaces over the same field F. The mapping f is a linear transformation of U into V *i.e.*,

$f: U \rightarrow V$ such that

$f(a\alpha + b\beta) = af(\alpha) = bf(\beta) \ \forall \ a, b \in F$ and $\forall \ \alpha, \beta \in U$.

Let S be a subspace of U. Then to prove that f (s) is a subspace of V.

Let $a, b \in F$ and $f(\alpha), f(\beta) \in f(S)$ where $\alpha, \beta \in S$.

Since S is a subspace of U, therefore

$a, b \in F$ and $\alpha, \beta \in S \Rightarrow a\alpha + b\beta \in S$

$\Rightarrow \quad f(a\alpha + b\beta) \in f(S)$

$\Rightarrow \quad af(\alpha) + bf(\beta) \in f(S). \quad [\because f(a\alpha + b\beta) = af(\alpha) + bf(\beta)].$

Thus $a, b \in F$ and $f(\alpha), f(\beta) \in f(S)$

$\Rightarrow \quad af(\alpha) + bf(\beta) \in f(S).$

Hence f (S) is a subspace of V.

Example 3(b):

Define linear transformation of a vector space V (F) into a vector space W(F). Show that the mapping

T: (a, b) → (a + 2, b + 3)

*of V_2 (**R**) into itself is not a linear transformation.*

Solution:

Linear Transformation. Definition. Let V(F) and W(F) be two vector spaces over the same field F. A mapping

$$T: V \rightarrow W$$

is called a linear transformation of V into W if

$T(a\alpha + b\beta) = a\,T(\alpha) + b\,T(\beta)\ \forall\ a, b, \in F$ and $\forall\ \alpha, \beta \in V$.

Now to show that the mapping

$$T: (a, b) \rightarrow (a + 2, b + 3)$$

of V_2 (**R**) into itself is not a linear transformation.

Take $\alpha = (1, 2)$ and $\beta = (1, 3)$ as two vectors of V_2(**R**) and $a = 1$, $b = 1$ as two elements of the field **R**.

Then $a\alpha + b\beta = 1\,(1, 2) + 1\,(1, 3) = (1, 2) + (1, 3)$

$= (2, 5)$.

By the definition of the mapping T, we have

$$T(a\alpha + b\beta) = T(2, 5) = (2 + 2, 5 + 3) = (4, 8). \quad ...(1)$$

Also $\quad T(\alpha) = T(1, 2) = (1 + 2, 2 + 3) = (3, 5)$

and $\quad T(\beta) = T(1, 3) = (1 + 2, 3 + 3) = (3, 6)$.

$\therefore\ aT(\alpha) + bT(\beta) = 1\,(3, 5) + 1\,(3, 6)$

$$= (3, 5) + (3, 6) = (6, 11). \quad ...(2)$$

From (1) and (2), we see that

$$T(a\alpha + b\beta) \neq aT(\alpha) + bT(\beta).$$

Hence T is not a linear transformation of V_2 (**R**) into itself.

Example 3(c):

*Show that the set S = {(1, 0, 0), (1, 1, 0), (1, 1, 1)} is a basis of $\mathbf{R}^3$ (**R**) where **R** is the field of real numbers. Hence find the coordinates of the vector (a, b, c) with respect to the above basis.*

Solution:

The dimension of the vector space $\mathbf{R}^3$(**R**) is 3. If the set S is linearly independent, then S will form a basis of $\mathbf{R}^3$(**R**) because S contains 3 vectors. Let x, y, z be scalars in R such that

$$x(1, 0, 0) + y(1, 1, 0) + z(1, 1, 1) = 0 = (0, 0, 0)$$

$\Rightarrow$ $(x + y + z, y + z, z) = (0, 0, 0)$

$\Rightarrow$ $x + y + z = 0, y + z = 0, z = 0$

$\Rightarrow$ $x = 0, y = 0, z = 0$

$\Rightarrow$ the set S is linearly independent.

$\therefore$ S is a basis of $\mathbf{R}^3(\mathbf{R})$.

Now to find the coordinates of (a, b, c) with respect to the ordered basis S. Let p, q, r be scalars in **R** such that

$$(a, b, c) = p(1, 0, 0) + q(1, 1, 0) + r(1, 1, 1)$$

$\Rightarrow$ $(a, b, c) = (p + q + r, q + r, r)$

$\Rightarrow$ $p + q + r = a, q + r = b, r = c$

$\Rightarrow$ $r = c, q = b - c, p = a - b.$

Hence the coordinates of the vector (a, b, c) are (p, q, r)

i.e., $(a - b, b - c, c)$.

Example 3(d):

A system S consisting of n vectors,

$e_1 = (1, 0, 0,..., 0), e_2 = (0, 1, 0,..., 0),..., e_n = (0, 0,..., 0, 1)$ is a basis of $V_n(F)$.

Solution:

First we should show that S is a linearly independent set of vectors. We have proved it in one of the previous examples.

Now we should prove that $L(S) = V_n(F)$. We have always $L(S) \subseteq V_n(F)$. So we should prove that $V_n(F) \subseteq L(S)$ *i.e.*, each vector in $V_n(F)$ is a linear combination of elements of S.

Let $\alpha = (a_1, a_3..., a_n)$ be any vector in $V_n(F)$. We can write

$(a_1, a_2,..., a_n) = a_1(1, 0,..., 0) + a_2(0, 1,..., 0) + ... + a_n(0, 0,..., 0, 1)$

i.e., $\alpha = a_1e_1 + a_2e_2 + ... + a_ne_n$.

Hence S is a basis of $V_n(F)$. We shall this particular basis the standard basis of $V_n(F)$.

Note: The set {(1, 0), (0, 1)} is a basis of $V_2(F)$. The set {(1, 0, 0), (0, 1, 0), (0, 0, 1)} is a basis of $V_3(F)$.

As a particular case a basis of F(F) is the set consisting of only the unit element of F.

Example 4:

Show that the infinite set

$$S = \{1, x, x^3, \ldots, x^n, \ldots\}$$

is a basis of the vector space F [x] of polynomials over the field F.

Solution:

First we should prove that S is a linearly independent set of vectors.

Now we should show that S spans F [x] *i.e.*, each polynomial in F [x] can be expressed as a linear combination of a finite number of elements of S.

Let $f(x) = a_0 + a_1x + a_2x^2 + \ldots + a_tx^t$ be a polynomial of degree t.

Then $f(x) = (a_0)1 + a_1x + a_2x^2 + \ldots a_tx_t$

Hence S is a basis of F [x].

Note: The vector space F [x] has no finite basis. If we take any finite set W of polynomials, we can find a polynomial of degree greater than that of each of them. Such a polynomial cannot at any cost be expressed as a linear combination of the elements of W.

Example 5:

Let V be the set of all pairs (x, y) of real numbers, and let F be the field of real numbers. Define

$$(x, y) + (x_1, y_1) = (x + x_1, y + y_1)$$

$$c\,(x, y) = (cx, y).$$

Show that with these operations V is not a vector space over the field of real numbers.

Solution:

The operation of addition of vectors defined by

$$(x, y) + (x_1, y_1) = (x + x_1, y + y_1),$$

obviously (V, +) is an abelian group.

$(a + b)\,\alpha = a\alpha + b\alpha\ \forall\ a, b \in F$ and $\alpha \in V$ fails.

Again, let $\alpha = (x, y)$ and a, b, $\in$ F. We have

$$(a + b)\,\alpha = (a + b)\,(x, y) = ((a + b)\,x, y)$$
$$= (ax + bx, y).$$

Also $a\alpha + b\alpha = a\,(x, y) + b\,(x, y) = (ax, y) + (bx, y)$

$$= (ax + bx, y + y) = (ax + bx, 2y).$$

Thus in general $(a + b)\,\alpha + a\alpha + b\alpha$

Hence V is not a vector space over the field.

Example 6:

Let R be the field of real numbers and let P_n be the set of all polynomials (of degree at most n) over the field R. Prove that P_n is a vector space over the field R.

Solution:

Here P_n is the set of all polynomials of degree at most n over the field R. The set P_n also includes the zero polynomial. Thus, we have

Thus $P_n = \{f(x): f(x) = a_0 + a_1x + a_2x^2 + ... + a_nx^n,$

Here $a_0, a_1, a_2, ..., a_n \in R\}$.

If $f(x) = a_0 + a_1x + a_2x^2 + ... + a_nx^n$

$g(x) = b_0 + b_1x + b_2x^2 + ... + b_nx^n$

be any two members of P_n, then

$$f(x) + g(x) = (a_0 + b_0) + (a_1 + b_1)\, x + ... + (a_n + b_n)\, x^n.$$

is also a member of P_n because it is also a polynomial of degree at most n over the field **R**.

Thus P_n is closed under for addition of polynomials.

Also we know that addition of polynomials is commutative as well as associative. The zero polynomial 0 is a member of P_n and is identity for addition of polynomials.

Also if $f(x) = a_0 + a_1x + ... + a_nx^n \in P_n$,

then $-f(x) = -a_0 - a_1x - ... - a_nx^n$ Î P_n because it is also a polynomial of degree at most n over the field **R**.

We have $-f(x) + f(x)$ = the zero polynomial.

$\therefore -f(x)$ is the inverse of $f(x)$ for addition of polynomials.

Hence P_n is an abelian group under for addition of composition.

Now if $f(x) = a_0 + a_1x + a_2x^2 + ... + a_nx^n$ is any member of P_n and $c \in$ **R**, we define scalar multiplication $cf(x)$ by the relation

$$cf(x) = ca_0 + (ca_1)\, x + (ca_2)\, x^2 + ... + (ca_n)\, x^n.$$

Obviously $cf(x) \in P_n$ because it is also a polynomial of degree at most n over the field **R**. Thus P_n is closed for scalar multiplication.

Now if $a, b \in$ **R** and $f(x), g(x) \in P_n$, we have

$a\,[f(x) + g(x)] = af(x) + ag(x)$,

$(a + b)\, f(x) = af(x) + bg(x)$

and $(ab)\, f(x) = a\,[bf(x)]$ as can be easily shown.

Also 1 f(x) = f(x) " f(x) $\in P_n$.

Hence P_n is a vector space over the field **R**.

Example 7(a):

Prove that the set of all vectors in a plane over the field of real numbers is a vector space.

Solution:

Let us consider V be the set of all vectors in a plane and let R be the field of real numbers whose elements will be scalars.

Further, let $\alpha, \beta \in V$. If $\alpha = \vec{AB}$ and $\beta = \vec{BC}$, then we define $\alpha + \beta = \vec{AB} + \vec{BC} = \vec{AC}$. Since $\vec{AC}$ is also a vector, therefore $\alpha, \beta \in V \Rightarrow \alpha + \beta \in V$ and thus V is closed for addition of vectors. We know that addition of vectors on the set V is commutative as well as associative. The zero vector $0 = \vec{AA}$ is identity for addition of vectors. If $\alpha = \vec{AB}$, then the vectors $-\alpha = \vec{BA}$ is the additive inverse of α because $-\alpha + \alpha = \vec{BA} + \vec{AB} = \vec{BB} = \vec{O}$

Hence (V, +) is an abelian group.

If $\alpha \in V$ and $m \in R$ *i.e.*, m is any scalar, then the scalar multiplication mα is defined as a vector whose direction is that of α or opposite to that α according as m is + ive or – ive and $|m\alpha| = |m| \cdot |\alpha|$.

Since $m \in \mathbf{R}$, $\alpha \in V \Rightarrow m\alpha \in V$ therefore V is closed for scalar multiplication 3.

Now if a, b $\in$ **R** and α, β, $\in$ V

a (α + β) = aα + aβ, (a + b) α = aα + bα and (ab) α = 0 (bα)

Also if α is any vector and 1 is the multiplicative identity of the field **R**, 1α is in the direction of the vector α and

$|1\alpha| = |1|.|\alpha| = 1.|\alpha| = |\alpha|$.

Hence V is a vector space over the field R.

Example 7(b):

The vector of all real valued continuous (differentiable or integrable) functions defined in some interval [0, 1].

Solution:

Let us consider V denote the set of all real valued continuous functions of x defined in the interval [0, 1].

Then V is a vector space over the field R of real numbers with vector addition and scalar multiplication then we have

$$(f + g)(x) = f(x) + g(x) \ \forall\ f, g \in V$$

and $$(af)(x) = af(x) \ \forall\ a \in R, \forall\ f \in V.$$

As in rings, we should first prove that V is an abelian group with respect to addition composition.

V is closed with respect to scalar multiplication since af is also a real valued continuous function in [0, 1]. Here, we observe that

1. If $a \in R$ and $f, g \in V$, then

$$[a(f + g)](x) = a[(f + g)(x)] = a[f(x) + g(x)] = af(x) + ag(x)$$
$$= (af)(x) + (ag)(x) = (af + ag)(x).$$

$\therefore\ a(f + g) = af + ag.$

2. If $a, b \in R$ and $f \in V$, then

$$[(a + b) f](x) = (a + b) f(x) = af(x) + bf(x) = (af)(x) + (bf)(x)$$
$$= (af + bf)(x).$$

$\therefore\ (a + b) f = af + bf.$

3. If $a, b \in R$ and $f \in V$, then

$$[(ab) f](x) = (ab) f(x) = a[bf(x)] = a[(bf)(x)] = [a(bf)](x):$$

$\therefore\ (ab) f = a(bf).$

4. If 1 is the unity element of R and $f \in V$, then

$$(1f)(x) = 1 f(x) = f(x).$$

$\therefore\ 1f = f.$

Hence V is a vector space over R.

Example 7(c):

*Show that the system of three vectors (1, 3, 2), (1, – 7, – 8), (2, 1, – 1) of V_3(**R**) is linearly dependent.*

Solution:

Let a, b, c be scalars *i.e.*, real numbers such that

$$a(1, 3, 2) + b(1, -7, -8) + c(2, 1, -1) = (0, 0, 0)$$

i.e. $$(a + b + 2c, 3a - 7b + c, 2a - 8b - c) = (0, 0, 0).$$

Then $$a + b + 2c = 0 \qquad ...(1)$$

$$3a - 7b + c = 0, \quad ...(2)$$

and $$2a - 8b - c = 0. \quad ...(3)$$

If the equations (1), (2), (3) posses a non-zero solution, then the given vectors are linearly dependent.

Adding (2) and (3), we get

$$5a - 15b = 0 \quad \Rightarrow a - 3b = 0. \quad ...(4)$$

Multiplying (3) by (2) and adding to (1), we get

$$5a - 15b = 0$$

$$\Rightarrow \quad a - 3b = 0. \quad ...(5)$$

The equations (4) and (5) are the same and give $a = 3b$.

Putting $a = 3b$ in (1), we get $\quad 2c + 4b = 0 \Rightarrow c = -2b$.

If we take $b = 1$, we get $a = 3$, $c = -2$.

Thus $a = 3$, $b = 1$, $c = -2$ is non-zero solution of the equations (1), (2) and (3). Hence the given set of vectors is linearly dependent.

Alternative Method. The coefficient matrix A of the system of equations (1), (2) and (3) is

$$A = \begin{bmatrix} 1 & 1 & 2 \\ 3 & -7 & 1 \\ 2 & -8 & -1 \end{bmatrix}.$$

$$\text{We have } \det A = \begin{vmatrix} 1 & 1 & 2 \\ 3 & -7 & 1 \\ 2 & -8 & -1 \end{vmatrix}$$

$$= \begin{vmatrix} 1 & 0 & 0 \\ 3 & -10 & -5 \\ 2 & -10 & -1 \end{vmatrix}, \text{ by } C_2 - C_1 \text{ and } C_3 - 2C_1$$

$$= 50 - 50 = 0.$$

$\therefore$ rank $A < 3$ *i.e.*, rank $A <$ the number of unknowns a, b, c in the equations (1), (2) and (3).

Therefore the equations (1), (2) and (3) must posses a non-zero solution. Hence the given vectors are linearly dependent.

Example 8:

Find whether the vectors $2x^3 + x^2 + x + 1$, $x^3 + 3x^2 + x - 2$ and $x^3 + 2x^2 - x + 3$ of **R** *[x], the vector space of all polynomials over the real number field, are linearly independent or not.*

Solution:

The zero vector of the vector space **R** [x] is the zero polynomial.

Let a, b, c be scalars (*i.e.*, real numbers) such that

$a\,(2x^3 + x^2 + x + 1) + b\,(x^2 + 3x^2 + x - 2) + c\,(x^3 + 2x^2 - x + 3) = 0$

i.e., zero polynomial.

Then $(2a+b+c)x^2+(a+3b+2c)x^2+(a+b-c)x+a-2b+3c=0$ ...(1)

Equating the coefficients of like powers of x on both sides of (1), we get

$$\begin{aligned} 2a + b + c &= 0, \\ a + 3b + 2c &= 0, \\ a + b - c &= 0, \\ \text{and} \quad a - 2b + 3c &= 0. \end{aligned} \qquad \text{...(2)}$$

The coefficients matrix A of the system of equations (2) is

$$A = \begin{bmatrix} 2 & 1 & 1 \\ 1 & 3 & 2 \\ 1 & 1 & -1 \\ 1 & -2 & 3 \end{bmatrix}$$

$$\sim \begin{bmatrix} 1 & 3 & 2 \\ 2 & 1 & 1 \\ 1 & 1 & -1 \\ 1 & -2 & 3 \end{bmatrix}, \text{ by } R_1 \leftrightarrow R_2$$

$$\sim \begin{bmatrix} 1 & 3 & 2 \\ 0 & -5 & -3 \\ 0 & -2 & -3 \\ 0 & -5 & 1 \end{bmatrix},$$

$$\sim \begin{bmatrix} 1 & 3 & 2 \\ 0 & 1 & 3/5 \\ 0 & -2 & -3 \\ 0 & -5 & 1 \end{bmatrix}, \text{ by } R_2 \to -\frac{1}{5}R_2$$

$$\sim \begin{bmatrix} 1 & 3 & 2 \\ 0 & 1 & 3/5 \\ 0 & 0 & -9/5 \\ 0 & 0 & 4 \end{bmatrix}, \text{ by } R_3 \to R_3 + 2R_2,\ R_4 \to R_4 + 5R_2$$

$$\sim \begin{bmatrix} 1 & 3 & 2 \\ 0 & 1 & 3/5 \\ 0 & 0 & 1 \\ 0 & 0 & 4 \end{bmatrix}, \text{ by } R_3 \to -5/9 R_3$$

$$\sim \begin{bmatrix} 1 & 3 & 2 \\ 0 & 1 & 3/5 \\ 0 & 0 & 1 \\ 0 & 0 & 0 \end{bmatrix}, \text{ by } R_4 \to R_4 - 4R_3$$

which is in echelon form.

∴ rank A = number of non-zero row in the its echelon form

= 3 = number of unknowns a, b, c in the system of equations (2).

Hence the system of equations (2) has the only solution

$$a = 0,\ b = 0,\ c = 0.$$

∴ the given set of vectors is linearly independent.

Example 9:

Determine whether the following set of vectors in V_3 (Q) is linearly dependent or independent, Q being the field of rational numbers:

$$\{(-1, 2, 1,), (3, 1, -2)\}$$

Solution:

Let a, b be scalars (*i.e.*, a, b, Î Q) such that

$$a(-1, 2, 1,) + b(3, 1, -2) = (0, 0, 0)$$

i.e., $(-a + 3b, 2a + b, a - 2b) = (0, 0, 0)$.

Then $-a + 3b = 0,$

$2a + b = 0,$ ...(1)

$a - 2b = 0,$

The coefficient matrix A of the system of equations (1) is

$$A = \begin{bmatrix} -1 & 3 \\ 2 & 1 \\ 1 & -2 \end{bmatrix}.$$

We have $\begin{vmatrix} -1 & 3 \\ 2 & 1 \end{vmatrix} = -1 - 6 = -7 \neq 0.$

Thus, there exists a 2-rowed minor of the matrix A which is not zero. Also the matrix A can have no minor of order greater than 2.

$\therefore$ rank A = 2 = the number of unknowns a and b.

Therefore the equations (1) have the only solution a = 0, b = 0. Hence the given get of vectors is linearly independent.

Note: If we do not want to use the concept of the rank of a matrix to discuss the solutions of the system of equations (1), we can directly say that solving the system of equations (1) we find that the only solution of the system of equations (1) is a = 0, b = 0.

Example 10:

Find a linearly independent subset T of the set

$S = \{\alpha_1, \alpha_2, \alpha_3, \alpha_4\}$

where $\alpha_1 = (1, 2, -1)$, $\alpha_2 = (-3, -6, 3)$,

$\alpha_3 = (2, 1, 3)$,

$\alpha_4 = (8, 7, 7) \in \mathbf{R}^3$

which spans the same space as S.

Solution:

First we observe that $\alpha_2 = -3\alpha_1$ so that the vectors α_1 and α_2 are linearly dependent.

$\therefore$ If $S_1 = \{\alpha_1, \alpha_3, \alpha_4\}$, then the subspace of $\mathbf{R}^3$ spanned by S_1 is the same as that spanned by S.

Now there exists no real number c such that $\alpha_3 = c\alpha_1$. Therefore the vectors α_1 and α_3 are linearly independent.

Let us now see whether the vector α_4 lies in the subspace of $\mathbf{R}^3$ spanned by the vectors α_1 and α_3 or not.

Let $\alpha_4 = a\alpha_1 + b\alpha_3$, where a, b $\in \mathbf{R}$.

Then (8, 7, 7) = a (1, 2, – 1) + b (2, 1, 3).

$$\therefore \quad a + 2b = 8.$$

$$2a + b = 7,$$

$$\text{and} \quad -a + 3b = 7.$$

Solving the first two of these three equations, we get a = 2, b = 3. These values of a and b also satisfy the third equation.

$$\therefore \quad \alpha_4 = 2\alpha_1 + 3\alpha_3.$$

Thus, the vector α_4 has been expressed as a linear combination of α_1 and α_3 so that the subspace of $\mathbf{R}^3$ spanned by the vectors α_1, α_3, and α_4 is the same as that spanned by the vectors α_1 and α_3.

Hence $T = \{\alpha_1, \alpha_3\}$ is a linearly independent subset of S which spans the same subspace of $\mathbf{R}^3$ as is spanned by S.

Example 11(a):

Prove that a set of vectors which contains the zero vector is linearly dependent.

Solution:

Let $S = \{\alpha_1, \alpha_2, ..., \alpha_n\}$ be a set of vectors of the vector space V(F).

Let α_r be equal to zero vector where $1 \leq r \leq n$.

To show that the set S of vectors in linearly dependent.

Obviously,

$0\alpha_1 + 0\alpha_2 + ... + a\alpha_r + 0\alpha_{r+1} + ... + 0\alpha_n = 0$ *i.e.*, zero vector ...(1)

for any non-zero scalar a *i.e.*, for any non-zero element a in the field F.

Since in the linear relation (1) among the vectors $\alpha_1, \alpha_2, ..., \alpha_n$, the scalar coefficient a is not zero, therefore the vectors $\alpha_1, \alpha_2, ..., \alpha_n$ are linearly dependent.

Example 11(b):

The set V of all m × n matrices with their elements as real numbers is a vector space over the field F of real numbers with respect to addition of matrices as addition of vectors and multiplication of a matrix by a scalar as scalar multiplication.

Solution:

We know that V is an abelian group w.r.t. addition of matrices and O the null matrix additive identity.

Again, if $a \in F$ and $\alpha \in V$ (*i.e.*, α is a matrix of the type $m \times n$ with elements as real numbers), then $a\alpha \in V$ because $a\alpha$ is also a matrix of the type $m \times n$ with elements as real numbers. Therefore V is closed with respect to scalar multiplication. Also from our study of matrices we observe that

(i) $a(\alpha + \beta) = a\alpha + a\beta \quad \forall\, a \in F$ and $\forall\, \alpha, \beta \in V$.

(ii) $(a + b)\alpha = a\alpha + b\alpha \quad \forall\, a, b, \in F$ and $\forall\, \alpha \in V$

(iii) $(ab)\alpha = a(b\alpha) \quad \forall\, a, b \in F$ and $\forall\, \alpha \in V$.

(iv) $1\alpha = \alpha\ \forall\, \alpha \in V$ where 1 is the unity element of the field F of real numbers.

Hence V (F) is a vector space.

Note: If V is the set of all m × n matrices with their elements as rational numbers and F is the field of real numbers, then V will not be closed with respect to scalar multiplication. For $\sqrt{3} \in F$ and if $\alpha \in V$, then $\sqrt{3}\, \alpha \notin V$ because the elements of the matrix $\sqrt{3}\, \alpha$ will not be rational numbers. Therefore V (F) will not be vector space.

Example 12:

Let $V = R^2 = \{(a_1, a_2): a_1, a_2 \in R\}$ *and* $F = R$.

Define the addition and scalar multiplication in R^2 *as follows:*

$$(a_1, a_2) + (b_1, b_2) = (a_1 + b_1, a_2 + b_2)$$

Show that $\mathbf{R}^2$ *is a vector space over* **R**.

Solution:

Do yourself.

Example 13:

The vector space of all polynomials over a field F.

Solution:

Let us denote F(x) is the set of polynomial in an indeterminate x over a field F. Then F[x] is a vector space over the field F with respect to addition of two polynomials as addition of vectors and the product of a polynomial by a constant polynomial (*i.e.* by an element of F) as scalar multiplication.

Let $f(x) = \Sigma\, a_i x^i = a_0 + a_1 x + a_2 x^2 + a_3 x^3 + \ldots$

$g(x) = \Sigma\, b_i x^i = b_0 + b_1 x + b_2 x^2 + b_2 x^2 + \ldots$

and $h(x) = \Sigma\, c_i x^i = c_0 + c_1 x + c_2 x^2 + c_3 x^2 + \ldots$

be any arbitrary members of F[x].

Equality of Two Polynomials. We define f(x) = g(x) if and only if $a_1 = b_1$ for each $l = 0, 1, 2, \ldots$.

Addition Composition in F[x]. We define

$$f(x) + g(x) = (a_0 + b_0) + (a_1 + b_1)\, x + (a_2 + b_2)\, x^2 + \ldots$$

$$= \Sigma\, (a_1 + b_1)\, x^i.$$

Since $a_0 + b_0$, $a_1 + b_1$, $a_2 + b_2, \ldots$ are all elements of F, therefore f(x)+ g(x) ∈ F[x] and thus F[x] is closed with respect to addition of polynomials.

Scalar Multiplication in F[x] Over F. If k is any scalar *i.e.*, k ∈ F, we define

$$kf\,(x0 = ka_0 + (ka_1)\, x + (ka_2)\, x^2 + (ka_3)\, x^2 + \ldots$$

$= \Sigma (ka_1) x^i.$

Since $ka_0, ka_1, ka_2,...$ are all elements of F, therefore $kf(x) \in F[x]$ and thus F[x] is closed with respect to scalar multiplication.

Now, we shall show that F[x] is a vector space for these two compositions.

Commutativity of Addition in F[x]. We have

$f(x) + g(x) = (a_0 + b_0) + (a_1 + b_1) x + (a_2 + b_2) x^2 + ...$

$= (b_0 + a_0) + (b_1 + a_1) x + (b_2 + a_2) x^2 + ...$

[$\because$ addition in the field F is commutative]

$= g(x) + f(x).$

Associativity of addition in F[x]. We have

$[f(x) + g(x)] + h(x) = \Sigma (a_i + b_i) x^i + \Sigma c_i x^i$

$= \Sigma [(a_i + b_i)c_i x^i + \Sigma [a_i + (b_i + c_i)] x^i$

$= \Sigma a_i x^i + \Sigma (b_i + c_i) x^i = f(x) + [g(x) + h(x)].$

Existence of additive identity in F[x]. Let 0 denote the zero polynomial over the field F *i.e.*,

$0 = 0 + 0x + 0x^2 + 0x^2 + ...$

Then $0 \in F[x]$ and $0 + f(x) = f(x)$.

$\therefore$ the zero polynomial 0 is the additive identity.

Existence of additive inverse of each member of F[x]. Let $- f(x)$ be the polynomial over the field F defined as

$- (x) = - a_0 + (- a_1) x + (- a_2) x^2 + ...$

$= - a_0 - a_1 x - a_2 x^2 - ... .$

Then $- f(x) \in F[x]$ and we get

$- f(x) + f(x) = 0$ *i.e.*, the zero polynomial.

$\therefore - f(x)$ is the additive inverse of f(x).

Thus F[x] is an abelian group with respect to addition of polynomials.

Now for the operation of scalar multiplication we make the following observations.

1. If $k \in F$, then

$k [f(x) + g(x)] = k [(a_0 + b_0) + (a_1 + b_1) x + (a_2 + b_2) x^2 + ...]$

$= k (a_0 + b_0) + k (a_1 + b_1) x + k (a_2 + b_2) x^2 + ...$

$= (ka_0 + kb_0) + (ka_1 + kb_1) x + (ka_2 + kb_2) x^2 + ...$

$= [ka_0 + (ka_1) x + (ka_2) x^2 + ...] + [kb_0 + (kb_1) x + (kb_2) x^2 + ...]$

$= k\,(a_0 + a_1x + a_2x^2 + ...) + k\,(b_0 + b_1x + b_2x^2 + ...)$

$= kf\,(x) + kg\,(x).$

2. If $k_1, k_2 \in F$, then

$$(k_1 + k_2)\, f(x) = (k_1 + k_2)\, a_0 + [(k_1 + k_2)\, a_1]\, x + [(k_1 + k_2)\, a_2]\, x^2 + ...$$

$$= (k_1a_0 + k_2a_0) + (k_1a_1 + k_2a_1)\, x + (k_1a_2 + k_2a_2)\, x^2 + ...$$

$$= [k_1a_0 + (k_1a_1)\, x + (k_1a_2)\, x^2 + ...] + [k_2a_0 + (k_2a_1)\, x + (k_2a_2)\, x^2 + ...]$$

$$= k_1\,(a_0 + a_1x + a_2x^2 + ...) + k_2\,(a_0 + a_1x + a_2x^3 + ...)$$

$$= k_1\, f(x) + k_2f(x).$$

3. If $k_1, k_2 \in F$, then

$$(k_1k_2)\, f(x) = (k_1k_2)\, a_0 + [(k_1k_2)\, a_1]\, x + [(k_1k_2)\, a_2]\, x^2 + ...$$

$$= k_1\,(k_2a_0) + [k_1\,(k_2a_1)]\, x + [k_1\,(k_2a_2)]\, x^2 + ...$$

$$= k_1\,[k_2a_0 + (k_2a_1)\, x + (k_2a_2\, x^2 + ...]$$

$$= k_1\,[k_2\, f(x)].$$

4. If 1 is the unity element of the field F, then

$$1\, f(x) = (1a_0) + (1a_1)\, x + (1a_2)\, x^2 + ...$$

$$= a_0 + a_1x + a_2x^2 + ... = f(x).$$

Hence F[x] is a vector space over the field F.

Example 14:

(a) One of the most important Banach algebras is the set $\mathscr{C}(X)$ of all bounded continuous complex functions defined on a topological space X. The case in which X is a compact Hausdorff space will have particular significance for our later work. If X has cnly one point, then $\mathscr{C}(X)$ can be identified with the simplest of all Banach algebras, the algebra of complex numbers.

(b) Consider the closed unit disc $D = \{z : |z| \leq 1\}$ in the complex plane. The subset of $\mathscr{C}(D)$ which consists of all functions analytic in the interior of D is obviously a subalgebra which contains the identity. A simple application of Morera's theorem from complex analysis shows that it is closed and is therefore a Banach subalgebra of $\mathscr{C}(D)$. This Banach algebra is called the disc algebra. It has a number of interesting properties, which are, of course, intimately related to the special character of its functions.

Example 15:

(a) If $G = \{g_1, g_2, \ldots, g_n\}$ is a finite group, then its group algebra $L_1(G)$ is the set of all complex functions defined on G. Addition and scalar multiplication are defined pointwise, and the norm by $\|f\| = \sum_{i=1}^{n} |f(g_i)|$. In order to see what underlies the definition of multiplication, it is convenient to regard a typical element f of $L_1(G)$ as a formal sum $\sum_{i=1}^{n} \alpha_i g_i$, where α_i is the value of f at g_i. With this interpretation, we use the given multiplication in G to define multiplication, in $L_1(G)$, as follows:

$$\left(\sum_{i=1}^{n} \alpha_i g_i\right)\left(\sum_{j=1}^{n} \beta_j g_j\right) = \sum_{k=1}^{n} \gamma_k g_k, \tag{1}$$

where $\quad \gamma_k = \sum g_i g_j = g_k \, \alpha_i \beta_j. \qquad (2)$

The meaning of the sum is (2) is that the summation is to be extended over all subscripts i and j such that $g_i g_j = g_k$. In effect, therefore, we formally multiply out the sums on the left of (1), and we then gather together all the resulting terms which contain the same element of G. With these ideas as an intuitive guide, we revert to out first point of view, in which the elements of $L_1(G)$ are functions, and we see that our definition of multiplication can be expressed in the following way. If two functions f and g in $L_1(G)$ are given, then their product, which is denoted by f * g and called their *convolution*, is that function whose value at g_k is

$$(f*g)(g_k) = \sum_{n} g_i g_j \; f(g_i) g(g_j)$$

$$= \sum_{j=1}^{n} f(g_k g_j^{-1}) g(g_j). \tag{3}$$

We note that if each element of G is identified with the function whose value is 1 at that element and 0 elsewhere, then G becomes a subset of $L_1(G)$. Further, multiplication in G agrees with convolution in $L_1(G)$, and the element of $L_1(G)$ which corresponds to the identity in G is an identity for $L_1(G)$. We conclude this description by observing that every element of G has norm 1, so that $\|1\| = 1$, and that the basic norm inequality for a Banach algebra is satisfied:

$$\|f * g\| = \sum_{k=1}^{n} |(f*g)(g_k)|$$

$$= \sum_{k=1}^{n} \left| \sum_{j=1}^{n} f(g_k g_j^{-1}) g(g_j) \right|$$

$$\leq \sum_{k=1}^{n} \sum_{j=1}^{n} \left|f\left(g_k g_j^{-1}\right)\right| \left|g\left(g_j\right)\right|$$

$$= \sum_{j=1}^{n} \sum_{k=1}^{n} \left|f\left(g_k g_j^{-1}\right)\right| \left|g\left(g_j\right)\right|$$

$$= \sum_{j=1}^{n} \left|g\left(g_j\right)\right| \sum_{k=1}^{n} \left|f\left(g_k g_j^{-1}\right)\right|$$

$$= \sum_{j=1}^{n} \left|g\left(g_j\right)\right| \|f\|$$

$$= \|f\| \sum_{j=1}^{n} \left|g\left(g_j\right)\right|$$

$$= \|f\| \|g\|.$$

(b) Let $(G) = \{\ldots, -2, -1, 0, 1, 2, \ldots\}$ be the additive group of integers. Its group algebra $L_1(G)$ is the set of all complex function f defined on G for which converges. The linear operations are defined pointwise, the norm by $\|f\| = \sum_{m=-\infty}^{\infty} |f(n)|$, and the convolution of f and g.

$$(f*g)(n) = \sum_{m=-\infty}^{\infty} f(n-m)g(m).$$

Just as in (a), G is contained in $L_1(G)$ in a natural way, and $L_1(G)$ is a Banach algebra. Any attempt to discuss the group algebra of a non-discrete topological group like the real line must clearly be based on an adequate theory of integration. It should also have available a theory of Banach algebras in which no identity is assumed to be present. These ideas constitute a rich and beautiful field of modern analysis. They are, however, outside the scope of this work. [1]

The Banach algebras described above are many and diverse, and there are yet others which we have not mentioned. Our attention in the following chapters will be centered on $\mathcal{C}(X)$'s and commutative C*-algebras, but the general theory we develop is equally applicable to all. It is worthy of notice that an arbitrary Banach algebra A can be regarded as a Banach subalgebra of $\mathcal{B}(A)$. In a sense, therefore, $\propto$ and its Banach subalgebras include all possible Banach algebras. To see this, we recall that $a \to M_\propto$, where $M_\propto(x) = \propto x$ is an isomorphism of A into $\mathcal{B}(A)$. It is easy to see that M_1 is the identity operator on A, so all that remains is to observe that $\|\alpha\| = \|M_\alpha\|$ for every a (proof: $\|M_\alpha(x)\| = \|\alpha x\| \leq \|\alpha\|\,\|x\|$ shows that $\|M_\alpha\| \leq \|\alpha\|$, and the fact that $\|\alpha\| \leq \|M_\alpha\|$ follows from

$$\|M_\alpha\| = \sup\left\{\|M_\alpha(x)\| : \|x\| \leq 1\right\} \geq \|M_\alpha(1)\| = \|\alpha\|).$$

The mapping $\alpha \to M\alpha$ is thus an isometric isomorphism of A onto a Banach subalgebra of $\mathscr{B}$(A), and it allows us to identify the abstract Banach algebra A with a concrete Banach algebra of operators on A.

Example 16(a):

A field K can be regarded as a vector space over any subfield F of K.

Solution:

We know that K is the set of vectors. Addition of vectors is the addition composition in the field K. Since K is a field, therefore (K, +) is an abelian group. Further the elements of the subfield F constitute the set of scalars. The composition of scalar multiplication is the multiplication composition in the field K. K is a field, therefore $a\alpha \in K \ \forall\ a \in F$ and $\forall\ \alpha \in K$ because both a and α are elements of K. If 1 is the unity element of K, then 1 is also the unity element of the subfield F.

(i) $a(\alpha + \beta) = a\alpha + a\beta \ \forall\ \alpha \in F$ and $\forall\ \alpha, \beta \in K$. (left distributive law).

(ii) $(a + b)\,\alpha = a\alpha + b\alpha \ \forall\ \alpha, \beta \in F$ and $\forall\ \alpha \in K$. (right distributive law).

(iii) $(ab)\,\alpha = a\,(b\alpha) \ \forall\ a, b \in F$ and $\forall\ \alpha \in K$. (associative law)

(iv) $1\alpha = \alpha \ \forall\ \alpha \in K$ and 1 is the unit element of the subfield F.

Since 1 is also the unity element of the field K, therefore $1\alpha = \alpha \ \forall\ \alpha \in K$. Hence K(F) is a vector space.

Example 16(b):

(a) If B is a non-trivial complex Banach space, then the set $\mathscr{B}$(B) of all operators on B is a Banach algebra. We assume that B is non-trivial in order to guarantee that the identity operator is an identity in the algebraic sense.

(b) If we consider a non-trivial Hilbert space H, then $\mathscr{B}$(H), is a Banach algebra. This is a special case of $\mathscr{B}$(B), and it is important to observe that additional structure is present here, namely, the adjoint operation $T \to T^*$.

(c) A subalgebra of $\mathscr{B}$(H) is said to be self-adjoint if it contains the adjoint of each of its operators. Banach subalgebras of $\mathscr{B}$(H)'s which are self-adjoint are called C*-algebras. We shall return to the subject of commutative C*-algebras.

(d) The weak operator topology on $\mathcal{B}(H)$ is the weak topology generated by all functions of the form $T \to (Tx, y)$; that is, it is the weakest topology with respect to which all these functions are continuous. It is easy to see from the inequality $|(Tx,y)-(T_0x,y)| \leq \|T - T_0\| \|x\| \|y\|$ that this topology is weaker than the usual norm topology, so that its closed sets are also closed in the usual sense. A C*-algebra with the further property of being closed in the weak operator topology is called a W*-algebra. Algebras of this kind are also called *rings of operators*, or *von Neumann algebras*.

Example 16(c):

The vector of all ordered n-tuples over a field F.

Solution:

Let F be a field. An ordered set $a = (a_1, a_2, a_3, ..., a_n)$ of n element of F is called an *n-tuple* over F. Let V be the totality of all ordered *n-tuples* over F *i.e.*, let

$$V = \{(a_1, a_2,...,a_n): a_1, a_2, a_3,...,a_n \in F\}.$$

Now, we shall give a vector space structure to V over the field F. Here this we define the following terms:

Equality of two n-tuples. Two elements $\alpha = (a_1, a_2,...,a_n)$ and $\beta = (b_1, b_2,..., b_n)$ of V are said to be equal if and only if

$a_i = b_i$ for each $i = i, 2,..., n$.

Addition composition in V. We define

$\alpha + \beta = (a_1 + b_1, a_2 + b_2,..., a_n + b_n)$

$\forall\ \alpha = (a_1, a_2..., a_n), \beta = (b_1, b_2,..., b_n) \in V.$

Since $a_1 + b_1, a_2 + b_2,..., a_n + b_n$ are all elements of F, therefore $\alpha + \beta \in V$ and thus V is closed with respect to addition of n-tuples.

Scalar multiplication in V over F. We define

$a\alpha = (aa_1, aa_2,..., aa_n)\ \forall\ \alpha \in F, \forall\ \alpha = (a_1, a_2,..., a_n \in V.$

Since $aa_1, aa_2,..., aa_n$ are all elements of F, therefore $a\alpha \in V$ and thus V is closed with respect to scalar multiplication.

Now we shall see that V is a vector space for these two compositions.

Associativity of addition in V. We have

$$(a_1, a_2,..., a_n) + [(b_1,..., b_n) + (c_1, c_2,..., c_n)]$$
$$= (a_1, a_2,..., a_n) + (b_1 + c_1, b_2 + c_2,..., b_n + c_n)$$
$$= (a_1 + [b_1 + c_1], a_2 + [b_2 + c_2],..., a_n + [b_n + c_n])$$

$= ([a_1 + b_1] + c_1, [a_2 + b_2] + c_2,..., [a_n + b_n] + c_n]$

$= (a_1 + b_1, a_2 + b_2,..., a_n + b_n) + (c_1, c_2,..., c_n)$

$= [(a_1, a_2,..., a_n) + (b_1, b_2,..., b_n)] + (c_1, c_2,..., c_n).$

Commutativity of addition in V. We have

$(a_1, a_2,..., a_n) + (b_1, b_2,..., b_n) = (a_1 + b_1, a_2 + b_2,..., a_n + b_n)$

$= (b_1 + a_1, b_2 + a_2,..., b_n + a_n) = (b_1, b_2,..., b_n) + (a_1, a_2,..., a_n).$

Existence of additive identity in V. We have

$(0, 0,..., 0) \in V$. Also if $(a_1, a_2,..., a_n) \in V$, then

$(a_1, a_2,..., a_n) + (0, 0,..., 0) = (a_1 + 0, a_2 + 0,..., a_n + 0)$

$= (a_1, a_2,..., a_n).$

$\therefore$ (0, 0,..., 0) is the additive identity in V.

$(a_1, a_2,..., a_n) \in V$, then $(- a_1, - a_2,..., - a_n) \in V$.

Also we have $(- a_1, - a_2,..., - a_n) + (a_1, a_2,..., a_n)$

$= (- a_1 + a_1, - a_2 + a_2,..., - a_n + a_n) = (0, 0,..., 0).$

$\therefore$ $(- a_1, - a_2,..., -a_n)$ is the additive inverse of $(a_1, a_2,..., a_n)$.

Thus V is an abelian group with respect to addition. Further we observe that

1. If $a \in F$ and $\alpha (a_1, a_2,..., a_n)$, $\beta = (b_1, b_2,..., b_n) \in V$, then $a (\alpha + \beta)$
$= a (a_1 + b_1, a_2 + b_2,..., a_n + b_n)$

$= (a [a_1 + b_1], a [a_2 + b_2],..., a [a_n + b_n])$

$= (aa_1, ab_1, aa_2 + ab_2,..., aa_n + ab_n)$

$= (aa_1, aa_2,..., aa_n) + (ab_1, ab_2,..., b_n)$

$= a(a_1, a_2,..., a_n) + a(b_1, b_2,..., b_n) = a\alpha + a\beta.$

2. If $a, b \in F$ and $a = (a_1, a_2,..., a_n) \in V$, then

$(a + b) \alpha = ([a + b] a_1, [a + b] a_2,..., [a + b] a_n)$

$= (aa_1 + ba_1, aa_2 + ba_2,..., aa_n + ba_n)$

$= (aa_1, aa_2,..., aa_n) + (ba_1, ba_2,..., ba_n)$

$= a (a_1, a_2,..., a_n) + b (a_1, a_2,..., a_n) = a\alpha + b\alpha.$

3. If $a, b, \in F$ and $\alpha = (a_1, a_2,..., a_n) \in V$, then

$(ab) \alpha = ([ab] a_1, [ab] a_2,..., [ab] a_n) = (a [ba_1], a [ba_2],..., a [ba_n])$

$= a (ba_1, ba_2,..., ba_n) = a [b (a_1, a_2,..., a_n)] = a (b\alpha).$

4. If 1 is unity element of F and $\alpha = (a_1, a_2,..., a_n) \in V$, then $1\alpha = (1a_1,$

$1a_2,..., 1a_n) = (a_1, a_2,..., a_n) = \alpha$.

Hence V is a vector space over F. *The vector space of all ordered n-tuples over F will be denoted by $V_n(F)$. Sometimes we also denote it by F^n.* Here the zero vector *i.e.*, 0 is the n-tuple (0, 0,..., 0).

Example 17:

Show that a field F may be considered as a vector space over F if scalar multiplication is identified with field multiplication.

Solution:

Let us consider (F, +, ·) be a field. Take F as the set of vectors and also as the set of scalars. Take the addition operation on the field F as the addition of vectors and the multiplication operation on the field F as the operation of scalar multiplication *i.e.*, multiplication of a vector by a scalar. Then F is a vector space over F as shown as follows:

Since (F, +, ·) is a field, therefore (F, +) is an abelian group.

Again, if a, b are any scalars *i.e.*, a, b, $\in$ F and α, β, are any vectors *i.e.*, α, β, $\in$ F then we have

$$a(\alpha + \beta) = a\alpha + a\beta \text{ and } (a + b)\alpha = a\alpha + b\alpha.$$

(by distribution law)

Also $(ab)\alpha = a(b\alpha)$ because the multiplication on the field F is associative.

Also if 1 is the unity element of the field F and α is any vector *i.e.*, $\alpha \in$ F, then $1\alpha = \alpha$.

Hence F (F) is a vector space.

Example 18:

Show that the complex field C is a vector space over the real field R.

Solution:

Let us consider C as the set of vectors and R as the set of scalars.

Take the addition of complex numbers as the addition of vectors. Then (C, +) is an abelian group.

Again, let a be any scalar *i.e.*, $a \in R$ and α be any vector *i.e.*, $\alpha \in C$. Now $a \in R \Rightarrow a \in C$. Thus, both a and α are in C. Regard the composition of scalar multiplication as the multiplication of a and α in the field. Since C is a field, therefore $a\alpha \in C$ and thus C is closed for scalar multiplication.

The real number 1 which is the unity element of the field R is also the unity element of the field C.

Now let α, β be any vectors *i.e.*, $\alpha, \beta \in C$ and a, b be any scalars *i.e.* $a, b \in R$. Since $R \subset C$, therefore, we have

$a, b \in R \Rightarrow a, b \in C$

We have $a(\alpha + \beta) = a\alpha + a\beta$ and $a(a + b)\ \alpha = a\alpha + b\alpha$. (distributive laws in the field C).

Also $(ab)\ \alpha = a\ (b\alpha)$ because the multiplication on the field C is associative.

Also $1\alpha = \alpha$ because 1 is also the multiplicative identity in the field C.

Hence C is a vector space over R.

Example 19:

How many elements are there in the vector space of polynomials of degree at most in which the coefficients are the elements of the field I (p) over the field I (p), p being a prime number?

Solution:

Here we know that the field I (p) is the field

$(\{0, 1, 2, ..., p - 1)\}, +_p, \times_p)$.

The number of distinct elements in the field I (p) is p.

If f(x) is a polynomial of degree at most n over the field I (p), then we have

$f(x) = a_0 + a_1x + a_2x^2 + ... + a_nx^n$, where $a_0, a_1, a_2, ..., a_n \in I\ (p)$.

Now in the polynomial f(x), the coefficient of each of the n + 1 terms $a_0, a_1x, a_2x^2, ..., a_nx^n$ can be filled in p ways because any of the p elements of the field I (p) can be filled there.

Thus, we can have $p \times p \times p \times ...$ upto (n + 1) times *i.e.*, p^{n+1} distinct polynomials of degree at most n over the field I (p). Hence if P_n is the vector space of polynomials of degree at most n in which the coefficients are the elements of the field I (p) over the field I (p), then P_n has p^{n+1} distinct elements.

Example 20:

*Prove that the set of all solutions (a, b, c) of the equation $a + b + 2c = 0$ is a subspace of the vector space V_3 (**R**).*

Solution:

Let $W = \{(a, b, c): a, b, c \in \mathbf{R}$ and $a + b + 2c = 0\}$.

To prove that W is a subspace of V_3 (**R**) or $\mathbf{R}^3$.

Let $\alpha = (a_1, b_1, c_1)$ and $\beta = (a_2, b_2, c_2)$ be any two elements of W. Then

$a_1 + b_1 + 2c_1 = 0$ ...(1)

and $a_2 + b_2 + 2c_2 = 0$ ...(2)

If a, b be any two elements of **R**, we have

$$a\alpha + b\beta = a(a_1, b_1, c_1) + b(a_2, b_2, c_2)$$
$$= (aa_1, ab_1, ac_1) + (ba_2, bb_2, bc_2)$$
$$= (aa_1 + ba_2, ab_1 + bb_2, ac_1 + bc_2).$$

Now $(aa_1 + ba_2) + (ab_1 + bb_2) + 2(ac_1 + bc_2)$

$$= a(a_1 + b_1 + 2c_1) + b(a_2 + b_2 + 2c_2)$$
$$= a.0 + b.0 \quad \text{[from (1) and (2)]}$$
$$= 0.$$

$\therefore\ a\alpha + b\beta\ (aa_1 + ba_2, ab_1 + bb_2, ac_1 + bc_2) \in W$

Thus $\Rightarrow \alpha, \beta \in W$ and $a, b \in R \Rightarrow a\alpha + b\beta \in W$.

Hence W is a subspace of $V_3(\mathbf{R})$.

Example 21:

Which of the following sets of vectors $\alpha = (a_1, a_2,\ldots, a_n)$ in R^n are subspaces of R^n $(n \geq 3)$?

(i) all α such that $a_1 \leq 0$;

(ii) all α such that a_3 is an integer;

(iii) all α such that $a_2 + 4a_2 = 0$.

Solution:

(i) Let $W = \{\alpha : \alpha \in \mathbf{R}^n \text{ and } a_1 \leq 0\}$.

If we take $a_1 = -3$, then $a_1 < 0$ and so

$$\alpha = (-3, a_2,\ldots, a_n) \in W.$$

Now if we take $a = -2$, then

$$a\alpha = (6, -2a_2,\ldots, -2a_2).$$

Since the first coordinate of $a\alpha$ is 6 which is > 0, therefore $a\alpha \notin W$.

Thus, $\alpha \in W$, $a \in R$ but $a\alpha \notin W$. Therefore W is not closed for scalar multiplication and so W is not a subspace of $\mathbf{R}^n$.

(ii) Let $W = \{\alpha : \alpha \in \mathbf{R}^n \text{ and } a_2 \text{ is an integer}\}$.

If we take $a_3 = 5$, then a_3 is an integer and so

$$a = (a_1, a_2, 5,\ldots, a_n) \in W.$$

Now if we take $a = \frac{1}{2}$, then

$$a\alpha = \left(\frac{1}{2}a_1, \frac{1}{2}a_2, 5/2,\ldots, \frac{1}{2}a_n\right).$$

Since the third coordinate of $a\alpha$ is 5/2 which is not in integer, therefore $a\alpha \notin W$.

Thus, $\alpha \in W$, $\alpha \in \mathbf{R}$ but $a\alpha \in W$. Therefore is not closed for scalar multiplication and so W is not a subspace of $\mathbf{R}^n$.

(iii) Let $\alpha = (a_1,..., a_n)$ and $\beta = (b_1,..., b_n)$ be any two members of W. Then $a_2 + 4a_2 = 0$ and $b_2 + 4b_3 = 0$.

If a, b, $\in \mathbf{R}$, then $a\alpha + b\beta = (aa_1 + bb_1,..., aa_n + bb_n)$.

We have $(aa_2 + bb_2) + 4\,(aa_3 + bb_3)$

$$= a\,(a_2 + 4a_3) + b\,(b_2 + 4b_3) + b\,(b_2 + 4b_3 = a, 0 + b.0 = 0.$$

Thus according to the definition of W, $a\alpha + b\beta \in W$.

In this way $\alpha, \beta, \in W$ and a, b, $\in \mathbf{R} \Rightarrow a\alpha + b\beta \in W$. Hence W is a subspace of $\mathbf{R}^n$.

Example 22(a):

*Show that the set W of the elements of the vector space V_3 **(R)** of the form $(x + 2y, y, -x + 3y)$, where $x, y \in R$ is a subspace of V_3 **(R)**.*

Solution:

Let $W = \{(x + 2y, y, -x + 3y): x, y \in \mathbf{R}\}$.

To prove that W is a subspace of V_3 (**R**).

Let $\alpha = (x_1 + 2y, y_1, -x_1 + 3y_1)$ and $\beta = (x_2 + 2y_2, -x_2 + 3y_2)$ be any two elements of W.

If a, b be any two elements of **R**, we have

$a\alpha + b\beta = a\,(x_1 + 2y_1, y_1, -x_1 + 3y_1) + b\,(x_2 + 2y_2, -x_2 + 3y_2)$

$-(ax_1 + 2ay_1, ay_1, -ax_1 + 3ay_1) + (bx_2 + 2by_2, by_2, -bx_2 + 3by_2)$

$= (ax_1 + 2ay_1 + bx_2 + 2by_2, ay_1 + by_2, -ax_1 + 3ay_1 - bx_2 + 3by_2)$

$= ([ax_1 + bx_2] + 2[ay_1 + by_2], ay_1 + by_2, -[ax_1 + bx_2] + 3\,[ay_1 + by_2]$

which is in W because it is of the form $(x + 2y, y, -x + 3y)$.

Here in place of y we have $ay_1 + by_2$ and in place of x we have

$ax_1 + bx_2$.

Thus $\alpha, \beta \in W$ and a, b $\in \mathbf{R} \Rightarrow a\alpha + b\beta \in W$.

Hence W is a subspace of V_2 (**R**).

Example 22(b):

In the vector space F[x] of all polynomials over the field F the infinite set $S = \{1, x, x^2, x^3, ...\}$ is linearly indepcndent.

Solution:

Let $S' = \{x^{m_1}, x^{m_2}, ..., x^{m_n}\}$ be any finite subset of S having n vectors. Here $m_1, m_2, ..., m_n$ are some non-negative integers. Let $a_1, a_2, ..., a_n$ be scalars such that

$$a_1 x^{m_1} + a_2 x^{m_2} + ... + a_n x^{m_n} = 0.$$

(*i.e.*, zero polynomial) ...(1)

By the definition of equality of two polynomials we have from (1)

$$a_1 = 0,\ a_2 = 0,\ ...,\ a_n = 0.$$

Thus every finite subset of S is linearly independent.

Therefore S is linearly independent.

Example 22(c):

Is the vector (2, – 5, 3) in the subspace of $\mathbf{R}^3$ *spanned by the vectors (1, – 3, 2), (2, – 4, – 1), (1, – 5, 7)?*

Solution:

Let $\alpha = (2, -5, 3)$, $\alpha_1 = (1, -3, 2)$, $\alpha_2 = (2, -4, -1)$, $\alpha_3 = (1, -5, 7)$. If a can be expressed as a linear combination of the vectors $\alpha_1, \alpha_2, \alpha_3$, then it will be in the subspace of $\mathbf{R}^3$ spanned by these vectors otherwise it will not be.

Let $\alpha = a_1\alpha_1 + a_2\alpha_2 + a_3\alpha_3$ where $a_1, a_2, a_2 \in \mathbf{R}$.

Then $(2, -5, 3) = a_1(1, -3, 2) + a_2(2, -4, -1) + a_3(1, -5, 7)$

$\Rightarrow (2, -5, 3) = (a_1 + 2a_2 + a_3, -3a_1 - 4a_2 - 5a_3, 2a_1 - a_2 + 7a_3)$.

$\therefore \quad a_1 + 2a_2 + a_3 = 2$ ...(1)

$-3a_1 - 4a_2 - 5a_3 = -5$ ...(2)

$2a_1 - a_2 + 7a_3 = 3$ ...(3)

Multiplying the equation (1) by 3 and adding to (2), we get

$$2a_2 - 2a_3 = 1 \quad \Rightarrow \quad a_2 - a_3 = \frac{1}{2}. \qquad ...(4)$$

Again multiplying the equation (1) by 2 and subtracting from (3), we get

$$-5a_2 + 5a_3 = -1 \quad \Rightarrow \quad a_2 - a_3 = 1/5. \qquad ...(5)$$

The relations (4) and (5) show that the above equations are inconsistent. Hence the vector α cannot be expressed as a linear combination of the vectors

$\alpha_1, \alpha_2, \alpha_3$. Therefore α is not in the subspace of $\mathbf{R}^3$ generated by the vectors $\alpha_1, \alpha_2, \alpha_3$.

Example 23:

Show that the vectors (1, 1, 2, 4), (2, – 1, – 5, 2) (1, – 1, – 4, 0) and (2, 1, 1, 6) are linearly dependent in $\mathbf{R}^4$.

Solution:

Let (1, 1, 2, 4) = a (2, – 1, – 5, 2) + b (1, – 1, – 4, 0) + c (1, 1, 1, 6).

Then $2a + b + 2c = 1$ ...(1)

$-a - b + c = 1$ ...(2)

$-5a - 4b + c = 2$ ...(3)

$22a + 0b + 6c = 4.$ ...(4)

Now we shall solve the simultaneous equations (1), (2), (3) and (4).

Adding (1) and (2), we get a + 3c = 2 which is the same equation as (4).

If we take c = 0, we get a = 2.

Putting a = 2 and c = 0 in (1), we get b = – 3.

We see that a = 2, b = 2 3, c = 0 satisfy all the four equations (1), (2), (3) and (4).

$\therefore$ (1, 1, 2, 4) = 2 (2, – 1, – 5, 2) – 3 (1, – 1, – 4, 0) + 0 (2, 1, 1, 6)

$\Rightarrow$ 1 (1, 1, 2, 4) – 2 (2, – 1, – 5, 2) + 3 (1, – 1, – 4, 0) – 0 (2, 1, 1, 6) = (0, 0, 0, 0), ...(1)

Since in the linear relation (1) among the four given vectors the scalar coefficients 1, – 2, 3, 0 are not all zero, therefore the given vectors are linearly dependent in $\mathbf{R}^4$.

Example 24:

Show that the set {1, x, x (1 – x)} is a linearly independent set of vectors in the space of all polynomials over the real number field.

Solution:

The zero vector of the vector space of all polynomials over the real number field is the zero polynomial.

Let a, b, c be scalars (*i.e.*, real numbers) such that

a(1) + bx + c [x (1 – x)] = 0 *i.e.*, zero polynomial.

We have $a(1) + bx + c\,(x - x^2) = 0$

$$\Rightarrow \quad a + (b + c)\,x - cx^2 = 0. \qquad ...(1)$$

Now two polynomials in x are said to be equal if the coefficients of like powers of x on both sides are equal. So by the definition of the equality of two polynomials, we have from (1).

$$a = 0,\ b + c = 0,\ -c = 0$$

$$\Rightarrow \quad c = 0,\ b = 0,\ a = 0.$$

Thus $\quad a(1) + bx + c\,[x\,(1 - x)] = 0$

$$\Rightarrow \quad a = 0,\ b = 0,\ c = 0.$$

$\therefore$ the vectors 1, x, x (1 – x) are linearly independent over the field of real numbers.

Example 25:

Show that the three vectors (1, 1, – 1), (2, – 3, 5) and (–2, 1, 4) of $\mathbf{R}^3$ *are linearly independent.*

Solution:

Let a, b, c be scalars *i.e.*, real numbers such that

$$a\,(1, 1, -1) + b\,(2, -3, 5) + c\,(-2, 1, 4) = (0, 0, 0)$$

i.e., $\quad (a + 2b - 2c,\ a - 3b + c,\ -a + 5b + 4c) = (0, 0, 0)$

i.e.

$$a + 2b - 2c = 0 \qquad ...(1)$$

$$a - 3b + c = 0 \qquad ...(2)$$

$$-a + 5b + 4c = 0 \qquad ...(3)$$

Now we shall solve the simultaneous equations (1), (2) and (3).

Multiplying (2) by 2 and adding to (1), we get

$$3a - 4b = 0 \qquad ...(4)$$

Again multiplying (1) by 2 and adding to (3), we get

$$a + 9b = 0$$

Multiplying (5) by 3 and subtracting from (4), we get

$$-31b = 0 \text{ or } b = 0.$$

Putting b = 0 in (5), we get a = 0.

Now putting a = 0, b = 0 in (1), we get c = 0.

Thus, a = 0, b = 0, c = 0 is the only solution of the equations (1), (2) and (3).

$$\therefore \quad a\,(1, 1, -1) + b\,(2, -3, 5) + c\,(-2, 1, 4) = (0, 0, 0)$$

$\Rightarrow \quad a = 0, b = 0, c = 0.$

Hence the vectors (1, 1, – 1), (2, – 3, 5), (– 2, 1, 4) of $\mathbf{R}^3$ are linearly independent.

Example 26:

If α, β, γ are linearly independent vectors of V(F) where F is the field of complex numbers, then so also are α + β, β + γ, γ + α.

Solution:

Let a, b, c be scalars such that

$$a\,(\alpha + \beta) + b\,(\beta + \gamma) + (\gamma + \alpha) = 0$$

i.e., $\quad (a + c)\,\alpha + (a + b) + \beta\,(b + c) + \gamma = 0.$ ...(1)

But α, β, γ are linearly independent. Therefore (1) implies

$$a + 0b + c = 0,\ a + b + 0c = 0,\ 0a + b + c = 0.$$

The coefficient matrix A of these equations is

$$A = \begin{bmatrix} 1 & 0 & 1 \\ 1 & 1 & 0 \\ 0 & 1 & 1 \end{bmatrix}.$$

We have rank A = 3 *i.e.*, the number of unknowns a, b, c. Therefore a = 0, b = 0, c = 0 is the only solution of the given equations.

Hence α + β, β + γ, γ + α are also linearly independent.

Example 27:

If α, β, γ are linearly independent vectors of V(F) where F is the field of complex numbers, then so also are

$$\alpha + \beta,\ \alpha - \beta,\ \alpha - 2\beta + \gamma.$$

Solution:

Let a, b, c be scalars such that

$(a + \beta) + b\,(\alpha - \beta) + c\,(\alpha - 2\beta + \gamma) = 0$ ...(1)

i.e., $\quad (a + b + c)\,\alpha + (a - b - 2c)\,\beta + c\gamma = 0.$ ...(2)

But α, β, γ are linearly independent. Therefore (2) implies

$$a + b + c = 0,\ a - b - 2c = 0,\ c = 0.$$

The only solution of these equations is c = 0, a = 0, b = 0.

Thus (1) implies a = 0, b = 0, c = 0.

Therefore the vectors α + β,

α – β, α – 2β + γ are linearly independent.

Example 28(a):

A system consisting of a single non-zero vector is always linearly independent.

Solution:

Let S = {α} be a subset of a vector space V and let α be not equal to zero vector. If a is any scalar, then

$$a\alpha = 0$$

$\Rightarrow$ $a = 0$ [Since a is not zero vector]

∴ the set S is linearly independent.

Example 28(b):

If the set $S = \{a_1, a_2, \ldots, a_n\}$

of vectors of V(F) is linearly independent, then of the vectors $\alpha_1, \alpha_2, \ldots, \alpha_n$ *can be zero vector.*

Solution:

Let α_r be equal to zero vector where $1 \leq r \leq n$. Then

$$0\alpha_1 + 0\alpha_2 + \ldots + a\alpha_r + 0\alpha_{r+1} + \ldots + 0\alpha_n = 0$$

for any $a \neq 0$ in F.

Since $a \neq 0$, therefore from this relation we conclude that S is linearly dependent. Thus we get a contradiction because it is given that S is linearly independent. Hence none of the vectors $\alpha_1, \alpha_2 \ldots, \alpha_n$ can be zero vector. We also conclude that *a set of vectors which contains the zero vector is necessarily linearly dependent.*

Example 29:

Every superset of a linearly dependent set of vectors is linearly dependent.

Solution:

Let $S = \{\alpha_1, \alpha_2, \ldots, \alpha_n\}$ be a linearly dependent set of vectors. Then there exist scalars $a_1, a_2, \ldots, a_n$ not all zero such that

$$a_1\alpha_1 + a_2\alpha_2 + \ldots + a_n\alpha_n = 0. \quad \ldots(1)$$

Now let $S' = \{\alpha_1, \alpha_2, \ldots, \alpha_n, \beta_1, \beta_2, \ldots, \beta_m)$ be a superset of S. Then we have from (1)

$$a_1\alpha_1 + a_2\alpha_2 + \ldots + a_n\alpha_n + 0b_1 + 0\beta_2 + \ldots + 0\beta_m = 0. \quad \ldots(2)$$

Since in the relation (2) the scalar coefficients are not all 0, therefore S' is linearly dependent.

From this we also conclude that *any subset of a linearly independent set of vectors is also linearly independent.*

Example 30:

Show that

$S = \{(1, 2, 4), (1, 0, 0), (0, 1, 0), (0, 0, 1)\}$

is a linearly dependent subset of the vector space $V_3(\mathbf{R})$ *where* **R** *is the field of real numbers.*

Solution:

We have

$1\ (1, 2, 4) + (-1)\ (1, 0, 0) + (-2)\ (0, 1, 0) + (-4)\ (0, 0, 1)$

$= (1, 2, 4) + (-1, 0, 0) + (0, -2, 0) + (0, 0, -4)$

$= (0, 0, 0)$ *i.e.*, zero vector.

Since in this relation the coefficients $1, -1, -2, -4$ are not all zero, therefore the given system S is linearly dependent.

Example 31:

Prove that if two vectors are linearly dependent, one of them is a scalar multiple of the other.

Solution:

Let α, β be two linearly dependent vectors of the vector space V. Then $\exists$ scalars a, b not both zero, such that

$$a\alpha + b\beta + 0.$$

If $a \neq 0$, then we get

$$a\alpha = -b\beta \Rightarrow a^{-1}(a\alpha) = a^{-1}(-b\beta)$$

$$\Rightarrow \quad (a^{-1}a)\,\alpha = [a^{-1}(-b)]\,\beta$$

$$\Rightarrow \quad 1\alpha = (-a^{-1}b)\,\beta$$

$$\Rightarrow \quad \alpha = -\frac{b}{a}\beta \Rightarrow \alpha \text{ is a scalar multiple of } \beta.$$

If $b \neq 0$, then we get

$$b\beta = -a\alpha$$

$$\Rightarrow \quad \beta = \left(-\frac{a}{b}\right)\alpha \Rightarrow \beta \text{ is a scalar multiple of } \alpha.$$

Thus one of the vectors α and β is a scalar multiple of the other.

Example 32:

In the vector $V_n(F)$, the system of n vectors

$e_1 = (1, 0, 0, ..., 0),$

$e_2 = (0, 1, ..., 0, 0),$

$e_n = (0, 0, ..., 0, 1)$

is linearly independent where 1 denotes the unity of the field F.

Solution:

If $a_1, a_2, a_3, ..., a_n$ be any scalars, then

$$a.e_1 + a_2e_2 + ... + a_ne_n = 0$$

$\Rightarrow a_1 (1, 0, 0, ..., 0) + a_2 (0, 1, 0, ..., 0) + ... + a_n (0, 0, ..., 1) = 0$

$\Rightarrow (a_1, a_2, ..., a_n) = (0, 0, ..., 0) \Rightarrow a_1 = 0, a_2 = 0, ..., a_n = 0.$

Therefore the given set of n vectors is linearly independent.

In particular {(1, 0, 0), (0, 1, 0), (0, 0, 1)} is a linearly independent subset of $V_3(F)$.

Example 33(a):

Let $\alpha_1, \alpha_2, \alpha_3$ be vectors of V(F), a, b, $\in$ F. Show that the set $\{\alpha_1, \alpha_2, \alpha_3\}$ is linearly dependent if the set $\{\alpha_1 + a\alpha_2 + b\alpha_3, \alpha_2, \alpha_3\}$ is linearly dependent.

Solution:

Since the set $\{\alpha_1 + a\alpha_2 + b\alpha_3, \alpha_2, \alpha_3\}$ is linearly dependent therefore exist scalars x, y, z not all zero such that

$$x (\alpha_1 + a\alpha_2 + b\alpha_3) + y\alpha_2 + z\alpha_3 = 0$$

i.e. $$x\alpha_1 + (xa + y) \alpha_2 + (xb + z) \alpha_3 = 0. \quad ...(1)$$

If in the relation (1), the coefficients x, xa + y, xb + z are not all zero, then the set $\{\alpha_1, \alpha_2, \alpha_3\}$ will also be linearly dependent.

If $x \neq 0$, then the problem is at once solved whatever y and z may be. However if x = 0, then at least one of y and z is not zero. Therefore at least one of xa + y and xb + z will not be zero since when x = 0 then xa + y and xb + z reduce to y and z respectively.

Hence in the relation (1) the scalar coefficients of $\alpha_1, \alpha_2, \alpha_3$ are not all zero. Therefore the set $\{\alpha_1, \alpha_2, \alpha_3\}$ is also linearly dependent.

Example 33(b):

If R is a ring, show that

$$Z(R) = \{x \in R : xy = yx \ \forall y \in R\}$$

is a subring of R. Further show that Z(R) is a field if R is a division ring.

Solution:

We have $0y = 0 = y0 \ \forall y \in R$. Therefore $0 \in Z(R)$ and so $Z(R) \neq \varnothing$.

Now let $z_1, z_2 \in Z(R)$. Then

$$z_1 y = yz_1 \text{ and } z_2 y = yz_2 \ \forall y \in R.$$

Now $\forall y \in R$, we have

$$(z_1 - z_2)\, y = z_1 y - z_2 y = yz_1 - yz_2 = y\,(z_1 - z_2)$$

and $(z_1 z_2)\, y = z_1\,(z_2 y) = z_1\,(yz_2)\, z_2 = (z_1 y)\, z_2 = (yz_1)\, z_2 = y\,(z_1 z_2)$.

$\therefore$ by definition of Z(R), both $z_1 - z_2$ and $z_1 z_2 \in Z(R)$.

Thus $z_1, z_2 \in Z(R) \Rightarrow z_1 - z_2 \in Z(R)$ and $z_1 z_2 \in Z(R)$.

$\therefore$ Z(R) is a subring of R.

Now suppose R is a division ring *i.e.*, R is a ring with unity and every non-zero element of R possesses multiplicative inverse. Then to prove that Z(R) is a field.

Z(R) is a commutative ring. Let $z_1, z_2 \in Z(R)$.

Now $z_1 \in Z(R) \Rightarrow z_1 y = yz_1 \ \forall y \in R$. Since $z_2 \in R$, therefore $z_1 z_2 = z_2 z_1$.

Thus $z_1 z_2 = z_2 z_1 \ \forall z_1, z_2 \in Z(R)$ and so Z(R) is a commutative ring.

The Ring Z(R) Possesses Multiplicative Identity: If 1 denotes the unity element of the division ring R, then $1y = y = y1 \ \forall y \in R$. Therefore $1 \in Z(R)$ and is also the unity element of Z(R). Thus Z(R) is also a ring with unity.

Each Non-zero Element of the Ring Z(R) Possesses Multiplicative Inverse: Let $0 \neq z \in Z(R)$ and let z^{-1} denote the multiplicative inverse of z in the division ring R. We shall show that $z^{-1} \in Z(R)$.

We have $z \in Z(R) \Rightarrow zy = yz \ \forall y \in R$

$\Rightarrow \ z^{-1}\,(zy)\, z^{-1} = z^{-1}\,(yz)\, z^{-1} \ \forall y \in R$

$\Rightarrow \ (z^{-1} z)\, yz^{-1} = (z^{-1} y)\, zz^{-1} \ \forall y \in R$

$\Rightarrow \ 1\,(yz^{-1}) = (z^{-1} y)\, 1 \quad \forall y \in R$

$\Rightarrow \ yz^{-1} = z^{-1} y \ \forall y \in R.$

$\therefore$ by the definition of Z(R), $z^{-1} \in Z(R)$.

Since $zz^{-1} = 1 = z^{-1}z$. where 1 is the unity element of the ring Z(R), therefore z^{-1} is also the multiplicative inverse of z in the ring Z(R).

Thus each non-zero element of the ring $\bar{Z}(R)$ is inversible.

Since Z(R) is a commutative ring with unity and each non-zero element of Z(R) is inversible, therefore Z(R) is a field.

Now $S_1 \cup S_2$ is not a subring of the ring of integers because $S_1 \cup S_2$ is not closed for addition.

We have $2 \in S_1 \cup S_2$ is because $2 \in S_1$ and $3 \in S_1 \cup S_2$ because $3 \in S_2$. But $2 + 3 = 5 \notin S_1 \cup S_2$ because neither $5 \in S_1$ nor $5 \in S_2$. Thus $S_1 \cup S_2$ is not closed for addition. Hence $S_1 \cup S_2$ is not a subring of the ring of integers.

However $S_1 \cup S_3 = S_1$ because $S_3 \subset S_1$. Thus $S_1 \cup S_2$ is a subring of the ring of integers.

From the above example it is obvious that the union of two subrings may or may not be a subring.

Example 34:

Prove or disprove that any subring of a non-commutative ring is non-commutative.

Solution:

A subring of a non-commutative ring may be commutative as is obvious from the following example.

Let M be the ring of all 2 × 2 matrices with elements as integers for addition and multiplication of matrices as the tow ring operations. Then M is a non-commutative ring because the operation of multiplication of matrices on the set M is not commutative.

Let S be the subset of M consisting of matrices of the type

$$\begin{bmatrix} a & 0 \\ 0 & 0 \end{bmatrix}, \text{ where a is any integer.}$$

Let $A = \begin{bmatrix} a_1 & 0 \\ 0 & 0 \end{bmatrix}, B = \begin{bmatrix} a_2 & 0 \\ 0 & 0 \end{bmatrix}$ be any two members of the set S.

The $A - B = \begin{bmatrix} a_1 - a_2 & 0 \\ 0 & 0 \end{bmatrix} \in S$

and $AB = \begin{bmatrix} a_1 a_2 & 0 \\ 0 & 0 \end{bmatrix} \in S$

∴ S is a subring of the ring M.

Now the subring S of the ring M is a commutative ring. For let

$$A = \begin{bmatrix} a_1 & 0 \\ 0 & 0 \end{bmatrix} \text{ and } B = \begin{bmatrix} a_2 & 0 \\ 0 & 0 \end{bmatrix}$$

be any two members of S. Then

$$AB = \begin{bmatrix} a_1a_2 & 0 \\ 0 & 0 \end{bmatrix} \text{ and } BA = \begin{bmatrix} a_2a_1 & 0 \\ 0 & 0 \end{bmatrix}$$

Since $a_1a_2 = a_2a_1$,

therefore AB = BA.

Thus, multiplication of matrices is a commutative operation on the set S. Hence the subring S of the non-commutative ring M is a commutative ring.

Example 35:

Show that the set of even integers forms a subring of the ring of integers.

Solution:

Let (I, +, .) be the ring of integers.

Let S = 2I = {..., –6, –4, –2, 0, 2, 4, 6, ...} be the set of even integers. Then $S \subset I$.

To prove that S is a subring of the ring of integers.

Let a = 2r and b 2s be any two elements of S where r and s are some integers.

Then $a - b = 2r - 2s = 2(r - s)$ which is an even integer and so $a - b \in S$.

Also $ab = (2r)(2s) = 2(2rs)$ which is also an even integer and so $ab \in S$.

Thus, $a, b \in S \Rightarrow a - b \in S$ and $ab \in S$. Hence the set of even integers S is a subring of the ring of integers

Example 36:

Show that the set of all 2-rowed matrices of the form

$$\begin{bmatrix} a & 0 \\ b & c \end{bmatrix}$$

where a, b, c are integers is a subring of the ring M of all 2-rowed matrices with integral entries.

Solution:

Let M be the set of all 2 × 2 matrices with elements as integers. Then M is a ring for addition and multiplication of matrices as the two ring operations.

Let S be the subset of M consisting of matrices of the type

$$\begin{bmatrix} a & 0 \\ b & c \end{bmatrix}, \text{ where a, b, c are integers.}$$

To prove that S is a subring of the ring M.

Let $A = \begin{bmatrix} a_1 & 0 \\ b_1 & c_1 \end{bmatrix}, B = \begin{bmatrix} a_2 & 0 \\ b_2 & c_2 \end{bmatrix}$ be any two members of the set S.

The $A - B = \begin{bmatrix} a_1 - a_2 & 0 \\ b_1 - b_2 & c_1 - c_2 \end{bmatrix}$ which is obviously a member of the set S.

Also $AB = \begin{bmatrix} a_1a_2 & 0 \\ b_1a_2 + c_1b_2 & c_1c_2 \end{bmatrix}$ which is also a member of the set S.

Thus A, B ∈ S ⇒ A – B ∈ S and AB ∈ S

Hence S is a subring of the ring M

Example 37:

*In V_3(**R**), where **R** is the field of real numbers. examine each of the following sets of vectors for linear dependence:*

(i) {(2, 1, 2), (8, 4, 8)}.

(ii) {(1, 2, 0), (0, 3, 1), (– 1, 0, 1)}

(iii) {(– 1, 2, 1), (3, 0, – 1), (– 5, 4, 3)}.

(iv) {(2, 3, 5), (4, 9, 25)}.

(v) (1, 2, 1), (3, 1, 5), (3, – 4, 7).

Solution:

(i) We have

4 (2, 1, 2) + (– 1) (8, 4, 8) = (8, 4, 8) + (– 8, – 4, – 8)

= (0, 0, 0) *i.e.*, the zero vector.

Since in this relation the scalar coefficients 4, – 1 are not both zero, therefore the given set is linearly dependent.

(ii) Let a, b, c be scalars *i.e.*, real numbers such that

$$a\,(1, 2, 0) + b\,(0, 3, 1) + c\,(-1, 0, 1) = (0, 0, 0)$$

i.e., $(a - c, 2a + 3b, b + c) = (0, 0, 0)$

i.e., $a + 0b - c, 2a + 3b + 0c = 0, 0a + b + c = 0.$

These equations will have a non-zero solution *i.e.*, a solution in which a, b, c are not all zero if the rank of the coefficient matrix is less than three *i.e.*, the number of unknowns a, b, c. If the rank is 3, then the zero solution a = 0, b = 0, c = 0 will be the only solution.

Coefficient matrix $A = \begin{bmatrix} 1 & 0 & -1 \\ 2 & 3 & 0 \\ 0 & 1 & 1 \end{bmatrix}.$

We have $|A| = 1\,(3 - 0) - 2\,(0 + 1) = 1 \neq 0.$

$\therefore$ Rank A = 3. Hence a = 0, b = 0, c = 0 is the only solution. Therefore the given system is linearly independent.

(iii) Let a, b, c be scalars such that

$$a\,(-1, 2, 1) + b\,(3, 0, -1) + c\,(-5, 4, 3) = (0, 0, 0)$$

i.e., $(-a + 3b - 5c, 2a + 0b + 4c, a - b + 3c) = (0, 0, 0)$

i.e., $-a + 3b - 5c = 0, 2a + 0b + 4c = 0, a - b + 3c = 0.$

The coefficient matrix A of these equations is

$$A = \begin{bmatrix} -1 & 3 & -5 \\ 2 & 0 & 4 \\ 1 & -1 & 3 \end{bmatrix}.$$

We have $|A| = -(0 + 4) - 2\,(9 - 5) + 1\,(12 - 0) = 0.$

$\therefore$ Rank A is < 3 *i.e.*, the number of unknowns a, b, c. Therefore the given system of equations will possess a non-zero solution. For example a = –2, b = 1, c = 1, is a non-zero solution. Hence the given system of vectors is linearly dependent.

(iv) Let a, b, be scalars *i.e.*, real numbers such that

$$a\,(2, 3, 5) + b\,(4, 9, 25) = (0, 0, 0)$$

i.e., $(2a + 4b, 3a + 9b, 5a + 25b) = (0, 0, 0)$

i.e., $2a + 4b = 0, 3a + 9b = 0, 5a + 25b = 0.$

The coefficients matrix A of these equations is

$$\begin{bmatrix} 2 & 4 \\ 3 & 9 \\ 5 & 25 \end{bmatrix}.$$

Obviously rank A = 2 *i.e.*, equal to the number of unknowns a and b.

Therefore these equations have the only solution a = 0, b = 0. Hence the given set of vectors is linearly independent.

(v) Let a, b, c be scalars *i.e.*, real numbers such that

$$a(1, 2, 1) + b(3, 1, 5) + c(3, -4, 7) = (0, 0, 0)$$

i.e., $(a + 3b + 3c, 2a + b - 4c, a + 5b + 7c) = (0, 0, 0)$

i.e., $a + 3b + 3c = 0,$ ...(1)

$2a + b - 4c = 0.$ ...(2)

$a + 5b + 7c = 0.$ ...(3)

Multiplying (1) by 2, we get

$2a + 6b + 6c = 0.$ ...(4)

Subtracting (4) from (2), we get

$$-5b - 10c = 0$$

$\Rightarrow$ $b + 2c = 0.$ ...(5)

Again subtracting (3) from (1), we get

$$-2b - 4c = 0,$$

$\Rightarrow$ $b + 2c = 0.$ ...(6)

The equations (5) and (6) are the same and give b = – 2c,

Putting b – 2c in (1), we get a = 3c. If we take c = 1, we get b = – 2 and a = 3. Thus b = 3, b = – 2, c = 1 is a non-zero solution of the equations (1), (2) and (3). Hence the given set of vectors is linearly dependent.

Example 38:

If F is the field of complex numbers, prove that the vectors (a_1, a_2) *and* (b_1, b_2) *in* $V_2(F)$ *are linearly dependent iff*

$$a_1b_2 - a_2b_2 = 0.$$

Solution:

Let x, y $\in$ F. Then

$$x(a_1, a_2) + y(b_1, b_2) = (0, 0)$$

$$\Rightarrow (xa_1 + by_1, xa_2 + yb_2) = (0, 0).$$

Therefore $a_1x + b_1y = 0$

and $a_2x + b_2y = 0.$

The necessary and sufficient condition for these equations to possess a non-zero solution is that

$$\begin{vmatrix} a_1 & b_1 \\ a_2 & b_2 \end{vmatrix} = 0 \text{ i.e., } a_1b_2 - a_2b_1 = 0.$$

Hence the given system is linearly dependent iff

$$a_1b_2 - a_2b_1 = 0.$$

Example 39(a):

If α_1 and α_2 are vectors of V(F), and a, b $\in$ F, show that the set $\{\alpha_1, \alpha_2, a\alpha_1 + b\alpha_2\}$ is linearly dependent.

Solution:

We have

$(-a)\,\alpha_1 + (-b)\,\alpha_2 + 1\,(a\alpha_1 + b\alpha_2)$

$= (-a + a)\,\alpha_1 + (-b + b)\,\alpha_2 = 0\alpha_1 + 0\alpha_2 = 0$

i.e., zero vector.

In the above linear combination the scalar coefficient 1 ¹ 0. Therefore whatever may be the scalars – a and – b, the given set of vectors is linearly dependent.

Example 39(b):

In the vector space $\mathbf{R}^3$, *let* $\alpha = (1, 2, 1)$, $\beta = (3, 1, 5)$, $\gamma = (3, -4, 7)$. *Show that the subspaces spanned by S $\{\alpha, \beta\}$ and T = $\{\alpha, \beta, \gamma\}$ are the same.*

Solution:

First we shall show that the vector g can be expressed as a linear combination of the vectors a and b. Let

$$(3, -4, 7) = a\,(1, 2, 1) + b\,(3, 1, 5).$$

Then $a + 3b = 3$, $2a + b = -4$, $a + 5b = 7$. Solving the first two equations we get $a = -3$, $b = 2$ and these satisfy the third equation also. Therefore we can write $\gamma = -3\alpha + 2\beta$.

Now $\quad S \subseteq T \Rightarrow L(S) \subseteq L(T)$.

Further let $\delta \in L(T)$. Then δ can be expressed as a linear combination of the vectors, α, β and γ. In this linear combination the vector γ can be replaced by $-3\alpha + 2\beta$.

Thus, δ can be expressed as a linear combination of the vectors α and β. Therefore d $\in$ L(S). Thus d $\in$ L(T) $\Rightarrow$ d $\in$ L(S). Therefore $L(T) \subseteq L(S)$.

Hence $\quad L(T) = L(S)$.

Example 40:

Consider the following polynomials over the ring

$(I_8, +_8, \times_8)$:

$f(x) = 2 + 6x + 4x^2$, $g(x) = 2x + 4x^2$, $h(x) = 2\ 4x$ *and find*

1. deg [f (x) + g (x)]
2. deg [f (x) g (x)]
3. deg [h (x) h (x)].

Solution:

1. We have $f(x) + g(x) = (2 + 6x + 4x^2) + (0 + 2x + 4x^2)$

   $= (2 +_8 0) + (6 +_8 2)\, x + (4 +_8 4)\, x^2 = 2 + 0x + 0x^2 = 2.$

   Thus f (x) + g (x) is a non-zero constant polynomial and so deg [f (x) + g (x)] = 0.

2. We have $f(x)\, g(x) = (2 + 6x + 4x^2)(2x + 4x^2)$

   $= (2 \times_8 2)\, x + [(2 \times_8 4) +_8 (6 \times_8 2)]\, x^2$

   $+ [(6 \times_8 4) +_8 (4 \times_8 2)]\, x^3 + (4 \times_8 4)\, x^4$

   $= 4x + (0 +_8 4)\, x^2 + (0 +_8 0)\, x^3 + 0x^4.$

   $= 4x + 4x^2 + 0x^2 + 0x^4 = 4x + 4x^2.$

   $\therefore$ deg [f (x) g (x)] = 2.

3. We have $h(x)\, h(x) = (2 + 4x)(2 + 4x)$

   $= (2 \times_8 2) + [(2 \times_8 4) +_8 (4 \times_8 2)]\, x + (4 \times_8 4)\, x^2$

   $= 4 + (0 +_8 0)\, x + 0x^2 = 4 + 0x + 0x^2 = 4.$

   Thus h (x) h (x) is a non-zero constant polynomial and so deg [h (x) h (x)] = 0.

Example 41:

The polynomial domain F [x] over a field F is not a field.

Solution:

In order to show that F [x] is not a field, we should show that there exists a non-zero element of F [x] which has no multiplicative inverse. Let f (x) be any member of F [x] such that deg f (x) is greater than zero. The inverse of f (x) cannot be the zero polynomial the product f (x) and the zero polynomial will be equal to the zero polynomial and not equal to the unity element of F [x] which is the polynomial $1 = 0x = 0x^2 + ...$ Suppose now g (x) is any non-zero polynomial. Then f being a field, we have deg [f (x) g (x)] = deg f (x) + deg g (x) > because deg f (x) > 0 and deg g (x) ≥ 0.

The degree of the unity element of F [x] is 0. Hence f (x) g (x) cannot be equal to the unity element of F [x]. Thus f (x) does not possess multiplicative inverse.

∴ F [x] is not field.

Note: The only inversible elements of F [x] are constant polynomials excluding the zero polynomial. No member of F [x] whose degree is greater then 0 is inversible.

Example 42:

Show that id a ring R has no zero divisors, then the ring R [x] has also no zero divisors.

Solution:

It is given that a ring R has no zero divisors and we have to show that the ring R [x] has also no zero divisors.

Let $f(x) = a_0 + a_1 x + a_2 x^2 + ... + a_m x^m,\ a_m \neq 0$

and $g(x) = b_0 + b_1 x + b_2 x^2 + ... + b_n x^n,\ b_n \neq 0$

be two non-zero elements of R [x].

Then f (x) g (x) cannot be the zero polynomial *i.e.*, the zero element of R [x]. Then reason is that at least on coefficient of f (x) g (x) namely $a_m b_n$ of x^{m+n} is ≠ 0 because a_m, b_n are non-zero elements of R and R is without zero divisors.

Thus, in R [x] the product of no two non-zero elements can be the zero element. Hence the ring R [x] has no zero divisors.

Example 43:

Prove that the relation of divisibility in an integral domain is reflexive and transitive.

Solution:

Let D be an integral domain with unity element 1.

If a ∈ D, then a is said to divide be ∈ D, if there exists an element c ∈ D such that b = ca. If a is a divisor of b, then symbolically we write a | b.

1. The relation of divisibility on D is reflexive. Let a be any element of D.

   We can write a = 1 a, where 1 ∈ D.

$\therefore$ a | a.

Thus a | a, $\forall$ a $\in$ D. Hence the relation of divisibility on D is reflexive.

2. The relation of divisibility on D is transitive *i.e.*, a | b and b | c $\Rightarrow$ a | c.

We have a | b $\Rightarrow$ b = ap for some p $\in$ D

and b | c $\Rightarrow$ c = bq for some q $\in$ D.

Now c = bq and b = aq $\Rightarrow$ c = (ap) q

$\Rightarrow$ c = a (pq) $\Rightarrow$ a | c since pq $\in$ D.

Hence the relation of divisibility on D is transitive.

Example 44:

Show that the set {1, x 1 + x + x^2} is a linearly independent set of vectors in the vector space of all polynomials over the real number field.

Solution:

Let a, b, c be scalars (real numbers) such that

$$a(1) + bx + c(1 + x + x^2) = 0. \text{ We have}$$

$$a(1) + bx + c(1 + x + x^2) = 0$$

$$\Rightarrow \quad (a + c) + (b + c)x + cx^2 = 0$$

$$\Rightarrow \quad a + c = 0, b + c = 0, c = 0$$

$$\Rightarrow \quad c = 0, b = 0, a = 0.$$

$\therefore$ the vectors 1, x, 1 + x + x^2 are linearly independent over the field of real numbers.

Example 45:

Show that the vectors (1, 1, 0, 0) (0, 1, – 1, 0), (0, 0, 0, 3) in $\mathbf{R}^4$ *are linearly independent.*

Solution:

Let a, b, c be scalars *i.e.*, real numbers such that

$$a(1, 1, 0, 0) + b(0, 1, -1, 0) + c(0, 0, 0, 3) = (0, 0, 0, 0) \quad ...(1)$$

Then $a + 0b + 0c = 0,$

$$a + b + 0c = 0,$$

$$0a - b + 0c = 0,$$

$$0a + 0b + 3c = 0.$$

The only solution of the above equations is

$$a = 0,\ b = 0,\ c = 0.$$

Thus the linear relation (1) among the three given vectors is possible only if $a = 0,\ b = 0,\ c = 0.$

Hence the three given vectors in $\mathbf{R}^4$ are linearly independent.

Example 46:

Is the vector (3, – 1, 0, – 1) in the subspace of $\mathbf{R}^4$ *spanned by the vectors (2, – 1, 3, 2), (– 1, 1, 1, – 3) and (1, 1, 9, – 5)?*

Solution:

Let $\alpha = (3, -1, 0, -1)$, $\alpha_1 = (2, -1, 3, 2)$,

$\alpha_2 = (-1, 1, 1, -3)$, $\alpha_3 = (1, 1, 9, -5)$.

If a can be expressed as a linear combination of the vectors $\alpha_1, \alpha_2, \alpha_3$, then it will be in the subspace of $\mathbf{R}^4$ spanned by these vectors otherwise it will not be.

Let $\alpha = a\alpha_1 + b\alpha_2 + c\alpha_3$ where $a, b, c \in \mathbf{R}$,

Then $(3, -1, 0, -1) = a\,(2, -1, 3, 2) + b\,(-1, 1, 1, -3) + c\,(1, 1, 9, -5).$

$$\therefore \quad 2a - b + c = 3, \qquad ...(1)$$

$$-a + b + c = -1, \qquad ...(2)$$

$$3a + b + 9c = 0, \qquad ...(3)$$

$$\text{and} \quad 2a - 3b - 5c = -1. \qquad ...(4)$$

Adding the equations (1) and (2), we get

$$a + 2c = 2$$

Again adding the equations (1) and (3) we get

$$5a + 10c = 3 \qquad ...(6)$$

Multiplying the equation (5) by 5, we get

$$5a + 10c = 10. \qquad ...(7)$$

The relations (6) and (7) show that the equations (1), (2), (3) and (4) are inconsistent *i.e.*, do not posses a common solution. Hence the vector α cannot be expressed as a linear combination of the vectors $\alpha_1, \alpha_2, \alpha_3$. Therefore α is not in the subspace of $\mathbf{R}^4$ generated by the vectors $\alpha_1, \alpha_2, \alpha_3$.

Example 47:

Show that the vectors (1, 2, 1), (2, 1, 0), (1, – 1, 2) form a basis of $\mathbf{R}^3$.

Solution:

We know that the set {(1, 0, 0), (0, 1, 0), (0, 0, 1) forms a basis for $\mathbf{R}^3$. Therefore dim $\mathbf{R}^3 = 3$. If we show that the set S = {(1, 2, 1,), (2, 1, 0), (1, – 1, 2)} is linearly independent, then this will also form a basis for $\mathbf{R}^3$.

We have

$$a_1 (1, 2, 1) + a_2 (2, 1, 0) + a_2 (1, -1, 2) = (0, 0, 0)$$

$$\Rightarrow \quad (a_1 + 2a_2 + a_3, 2a_1 + a_2 - a_3, a_1 + 2a_3) = (0, 0, 0).$$

$$\therefore \quad a_1 + 2a_2 + a_3 = 0 \qquad ...(1)$$

$$2a_1 + a_2 - a_3 = 0 \qquad ...(2)$$

$$a_1 + 2a_3 = 0 \qquad ...(3)$$

Now, we shall solve these equations to get the values of a_1, a_2, a_3. Multiplying the equation (2) by 2, we get

$$4a_1 + 2a_2 - 2a_3 = 0. \qquad ...(4)$$

Subtracting (4) from (1), we get

$$-3a_1 + 3a_3 = 0$$

$$\Rightarrow \quad -a_1 + a_3 = 0. \qquad ...(5)$$

Adding (3) and (5), we get $3a_3 = 0 \Rightarrow a_3 = 0$. Putting $a_3 = 0$ in (3), we get $a_1 = 0$. Now putting $a_3 = 0$ and $a_1 = 0$ in (1), we get $a_2 = 0$.

Thus, solving the equations (1), (2) and (3), we get $a_1 = 0$, $a_2 = 0$, $a_3 = 0$. Therefore the set S is linearly independent. Hence it forms a basis of $\mathbf{R}^3$.

Example 48:

Select a basis, if any, of $\mathbf{R}^3(\mathbf{R})$ from the set

$\{\alpha_1, \alpha_2, \alpha_3, \alpha_4\}$, where $\alpha_1 = (1, -3, 2)$, $\alpha_2 = (2, 4, 1)$, $\alpha_3 = (3, 1, 3)$, $\alpha_4 = (1, 1, 1)$.

Solution:

Let $S = \{\alpha_1, \alpha_2, \alpha_3, \alpha_4\}$.

If any three vectors in S are linearly independent, then they will form a basis of the vector space $\mathbf{R}^3$ (**R**),

First consider the set $S_1 = \{\alpha_1, \alpha_2, \alpha_3\}$. Let us see whether the vectors in the set S_1 are linearly independent or not

The determinant of order 3 whose columns consist of the coordinates of the vectors α_1, α_2, α_3 is

$$= \begin{vmatrix} 1 & 2 & 3 \\ -3 & 4 & 1 \\ 2 & 1 & 3 \end{vmatrix}$$

$$= \begin{vmatrix} 1 & 0 & 0 \\ -3 & 10 & 10 \\ 2 & -3 & -3 \end{vmatrix}, \text{ by } C_2 - 2C_1 \text{ and } C_3 - 3C_1$$

$= -30 + 30 = 0.$

$\therefore$ the vectors $\alpha_1, \alpha_2, \alpha_3$ are linearly dependent and so they do not form a basis of $\mathbf{R}^3(\mathbf{R})$.

Now consider the set $S_2 = \{\alpha_1, \alpha_2, \alpha_4\}$.

The determinant of order 3 whose columns consist of the coordinates of the vectors $\alpha_1, \alpha_2, \alpha_4$ is

$$= \begin{vmatrix} 1 & 2 & 1 \\ -3 & 4 & 1 \\ 2 & 1 & 1 \end{vmatrix}$$

$$= \begin{vmatrix} 1 & 0 & 0 \\ -3 & 10 & 4 \\ 2 & -3 & -1 \end{vmatrix}, \text{ by } C_2 - 2C_1 \text{ and } C_2 - C_1$$

$= -10 + 12 = 2$ *i.e.*, $\neq 0$.

$\therefore$ the vectors $\alpha_1, \alpha_2, \alpha_4$ are linearly independent.

Since $S_2 = \{\alpha_1, \alpha_2, \alpha_4\}$ is a linearly independent subset of $\mathbf{R}^3$ containing three vectors, therefore it is a basis of $\mathbf{R}^3$.

Example 49:

Show that the set $S = \{1, x, x^2, ..., x^n\}$ of $n + 1$ polynomials in x is a basis of the vector space $P_n(\mathbf{R})$, of all polynomials in x (of degree at most n) over the field of real numbers.

Solution:

$P^n(\mathbf{R})$ is the vector space of all polynomials in x (of degree at most n) over the field $\mathbf{R}$ of real numbers.

$S = \{1, x, x^2, ..., x^n)$ is a subset of P_n consisting of $n + 1$ polynomials. To prove that S is a basis of the vector space $P_n(\mathbf{R})$.

First we show that the vectors in the set S are linearly independent over he field **R**.

The zero vector of the vector space $P_n(\mathbf{R})$ is the zero polynomial. Let $a_0, a_1, a_2, \ldots, a_n \in R$ be such that

$a_0 (1) + a_1x + a_2x^3 + \ldots + a_nx^n = 0$ *i.e.*, zero polynomial.

Now by the definition of the equality of two polynomials, we have

$$a_0 + a_1x + a_2x^2 + \ldots a_nx^n = 0$$

$$\Rightarrow \quad a_0 = 0, a_1 = 0, a_2 = 0, \ldots, a_n = 0.$$

$\therefore$ the vectors $1, x, x^2, \ldots, x^n$ of the vector space $P_n(\mathbf{R})$ are linearly independent.

Now we shall show that the set S generates the whole vector space $P_n(\mathbf{R})$

Let $\alpha = a_0 = a_1x + a_2x^2 + \ldots a_nx^n$ be any arbitrary member of P_n, where $a_0, a_1, \ldots, a_n \in \mathbf{R}$. The α is a linear combination of the polynomials $1, x, x^2, \ldots, x^n$ over the field **R**. Therefore S generates $P_n(\mathbf{R})$.

Since S is a linearly independent subset of $P_n(\mathbf{R})$ and it also generates $P_n(\mathbf{R})$, therefore it is a basis of $P_n(\mathbf{R})$.

Example 50(a):

The set W of ordered triads $(a_1, a_2, 0)$ where $a_1, a_2 \in F$ is a subspace of V_3 (F).

Solution:

Let us consider $\alpha = (a_1, a_2, 0)$ and $\beta = (b_1, b_2, 0)$ be any two elements of W. Then $a_1, a_2, b_1, b_2 \in F$. If a, b be any two elements of F, we have

$a\alpha + b\beta = a\,(a_1, a_2, 0) + b\,(b_1, b_2, 0) = (aa_1, aa_2, 0) + (bb_1, bb_2, 0)$

$= (aa_1 + bb_1, aa_2 + bb_2, 0) \in W$ since $aa_1 + bb_1, aa_2 + bb_2 \in F$ and the last co-ordinate of this triad is zero.

Hence W is a subspace of V_3 (F).

Example 50(b):

Let V be the vector space of all polynomials in an indeterminate x over a field F. Let W be a subset of V consisting of all polynomials of degree $\leq$ n. Then W is a subspace of V.

Solution:

Let α and β be any two elements of W. Then α, β, are polynomials over F of degree $\leq$ n. If a, b are any two elements of F, then $a\alpha + b\beta$ will also be a polynomial of degree $\leq$ n. Therefore $a\alpha + b\beta \in W$. Hence W is a subspace of V.

Example 51:

*Let **R** be the field of real numbers. Which of the following are subspaces of V_3 (**R**) ?*

(i) $\{(x, 2y, 3z): x, y, z \in \mathbf{R}\}$

(ii) $\{(x, x, x): x \in \mathbf{R}\}$.

(iii) $[(x, y, z)$: x, y, z are rational numbers}?

Solution:

(i) Let $W = \{(x, 2y, 3z): x, y, z \in \mathbf{R}\}$.

Let $\alpha = (x_1, 2y_1, 3z_1)$ and $\beta = (x_2, 2y_2, 3z_2)$ be any two elements of W. Then $x_1, y_1, z_1, x_2, y_2, z_2$ are real numbers. If a, b are any two real numbers, then

$$a\alpha + b\beta = a\,(x_1, 2y_1, 3a_1) + b\,(x_2, 2y_2, 3z_2)$$
$$= (ax_1 + bx_2, 2ay_1 + 2by_2, 3az_1 + 3bz_2)$$
$$= (ax_1 + bx_2, 2\,[ay_1 + by_2], 3\,[az_1 + bz_2])$$

$\in W$ since $ax_1 + bx_2$, $ay_1 + by_2$, $az_1 + bz_2$ are real numbers.

Thus $a, b \in \mathbf{R}$ and $\alpha, \beta \in W \Rightarrow a\alpha + b\beta \in W$.

$\therefore$ W is a subspace of V_3 (**R**).

(ii) Let $W = \{(x, x, x): x \in \mathbf{R}\}$.

Let $\alpha = (x_1, x_1, x_1)$ and $\beta = (x_2, x_2, x_2)$ be any two elements of W. Then x_1, x_2, are real numbers. If a, b, are any real numbers, then

$$a\alpha + b\beta = a\,(x_1, x_1, x_1) + b\,(x_2, x_2, x_2)$$
$$= (ax_1 + bx_2, ax_1 + bx_2, ax_1 + bx_2) \in W$$

since $ax_1 + bx_2 \in \mathbf{R}$.

(iii) Let W = {(x, y, z): x, y, z are rational numbers}.

Now $\alpha = (3, 4, 5)$ is an element of W. Also $a = \sqrt{3}$ is an element of R. But $a\alpha = \sqrt{3}\,(3, 4, 5) = (3\sqrt{3}, 4\sqrt{3}, 5\sqrt{3}) \notin W$ since $3\sqrt{3}, 4\sqrt{3}, 5\sqrt{3}$ are not rational numbers.

Therefore W is not closed under scalar multiplication. Hence W is not a subspace of V_2 (**R**).

Example 52:

If a_1, a_2, a_3 are fixed elements of a field F, then the set W of all ordered triads (x_1, x_2, x_3) of elements of F, such that

$a_1x_1 + a_2x_2 + a_2x_2 = 0$ is a subspace of V_2 (F).

Solution:

Let $\alpha = (x_1, x_2, x_3)$ and $\beta = (y_1, y_2, y_3)$ be any two elements of W. Then $x_1, x_2, x_3, y_1, y_2, y_3$ are elements of F and are such that

$$a_1x_1 + a_2x_2 + a_3x_3 = 0 \quad ...(1)$$

$$a_1y_1 + a_2y_2 + a_3y_3 = 0 \quad ...(2)$$

If a, b, be any two elements of F, we have

$$a\alpha + b\beta = a(x_1, x_2, x_3) + b(y_1, y_2, y_3)$$

$$= (ax_1, ax_2, ax_3) + (by_1, by_2, by_3) = (ax_1 + by_1, ax_2 + by_2, ax_3 + by_3).$$

Now $a_1(ax_1 + by_1) + a_2(ax_2 + by_2) + a_3(ax_3 + by_3)$

$= a(a_1x_1 + a_2x_2 + a_3x_3) + b(a_1y_1 + a_2y_2 + a_3y_3)$

$= a0 + b0$ [by (1) and (2)]

$= 0.$

$\therefore a\alpha + b\beta = (ax_1 + by_1, ax_2 + by_2, ax_3 + by_3) \in W.$

Hence W is a subspace of V_3 (F).

Example 53:

Prove that any finite set S of vectors, not all the zero vectors, contains a linearly independent subset T which spans the same space as S.

Solution:

Let S be a finite set of vectors belonging to a vector space V(F). Assume that no member of S is zero vector for if any member of S is zero vector, we can omit it from S without affecting the subspace spanned by S.

Let $S = \{\alpha_1, \alpha_2, ..., \alpha_m\}$.

If S is linearly independent, then S itself is the required linearly independent subset T of S which spans the same subspace of V as S.

If S is linearly dependent, then there exists, a vector $\alpha_1 \in S$ which can be expressed as a linear combination of the preceding vectors $\alpha_1, \alpha_2, ..., \alpha_{i-1}$.

If we omit this vector α_i from S, then the remaining subset S' of S containing m – 1 vectors spans the same subspace of V as S.

If S' is linearly independent, then S' will be the required linearly independent subset of S which spans the same subspace of V as S. If S' is linearly dependent, then proceeding as above we shall get a new subset of S.

Continuing this process we shall, after a finite number of steps, obtain a linearly independent subset of S which spans the same space as S.

At the most it may happen that we shall be left with a subset of S which contains only one non-zero vector and which spans the same space as S. We know that a set containing a single non-zero vector is definitely linearly independent.

Hence any finite set S of vectors, not all the zero vectors, definitely contains a linearly independent subset T which spans the same space as S.

Example 54:

Let V be a vector space. Let W be a subspace of V generated by the vectors $\alpha_1, \ldots, \alpha_3$. Prove that W is spanned by a linearly independent subset of $\alpha_1, \ldots, \alpha_3$.

Solution:

W is a subspace of V generated by a finite set

$$S = \{\alpha_1, \ldots, \alpha_2\}.$$

To show that there exists a linearly independent subset T of S which also spans W.

Example 55:

If W is a subspace of a finite dimensional vector space V, prove that any basis of W can be extended to form a basis of V.

Solution:

Let V (F) be a finite dimensional vector space of dimension n. Let $\{\beta_1, \beta_2, \ldots, \beta_n\}$ be a basis of V.

Let W be subspace of V. Then W itself is finite dimensional and dim $W \leq n$. Let dim W = m and let $S = \{\alpha_1, \alpha_2, \ldots, \alpha_m\}$ be a basis of W.

Then S is a linearly independent subset of V. To show that S can be extended to form a basis of V.

Example 56:

If n vectors span a vector space V containing r linearly independent vectors, then show that $n \geq r$.

Solution:

Suppose a subset S of V containing n vectors spans V. Then there exists a linearly independent subset T of S which also spans V. This subset T of S will form a basis of V. Suppose the number of vectors in T is *m*. Then dim $V = m \leq n$.

Since dim V = m, therefore any subset of V containing more than m vectors will be linearly dependent.

Hence if S_1 is a linearly independent subset of V and S_1 contains r vectors, we must have

$$r \leq m \quad \Rightarrow r \leq n. \qquad [\because m \leq n]$$

Example 57:

Show that a finite subset W of a vector space V(F) is linearly dependent if and only if some element of W can be expressed as a linear combination of the others.

Solution:

Let $W = \{\alpha_1, \alpha_2, ..., \alpha_n\}$ be a finite subset of a vector space V(F).

First suppose that W is linearly dependent. Then to show that some vector of W can be expressed as a linear combination of the others.

Since the vectors $\alpha_1, \alpha_2, ..., \alpha_n$ are linearly dependent, therefore there exist scalars $a_1, a_2, ..., a_n$ not all zero such that

$$a_1\alpha_1 + a_2\alpha_2 + ... + a_n\alpha_n = 0. \qquad ...(1)$$

Suppose $a_r \neq 0$, $1 \leq r \leq n$.

Then from (1), we have

$$a_r\alpha_r = -a_1\alpha_1 - a_2\alpha_2 - ... - a_{r-1}\alpha_{r-1} - a_{r+1}\alpha_{r+1} - ... - a_n\alpha_n$$

$$\Rightarrow \quad a_r^{-1}(a_r\alpha_r) = a_r^{-1}(-a_1\alpha_1 - a_2\alpha_2 - ... - a_{r-1}\alpha_{r-1} - a_{r+1}\alpha_{r+1} - ... - a_n\alpha_n)$$

$$\Rightarrow \quad \alpha_r = (-a_r^{-1}a_1)\alpha_1 + (-a_r^{-1}a_2)\alpha_2 + ... + (-a_r^{-1}a_{r-1})\alpha_{r-1}$$
$$+ (a_r^{-1}a_{r+1})\alpha_{r+1} + ... + (-a_r^{-1}a_n)\alpha_n$$

Thus, α_r is a linear combination of the other vectors of the set W.

Conversely suppose that some vector of W is a linear combination of the others.

Then to prove that W is linearly dependent.

Without loss of generality suppose α_1 is a linear combination of the vectors $\alpha_2, ..., \alpha_n$.

Let $\quad \alpha_1 = b_2\alpha_2 + b_3\alpha_3 + ... + b_n\alpha_n$,

where $\quad b_2, b_2, ..., b_n \in F$.

Then $\quad \alpha_1 - b_2\alpha_2 - b_3\alpha_3 - ... - b_n\alpha_n = 0. \qquad ...(2)$

Since in the linear relation (2) among the vectors $\alpha_1, \alpha_2, ..., \alpha_n$, the scalar coefficient of α_1 is 1 which is $\neq 0$, therefore the vectors $\alpha_1, \alpha_2, ..., \alpha_n$ are linearly dependent.

Example 58:

Determine whether or not the following vectors form a basis of $\mathbf{R}^3$*:*

(1, 1, 2), (1, 2, 5), (5, 3, 4).

Solution:

We know that dim $\mathbf{R}^3 = 3$. If the given set of vectors is linearly independent, it will form a basis of $\mathbf{R}^3$ otherwise not. We have

$$a_1 (1, 1, 2) + a_2 (1, 2, 5) + a_2 (5, 3, 4) = (0, 0, 0)$$

$$\Rightarrow (a_1 + a_2 + 5a_3, a_1 + 2a_2 + 3a_3, 2a_1 + 5a_2 + 4a_2) = (0, 0, 0).$$

$$\therefore \quad a_1 + a_2 + 5a_3 = 0 \quad ...(1)$$

$$a_1 + 2a_2 + 3a_3 = 0 \quad ...(2)$$

$$2a_1 + 5a_2 + 4a_3 = 0. \quad ...(3)$$

Now, we shall solve these equations to get the values of a_1, a_2, a_3. Subtracting (2) from (1), we get

$$-a_2 + 2a_3 = 0. \quad ...(4)$$

Multiplying (1) by 2, we get

$$2a_1 + 2a_2 + 10a_2 = 0. \quad ...(5)$$

Subtracting (5) from (3), we get

$$3a_2 - 6a_3 = 0$$

$$\Rightarrow \quad a_2 - 2a_3 = 0. \quad ...(6)$$

We see that the equations (4) and (6) are the same and give $a_2 = 2a_3$. Putting $a_2 = 2a_3$ in (1), we get $a_1 = -7a_3$. If we put $a_2 = 1$ we get $a_2 = 2$ and $a_1 = -7$. Thus, $a_1 = -7$, $a_2 = 2$, $a_3 = 1$ is a non-zero solution of the equations (1), (2) and (3). Hence the give set is linearly dependent so it does not form a basis of $\mathbf{R}^2$.

Example 59:

Show that the set {(1, i, 0), (2i, 1, 1), (0, 1 + i, 1 – i)} is a basis for $V_3(C)$*.*

Solution:

We know that dim $V_3(C)$ or $C^3 = 3$. If the given set containing three vectors is linearly independent, it will form a basis of V_3 (C) otherwise not.

Let a, b, c $\in$ C be such that

$$a (1, i, 0) + b (2i, 1, 1) + c (0, 1, + i, 1 - i) = (0, 0, 0)$$

$$\Rightarrow (a + 2ib + 0c, ai + b + c [1 + i], 0a + b + c [1 - i]) = (0, 0, 0).$$

$$\therefore \quad a + 2ib = 0, \quad ...(1)$$

$$ai + b + c\,(1 + i+ = 0, \qquad ...(2)$$

and $\quad b + c\,(1 - i) = 0. \qquad ...(3)$

Now we shall solve these equations to get the values of a, b, c.

Multiplying (1) y – i and adding to (2), we get

$$3b + c\,(1 + i) = 0. \qquad ...(4)$$

Multiplying (3) by 3 and subtracting from (4), we get

$$c\,(1 + i) - 3c\,(1 - i) = 0$$

$\Rightarrow \quad c\,(1 + i - 3 + 3i) = 0 \Rightarrow c\,(-2 + 4i) = 0 \Rightarrow c = 0.$

Putting c = 0 in (3), we get b = 0

Putting b = 0 in (1), we get a = 0.

Thus, the only solution of the equations (1), (2) and (3) is a = 0, b = 0, c = 0. Therefore the three given vectors are linearly independent and so they form a basis of $V_3(C)$.

Example 60:

For the 3-dimensional space $\mathbf{R}^3$ *over the field of real numbers* $\mathbf{R}$, *determine if the set {(2, – 1, 0), (3, 5, 1), (1, 1, 2)} is a basis.*

Solution:

We have dim $\mathbf{R}^3 = 3$. If the given set containing three vectors is linearly independent, it will form a basis of $\mathbf{R}^3$ otherwise not.

Let a, b, c $\in \mathbf{R}$ be such that

$$a\,(2, -1, 0) + b\,(3, 5, 1) + c\,(1, 1, 2) = (0, 0, 0)$$

$\Rightarrow \quad (2a + 3b + c, -a + 5b + c, 0a + b + 2c) = (0, 0, 0).$

$\therefore \quad 2a + 3b + c = 0, \qquad ...(1)$

$$-a + 5b + c = 0, \qquad ...(2)$$

and $\quad b + 2c = 0. \qquad ...(3)$

Now we shall solve these equations to get the values of a, b, c.

Multiplying (2) by 2 and adding to (1), we get

$$13b + 3c = 0. \qquad ...(4)$$

Multiplying (3) by 13 and then subtracting (4) from it, we get

$$23c = 0 \Rightarrow c = 0.$$

Putting c = 0 in (3), we get b = 0.

Putting b = 0, c = 0 in (1), we get a = 0.

Thus, the only solution of the equations (1), (2) and (3) is a = 0, b = 0, c = 0. Therefore the three given vectors are linearly independent and so they form a basis of $\mathbf{R}^3$.

Example 61:

Show that a system X consisting of the vectors

$\alpha_1 = (1, 0, 0, 0)$, $\alpha_2 = (0, 1, 0, 0)$, $\alpha_3 = (0, 0, 1, 0)$ and $\alpha_4 = (0, 0, 0, 1)$ is a basis set of $\mathbf{R}^4$ (**R**).

Solution:

First we show that the set X is a linearly independent set of vectors.

If a_1, a_2, a_3, a_4 be any scalars *i.e.*, elements of the field R, then

$$a_1\alpha_1 + a_2\alpha_2 + a_3\alpha_3 + a_4\alpha_4 = \text{zero vector}$$

$$\Rightarrow a_1 (1, 0, 0, 0) + a_2 (0, 1, 0, 0) + a_3 (0, 0, 1, 0) + a_4 (0, 0, 0, 1) = (0, 0, 0, 0)$$

$$\Rightarrow (a_1, a_2, a_3, a_4) = (0, 0, 0, 0)$$

$$\Rightarrow a_1 = 0, a_2 = 0, a_3 = 0, a_4 = 0.$$

Therefore the given set X of four vectors is linearly independent.

Now we shall that X generates R^4 i.e., each vector of R^4 can be expressed as a linear combination of the vectors of X.

Let (a, b, c, d) be any vector in R^4. We can write

$$(a, b, c, d) = a(1, 0, 0, 0) + b(0, 1, 0, 0) + c(0, 0, 1, 0) + d(0, 0, 0, 1)$$
$$= a\alpha_1 + b\alpha_2 + c\alpha_3 + d\alpha_4.$$

Thus (a, b, c, d) has been expressed as a linear combination of the vectors of X and so X generates R^4.

Since X is a linearly independent subset of R^4 and it also generates R^4, therefore it is a basis of R^4.

Example 62:

If V is a finite dimensional vector space and f is an isomorphism of V into V, prove that f must map V onto V.

Solution:

Let V (F) be a finite dimensional vector space of dimension n. Let f be an isomorphism of V into V *i.e.* f is a linear transformation and f is one-one. To prove that f is onto V.

Let S = $\{\alpha_1, \alpha_2, ..., \alpha_n\}$ be a basis of V. We shall first prove that

$S' = \{f(\alpha_1), f(\alpha_2), ..., f(\alpha_n)\}$ is also a basis of V. We claim that S' is linearly independent. The proof is as follows:

Let $a_1 f(\alpha_1) + a_2 f(\alpha_2) + ... + a_n f(\alpha_n) = 0$ (zero vector of V)

$\Rightarrow f(a_1\alpha_1 + a_2\alpha_2 + ... + a_n\alpha_n) = 0$ [$\because$ f is a linear transformation]

$\Rightarrow a_1 a_1 + a_2 a_2 + ... + a_n \alpha_n = 0$ [$\because$ f is one-one and f (0) = 0]

$\Rightarrow a_1 = 0, a_2 = 0, ..., a_n = 0$

since $\alpha_1, \alpha_2, ..., \alpha_n$ are linearly independent.

$\therefore$ S' is linearly independent.

Now V is of dimension n and S' is a linearly independent subset of V containing n vectors. Therefo:e S' must be a basis of V.

Therefore each vector in V can be expressed as a linear combination of the vectors belonging to S'.

Now we shall show that f is onto V. Let α be any element of V. Then there exist scalars $c_1, c_2, ... c_n$ such that

$$\alpha = c_1 f(\alpha_1) + c_2 f(\alpha_2) + ... + c_n f(\alpha_n)$$
$$= f(c_1\alpha_1 + c_2\alpha_2 + ... + c_n\alpha_n).$$

Now $c_1\alpha_1 + c_2\alpha_2 + ... + c_n\alpha_n \in V$ and the f image of this element is a. Therefore f is onto V. Hence f is an isomorphism of V onto V.

Example 63:

If V is finite dimensional and f is a homomorphism of V into itself which is not onto prove that there is some $\alpha \neq 0$ in V such that $f(\alpha) = 0$.

Solution:

If f is a homomorphism of V into itself, then f (0) = 0. Suppose there is no non-zero vector a in V such that $f(\alpha) = 0$.

Then f is one-one. Because

$$f(\beta) = f(\gamma) \Rightarrow f(\beta) - f(\gamma) = 0$$
$$\Rightarrow f(\beta - \gamma) = 0 \quad [\because \text{ f is linear transformation}]$$
$$\Rightarrow \beta - \gamma = 0 \Rightarrow \beta = \gamma.$$

Now V is finite dimensional and f is a linear transformation of V into itself. Since f is one-one, therefore f must be onto V. But it is given that f is not onto. Therefore our assumption is wrong. Hence there will be a non-zero vector α in V such that

$$f(\alpha) = 0.$$

Example 64:

The mapping $f: V_3(F) \rightarrow V_2(F)$ *defined by*

$$f(a_1, a_2, a_3) = (a_1, a_2)$$

is a homomorphism of $V_3(F)$ *onto* $V_2(F)$. *What is the kernel of this homomorphism ?*

Solution:

Let $\alpha = (a_1, a_2, a_3)$ and $\beta = (b_1, b_2, b_3)$ be any two elements of $V_3(F)$. Also let a, b be any two elements of F. We have

$$f(a\alpha + b\beta) = f[a(a_1, a_2, a_3) + b(b_1, b_2, b_3)]$$

$$= f[(aa_1 + bb_1, aa_2 + bb_2, aa_2 + bb_3)] = (aa_1 + bb_1, aa_2 + bb_2)$$

$$= a(a_1, a_2) + b(b_1, b_2) = af(a_1, a_2, a_2) + bf(b_1, b_2, b_3)$$

$$= af(\alpha) + bf(\beta).$$

$\therefore$ f is a linear transformation.

To show that f is onto $V_2(F)$. Let (a_1, a_2) be any element of $V_2(F)$. Then $(a_1, a_2, 0) \in V_2(F)$ and we have $f(a_1, a_2, 0) = (a_1, a_2, 0) = (a_1, a_2)$. Therefore f is onto $V_2(F)$.

Therefore f is a homomorphism of $V_3(F)$ onto $V_2(F)$. If W is the kernel of this homomorphism then $W = \{(0, 0, a): a \in F\}$.

We have $\forall\, a \in F$, $f(0, 0, a) = (0, 0)$ = the zero vector of $V_2(F)$.

Also if $f(a_1, a_2, a_3) = (0, 0)$ then $f(a_1, a_2, a_3) = (a_1, a_2) = (0, 0)$ implies $a_1 = 0$, $a_2 = 0$. Therefore $(a_1, a_2, a_3) \in W$.

Hence W is the kernel of f.

Example 65:

Show that the three vectors (1, 1, – 1), (2, – 3, 5) and (–2, 1, 4) of $\mathbf{R}^3$ *are linearly independent.*

Solution:

Let a, b, c be scalars *i.e.*, real numbers such that

$$a(1, 1, -1) + b(2, -3, 5) + c(-2, 1, 4) = (0, 0, 0)$$

i.e., $(a + 2b - 2c, a - 3b + c, -a + 5b + 4c) = (0, 0, 0)$

i.e.

$$a + 2b - 2c = 0 \quad ...(1)$$

$$a - 3b + c = 0 \quad ...(2)$$

$$-a + 5b + 4c = 0 \quad ...(3)$$

Now we shall solve the simultaneous equations (1), (2) and (3).

Multiplying (2) by 2 and adding to (1), we get

$$3a - 4b = 0 \qquad ...(4)$$

Again multiplying (1) by 2 and adding to (3), we get

$$a + 9b = 0$$

Multiplying (5) by 3 and subtracting from (4), we get

$$-31b = 0 \text{ or } b = 0.$$

Putting b = 0 in (5), we get a = 0.

Now putting a = 0, b = 0 in (1), we get c = 0.

Thus, a = 0, b = 0, c = 0 is the only solution of the equations (1), (2) and (3).

$\therefore \qquad a(1, 1, -1) + b(2, -3, 5) + c(-2, 1, 4) = (0, 0, 0)$

$\Rightarrow \qquad a = 0, b = 0, c = 0.$

Hence the vectors (1, 1, – 1), (2, – 3, 5), (– 2, 1, 4) of R^3 are linearly independent.

Example 66:

How many elements are there in the vector space of polynomials of degree at most in which the coefficients are the elements of the field I (p) over the field I (p), p being a prime number?

Solution:

Here we know that the field I (p) is the field

$(\{0, 1, 2, ..., p - 1)\}, +_p, \times_p)$.

The number of distinct elements in the field I (p) is p.

If f(x) is a polynomial of degree at most n over the field I (p), then we have

$f(x) = a_0 + a_1x + a_2x^2 + ... + a_nx^n$, where $a_0, a_1, a_2, ..., a_n \in I(p)$.

Now in the polynomial f(x), the coefficient of each of the n + 1 terms $a_0, a_1x, a_2x^2, ..., a_nx^n$ can be filled in p ways because any of the p elements of the field I (p) can be filled there.

Thus, we can have p × p × p × ... upto (n + 1) times *i.e.*, p^{n+1} distinct polynomials of degree at most n over the field I (p). Hence if P_n is the vector space of polynomials of degree at most n in which the coefficients are the elements of the field I (p) over the field I (p), then P_n has p^{n+1} distinct elements.

Example 67:

Show that the complex field C is a vector space over the real field R.

Solution:

Let us consider C as the set of vectors and R as the set of scalars.

Take the addition of complex numbers as the addition of vectors. Then (C, +) is an abelian group.

Again, let a be any scalar *i.e.*, $a \in R$ and α be any vector *i.e.*, $\alpha \in C$. Now $a \in R \Rightarrow a \in C$. Thus, both a and α are in C. Regard the composition of scalar multiplication as the multiplication of a and α in the field. Since C is a field, therefore $a\alpha \in C$ and thus C is closed for scalar multiplication.

The real number 1 which is the unity element of the field R is also the unity element of the field C.

Now let α, β be any vectors *i.e.*, $\alpha, \beta \in C$ and a, b be any scalars *i.e.* $a, b \in R$. Since $R \subset C$, therefore, we have

$$a, b \in R \Rightarrow a, b \in C$$

We have $a(\alpha + \beta) = a\alpha + a\beta$ and $a(a + b)\, \alpha = a\alpha + b\alpha$. (distributive laws in the field C).

Also $(ab)\,\alpha = a\,(b\alpha)$ because the multiplication on the field C is associative.

Also $1\alpha = \alpha$ because 1 is also the multiplicative identity in the field C.

Hence C is a vector space over R.

Example 68:

The vector of all real valued continuous (differentiable or integrable) functions defined in some interval [0, 1].

Solution:

Let us consider V denote the set of all real valued continuous functions of x defined in the interval [0, 1].

Then V is a vector space over the field R of real numbers with vector addition and scalar multiplication then we have

$$(f + g)\,(x) = f(x) + g(x)\ \forall\ f, g \in V$$

and $$(af)\,(x) = af(x)\ \forall\ a \in R, \forall\ f \in V.$$

As in rings, we should first prove that V is an abelian group with respect to addition composition.

V is closed with respect to scalar multiplication since af is also a real valued continuous function in [0, 1]. Here, we observe that

1. If $a \in R$ and $f, g \in V$, then

   $[a\,(f + g)]\,(x) = a\,[(f + g)\,(x)]$

$= a\,[f(x) + g(x)] = af(x) + ag(x)$

$= (af)(x) + (ag)(x) = (af + ag)(x).$

$\therefore\ a(f + g) = af + ag.$

2. If $a, b \in R$ and $f \in V$, then

$[(a + b) f](x) = (a + b) f(x) = af(x) + bf(x) = (af)(x) + (bf)(x)$

$= (af + bf)(x).$

$\therefore\ (a + b) f = af + bf.$

3. If $a, b \in R$ and $f \in V$, then

$[(ab) f](x) = (ab) f(x) = a\,[bf(x)] = a\,[(bf)(x)] = [a(bf)](x):$

$\therefore\ (ab) f = a(bf).$

4. If 1 is the unity element of R and $f \in V$, then

$(1f)(x) = 1\,f(x) = f(x).$

$\therefore\ 1f = f.$

Hence V is a vector space over R.

2

INNER PRODUCT SPACES

INTRODUCTION

Throughout this chapter we shall deal only with real or *complex* vector spaces. Thus, if V is the vector space over the field F, then F will not be an arbitrary field. In this chapter, F will be either the field R of real numbers or the field C of complex numbers.

Before defining inner product and inner product spaces, we shall just give some important properties of complex numbers.

Let $z \in C$ *i.e.*, let 2 be a complex number. Then $2 = x + iy$ where $x, y \in R$ and $i = \sqrt{-1}$. Here x is called the real part of z and y is called the imaginary part of z. We write x = Re z, and y = Im z. The *modulus* of the complex number $z = x + iy$ is the non-negative real number $\sqrt{(x^2 + y^2)}$ and is denoted by $|z|$. Also if $z = x + iy$ is a complex number, then the complex number $\overline{z} = x - iy$ is called the *conjugate complex* of z. If $z = \overline{z}$, then $x + iy = x - iy$ and therefore $y = 0$. Thus $z = \overline{z}$ implies that z is real. Obviously we have

(i) $z + \overline{z} = 2x = 2 \text{ Re } z$,

(ii) $z + \overline{z} = 2iy = 2i \text{ Im } z$,

(iii) $z\overline{z} = x^2 + y^2 = |z|^2$,

(iv) $|z| = 0 \Leftrightarrow x = 0, y = 0$

i.e., $|z| = 0 \Leftrightarrow z = 0$,

(v) $(\overline{z}) = z$,

(vi) $|\overline{z}| = |z| = |z|$, and

(vii) $|z| = \sqrt{(x^2 + y^2)} \geq x$ *i.e.*, $|z| \geq \text{Re } z$.

If z_1 and z_2 are two complex numbers, then

(i) $|z_1 + z_2| \leq |z_1| + |z_2|$

(ii) $\overline{z_1 + z_2} = \overline{z}_1 + \overline{z}_2$

(iii) $\overline{z_1 z_2} = \overline{z}_1 \overline{z}_2$ and

(iv) $\overline{z_1 - z_2} = \overline{z}_1 - \overline{z}_2$

INNER PRODUCT SPACES

Definition : *Let V (F) be a vector space where F is either the field of real numbers or the field of complex numbers. An inner product on V is a function from V × V into F which assigns to each ordered pair of vectors α, β in V a scalar (α, β) in such a way that*

(i) $(\alpha, \beta) = \overline{(\beta, \alpha)}$ [Here $\overline{(\beta, \alpha)}$ denotes the conjugate complex of the number (β, α)].

(2) $(a\alpha + b\beta, \gamma) = \alpha(\alpha, \gamma) + b(\beta, \gamma)$

(3) $(\alpha, \alpha) \geq 0$ and $(\alpha, \alpha) = 0 \Rightarrow \alpha = 0$

for any $\alpha, \beta, \gamma \in V$ and $a, b \in F$.

Also the vector space V is then said to be an *inner product space* with respect to the specified inner product defined on it.

It should be noted that in the above definition (α, β) does not denote the ordered pair of the vectors α and β. But it denotes the inner product of the vectors α and β. It is an element of V which has been assigned by the function (named as inner product) to the vectors α and β, Sometimes the inner product of the ordered pair of vectors α, β is also written as (α | β) If F=R, then (α, β) is a real number and if F = C, then (α, β) is a complex number.

If F is the field of real numbers, then the complex conjugate appearing in (1) is superfluous and (1) should be read as (α, β) = (β, α). If F is the field of complex numbers, then from (1), we have $(\alpha, \alpha) = \overline{(\alpha, \alpha)}$ and therefore (α, α) is real Thus (α, α) is always real whether F = R or F = C Therefore the inequality given in (3) makes sense.

If V (F) is an inner product space, then it is called a *Euclidean* space if F is the field of real numbers. Also it is called a Unitary space if F is the field of complex numbers.

Notes:

1. The properly (3) in the definition of inner product is called non-negativity. The property (2) is called the *linearity property*. If F=R,

then the property (.1) is called *symmetry* and if F = C, then it is called conjugate symmetry.

2. If in an inner product space V (F), the vector α is 0, then $(\alpha, \alpha) = 0$.

We have $(0, 0) = (00,0)$ [$\because$ 00 = 0 in V]

$= 0\ (0,0)$ [by linearity property of inner product]

$= 0$ [$\because$ (0, 0) $\in$ F and therefore 0(0, 0) = 0]

NORM OR LENGTH OF A VECTOR IN AN INNER PRODUCT SPACE

Consider the vector space V_3 (R) with standard inner product defined on it. If $\alpha = (a_1, a_2, a_3) \in V_3$ (R), we have

$$(\alpha, \alpha) = a_1^2 + a_2^2 + a_3^2$$

Now we know that in the three dimensional Euclidcan space $\sqrt{(a_1^2 + a_2^2 + a_3^2)}$ is the length of the vector $\alpha = (a_1, a_2, a_3)$. Taking motivation from this fact, we make the following definition.

Definition : Let V be an inner product space. If $\alpha \in V$, then the norm or the length of the vector α, written as $\| \alpha \|$, is defined as she positive square root of (α, α) *i.e.*,

$$\| \alpha \| = \sqrt{(\alpha, \alpha)}.$$

Unit Vector

Definition : Let V he an inner product space. If $\alpha \in$ V is such that \\ a ||=l, then a is called a unit vector. Thus in an inner product space a vector is called a unit vector if its length is 1

Theorem:

In an inner product space V(F), prove that

(i) $(a\alpha - b\beta, \gamma) = a(\alpha, \gamma) - b\ (\beta, \gamma)$

(ii) $(\alpha, a\beta + b\gamma) = \bar{a}\ (\alpha, \beta) + \bar{b}\ (\alpha, \gamma)$.

Proof:

(i) We have

$$(a\alpha - b\beta, \gamma) = (a\alpha + (-b)\ \beta, \gamma)$$

$= a\ (\alpha, \gamma) + (-b)\ (\beta, \psi)$ [by linearity property]

$= a\ (\alpha, \gamma) - b\ (\beta, -\gamma)$.

(ii) $(\alpha, a\beta + b\gamma) = \overline{(a\beta + b\gamma, \alpha)}$ [by conjugate symmetry]

$= \overline{a(\beta, \alpha) + b(\gamma, \alpha)}$ [by linearity property]

$= \bar{a}\ (\alpha, \beta) + \bar{b}\ (\alpha, \gamma)$

Notes:

1. If F = R, then the result (ii) can be simply read as

   $(\alpha, \alpha\beta + b\gamma) = a\ (\alpha, \beta) + b\ (\alpha, \gamma)$.

2. Similarly it can be proved that

$$(\alpha, a\beta - b\gamma) = \bar{a}(\alpha, \beta) - \bar{b}\ (\alpha, \gamma).$$

Also $(\alpha, \beta + y) = (\alpha, 1\beta + 1\gamma) = \bar{1}\ (\alpha, \beta) + \bar{1}\ (\alpha, \gamma)$

$= (\alpha, \beta) + (\alpha, \gamma)$.

ORTHOGONALITY

Definition : Let α and β be vectors in an inner product space V. Then a is said to he orthogonal to β if

$$(\alpha, \beta) = 0.$$

The relation of orthogonality in an inner product space is *symmetric*. We have

α is orthogonal to $\beta \Rightarrow (\alpha, \beta) = 0$

$\Rightarrow \overline{(\alpha, \beta)} = \bar{0}$

$\Rightarrow (\alpha, \beta) = 0$

$\Rightarrow \beta$ is orthogonal to α.

So we can say that two vectors α and β in an inner product space are orthogonal if $(\alpha, \beta) = 0$.

Notes:

1. If α is orthogonal to β, then every scalar multiple of α is orthogonal to β. Let k be any scalar. Then

   $(k\alpha, \beta) = k\ (\alpha, \beta) = k0 = 0$ [$\because (\alpha, \beta) = 0$]

   Therefore $k\alpha$ is orthogonal to β.

2. The zero vector is orthogonal to every vector. For every vector α in V, we have $(0, \alpha) = 0$.

3. The zero vector is the only vector which is orthogonal to itself.

We have

α is orthogonal to $\alpha \Rightarrow (\alpha, \alpha) = 0$

$\Rightarrow \alpha = 0$, by def. of an inner product space.

Definition : A vector a is said to be orthogonal to a set S if it is orthogonal to each vector in S. Similarly two subspaces are called orthogonal if every vector in each is orthogonal to every vector in the other.

ORTHOGONAL SET

Definition : Let S be a set of vectors in an inner product space V. Then S is said to be an orthogonal set provided that any two distinct vectors in S are orthogonal.

Theorem:

Let $S = \{\alpha_1, ..., \alpha_m\}$ be an orthogonal set of non-zero vectors in an inner product space V. If α vector β in V is in the linear span of S, then

$$\beta = \sum_{k=1}^{m} \frac{(\beta, \alpha_k)}{\| \alpha_k \|^2} \alpha_k.$$

Proof:

Since B $\in$ L(S), therefore b can be expressed as a linear combination of the vectors in S. Let

$$\beta = c_1\alpha_1 + ... c_m\alpha_m = \sum_{j=1}^{m} c_j\alpha_j$$

We have for each k where $1 \le k \le m$.

$$(\beta, \alpha_k) = \left(\sum_{j=1}^{m} c_j\alpha_j, \alpha_k \right)$$

$$= \sum_{j=1}^{m} c_j \left(\alpha_j, \alpha_k \right) \text{ [by linearity property of inner product]}$$

$= c_k \{\alpha_k, \alpha_k\}$ [On summing with respect to j. Note that 5 is an orthogonal set of non-zero vectors and so $(\alpha_k, \alpha_k) = 0$ if $j \ne k$]

Now $\alpha_k \ne 0$. Therefore $(\alpha_k, \alpha_k) \ne 0$ Thus

$\| \alpha_k \|^2 \ne 0.$

$$\therefore \quad c_k = \frac{(\beta, \alpha_k)}{\| \alpha_k \|^2}, 1 \le k \le m$$

Putting these values of $c_1, \ldots, c_m$ in (1), we get

$$\beta = \sum_{k=1}^{m} \frac{(\beta, \alpha_k)}{\|\alpha_k\|^2} \alpha_k$$

LINEAR FUNCTIONALS AND ADJOINTS

Theorem 1:

Let V be a finite dimensional inner product space, and f a linear functional on V. Then there exists a unique vector β in V such that f(α) = (α, β) for all α in V.

Proof:

Suppose $B = \{\alpha_1, \alpha_2, \ldots, \alpha_n\}$ is an orthonormal basis for V and f is linear functional on V. Let

$$\beta = \sum_{j=1}^{n} \overline{f(\alpha_j)\alpha_j} \qquad \ldots(1)$$

Then, β is a vector in V. Let β be a function from V to F defined by

$$g(\alpha) = (\alpha, \beta) \ \forall \ \alpha \in V \qquad \ldots(2)$$

We claim that g is a linear functional on V. If $\alpha, b \in F$ and $\gamma_1, \gamma_2 \in V$, then we have

$$g(a\gamma_1 + b\gamma_2) = (a\gamma_1 + b\gamma_2, \beta\} \qquad \text{[trom (2)]}$$

$$= a(\gamma_1, \beta) + b(\gamma_2, \beta)$$

$$= ag(\gamma_1) + bg(\gamma_2). \qquad \text{[from (2)]}$$

Thus g is a linear functional on V.

Now we shall show that g = f. If $\alpha_k \in B$, then

$$g(\alpha_k) = (\alpha_k, \beta)$$

$$= \left(\alpha_k, \sum_{j=1}^{n} \overline{f(\alpha_j)\alpha_j}\right) \qquad \text{[Putting } \beta \text{ from (1)]}$$

$$= \sum_{j=1}^{n} \overline{\left[\overline{f(\alpha_j)}\right]} \left(\alpha_k, \alpha_j\right) = \sum_{j=1}^{n} f(a_j) \left(\alpha_k, \alpha_j\right)$$

$$= f(\alpha_k)$$

[On summing with respect to j. Remember that $(\alpha_k, \alpha_j) = 1$ if j = k and $(\alpha_k, \alpha_j) = 0$ if $j \neq k$].

Thus g and f agree on a basis for V. Therefore g = f.

Therefore corresponding to a linear functional f on V there exists a vector β in such that $f(\alpha) = (\alpha, \beta) \ \forall \ \alpha \in V$.

Now to show that β is unique.

Let γ be a vector in V such that $f(\alpha) = (\alpha, \gamma) \ \forall \ \alpha \in V$,

Then $(\alpha, \beta) = (\alpha, \gamma) \ \forall \ \alpha \in V$

$\Rightarrow (\alpha, \beta) - (\alpha, \gamma) = 0 \ \forall \ \alpha \in V$

$\Rightarrow (\alpha, \beta - \gamma) = 0 \ \forall \ \alpha \in V$

$\Rightarrow (\beta - \gamma, \beta - \gamma) = 0$ [taking $\alpha = \beta - \gamma$]

$\Rightarrow \beta - \gamma = 0$

$\Rightarrow \beta = \gamma$.

Thus β is unique. Hence the theorem.

POSITIVE OPERATORS

Definition : *A linear operator T on an inner product space V is called positive, in symbols $T > 0$, if it is self-adjoint and if $(T\alpha, \alpha) > 0$ whenever $\alpha \neq 0$.*

If $\alpha = 0$, then $(T\alpha, \alpha) = 0$. Thus if T is positive, then $(T\alpha, \alpha) \leq 0$ for all α and $(T\alpha, \alpha) = 0 \Rightarrow \alpha = 0$. Also if T is self-adjoint and if $(T\alpha, \alpha) \geq 0$ for all α, and $(T\alpha, \alpha) = 0 \Rightarrow \alpha = 0$, then T is positive.

If V is a complex inner product space, then by theorem 10 of 4, $(T\alpha, \alpha) \geq 0$ for every, a implies that T must be self-adjoint Therefore a linear operator T on a complex inner product space is positive if and only if $(T\alpha, \alpha) > 0$ whenever $\alpha \neq 0$.

Non-Negative Operator

Definition : *A linear operator T on an inner product space V is called non-negative, in symbols $T \geq 0$, if it is self-adjoint and if $(T\alpha, \alpha) \geq 0$ for all α in V.*

Every positive operator is also a non-negative operator. If T is a non-negative operator, then $(T\alpha, \alpha) = 0$ is possible even if $\alpha \neq 0$. Therefore a non-negative operator may or may not be a positive operator.

If S and T are two linear operators on an inner product space V, then we define $S > T$ (or $T < S$) if $S - T > 0$.

Note : Some authors call a positive operator by the name 'strictly positive' or 'positive definite'. Also they use the phrase 'positive operator' in place of 'non-negative operator'.

Theorem:

Let V be an inner product space, and let T be a linear operator on V. Let p be the function defined on ordered pairs of vectors β, β in V by

$$p(\beta, \beta) = (T\beta, \beta).$$

Show that the function p is an inner product on V if and only if T is a positive operator.

Proof:

The function p obviously satisfies linearity property.

If a, b ∈ F and $\alpha_1, \alpha_2 \in V$, then

$p(a\alpha_1 + b\alpha_2\ \beta) = (T(a\alpha_1 + b\alpha_2), \beta) = (aT\alpha_1 + bT\alpha_2, \beta)$

$= a(T\alpha_1, \beta) + b(T\alpha_2, \beta) = ap(\alpha_1, \beta) + bp(\alpha_2, \beta)$.

∴ the function p satisfies linearity property.

Now the function p will be an inner product on V if and only

if $p(\alpha, \beta) = \overline{p(\beta, \alpha)}$

and $p(\alpha, \alpha) > 0$ if $\alpha \neq 0$.

We have $p(\alpha, \beta) = (T\alpha, \beta)$

and p $\overline{p(\beta, \alpha)} = \overline{(T\beta, \alpha)} = (\alpha, T\beta)$.

Also $p(\alpha, \alpha) = (T\alpha, \alpha)$.

∴ the function p will be an inner product on V if and only if

(i) $(T\alpha, \beta) = (\alpha, T\beta)\ \forall\ \alpha, \beta \in V$ *i.e.* T is self-adjoint.

(ii) $(T\alpha, \alpha) > 0$, if $\alpha \in 0$.

Hence the function p will be an inner product on V if and only if the linear operator T is positive.

Now we shall show that if V is finite-dimensional, then every inner product on V is of the type just described.

UNITARY OPERATORS

Definitions : Let U and V be two inner product spaces over the same field F and let T be a linear transformation from U into V. We say that

(i) T preserves inner products if $(T\alpha, T\beta) = (\alpha, \beta)$ for all α, β in U.

(ii) T preserves norms if $\| T\alpha \| = \| \alpha \|\ \forall\ \alpha \in U$.

(iii) T is an isometry if T preserves distances, *i.e.*, if

$\| T\alpha - T\beta \| = \| \alpha - \beta \|$ for all α, β in U.

Note that d $(T\alpha, \beta) = \| T\alpha - T\beta \|$.

Theorem:

Let U and V be two inner product spaces over the same field F and let T be a linear operator from U into V. Then the following three conditions on Tare equivalent:

(i) T preserves inner products.

(ii) T preserves norms.

(iii) T is an isometry.

Proof:

(i) ⇒ (ii).

It is given that T preserves inner products. Therefore

$(T\alpha, T\beta) = (\alpha, \beta)$ for all α, β in U. Taking $\beta = \alpha$, we get

$(T\alpha, T\alpha) = (\alpha, \alpha)$ *i.e.* $\| T\alpha \|^2 = \| \alpha \|^2$. Thus $\| T\alpha \| = \| \alpha \|$ for every α in U. Hence T preserves norms.

(ii) ⇒ (iii).

It is given that $\| T\alpha, \| = \| \alpha \|$ for every α in U. Therefore for all α, β in U, we have

$$\| T(\alpha - \beta) \| = \|\alpha - \beta\| \qquad \text{(taking } \alpha\text{–}\beta \text{ in place of } \alpha]$$

$$\Rightarrow \quad |T\alpha - \beta\| = \| = \|\alpha - \beta\|$$

⇒ T is an isometry.

(iii) ⇒ (i).

It is given that $\|T\alpha - T\beta\| = \|\alpha - \beta\|$ for all α, β in U, taking $\beta = 0$, we see that

$$\| T\alpha \| = \| \alpha \| \text{ for every } \alpha \text{ in U}$$

$$\Rightarrow \quad (T\alpha, T\alpha) = (\alpha, \alpha) \text{ for all } \alpha \text{ in U.}$$

Now let $\alpha, \beta \in$ U. Then

$$(T(\alpha + \beta), T(\alpha - \beta)) = (\alpha + \beta, \alpha + \beta)$$

$$\Rightarrow \quad (T\alpha + T\beta, T\alpha + T\beta) = (\alpha + \beta, \alpha + \beta)$$

$$\Rightarrow \quad (T\alpha + T\alpha) + (T\alpha + T\beta) + (T\beta, + \alpha + \beta)$$

$$= (\alpha, \alpha) + (\alpha, \beta) + (\beta, \alpha) + (\beta, \beta)$$

$$\Rightarrow \quad (T\alpha, T\beta) + (T\beta, T\alpha) = (\alpha, \beta) + (\beta, \alpha)$$

$$[\because (T\alpha, T\alpha) = (\alpha, \alpha) \text{ and } (T\beta, T\beta) = (\beta, \beta)]$$

Thus if a, b are any vectors in U, then

$(T\alpha, T\beta) + (T\beta, T\alpha) = (\alpha, \beta) + (\beta, \alpha)$

If F is the field of real numbers, then

$(T\beta, T\alpha) + (T\alpha, T\beta) = (\beta, \alpha) + (\alpha, \beta)$

$\therefore$ (1) gives $2\,(T\alpha, T\beta) = 2\,(\alpha, \beta)$

$\Rightarrow$ $(T\alpha, T\beta) = (\alpha, \beta)$

If F is the field of complex numbers, then $\beta \in U \Rightarrow i\beta \in U$

So replacing β by $i\beta$ in (1), we get

$$(T\alpha, Ti\beta) + (Ti\beta, T\alpha) = (\alpha, i\beta) + (i\beta, \alpha)$$

$$\Rightarrow \quad (T\alpha, iT\beta) + (iT\beta, T\alpha) = \bar{i}\,(\alpha, \beta) + i\,(\beta, \alpha)$$

$$\Rightarrow \quad \bar{i}\,(T\alpha, T\beta) + i\,(T\beta, T\alpha) = -i\,(\alpha, \beta) + i\,(\beta, \alpha)$$

$$\Rightarrow \quad -i\,(T\alpha, T\beta) + i\,(T\beta, T\alpha) = -i\,(\alpha, \beta) + i\,(\beta, \alpha)$$

$$\Rightarrow \quad -\,(T\alpha, T\beta) + i\,(T\beta, T\alpha) = -\,(\alpha, \beta) + (\beta, \alpha) \qquad ...(2)$$

Subtracting (2) from (1), we get

$$2\,(T\alpha, T\beta) = 2\,(\alpha, \beta)$$

$$\Rightarrow \quad (T\alpha, T\beta) = 2\,(\alpha, \beta) \text{ for all } \alpha, \beta \text{ in U.}$$

Hence the theorem.

INNER PRODUCT SPACE ISOMORPHISM

Definition: Let U and V be inner product spaces over the same field F, and let T be a linear transformation from U into V. Then T is said to be an inner product space isomorphism if

(i) T is invertible *i.e.* T is one-one and onto,

(ii) T preserves inner products.

Also the inner product spaces U and V are then said to be isomorphic and we write $U \cong V$.

If T preserves inner products, then $\| T\alpha \| = \| \alpha \|$ for every α in U *i.e.* $(T\alpha, T\alpha) = (\alpha, \alpha)$ for every α in U. So

$$T\alpha = 0 \Rightarrow (\alpha, \alpha) = 0 \Rightarrow \alpha = 0$$

$\therefore$ T is non-singular *i.e.* T is one-one.

Therefore an inner product space isomorphism from U onto V can also be defined as a linear transformation from U onto V which preserves inner products.

PRODUCT OF TWO METRIC SPACES

Theorem:

Let (X, d) and (Y, e) be two metric spaces. Define δ by setting $\delta((x_1, y_1), (x_2, y_2)) = \sqrt{\{d^2(x_1, x_2) + e^2(y_1, y_2)}$. Then δ is a metric for $X \times Y$ and the metric topology is the same as the product topology.

Proof:

First we prove that d is a metric for $X \times Y$. Evidently $\delta((x_1, y_1). (x_2, y_2)) \geq 0$ for all $(x_1, y_1), (x_2, y_2) \in X \times Y$

Also $\quad \delta((x_1, y_1), ((x_2, y_2)) = 0$

$\Leftrightarrow \quad d^2(x_1, x_2) + \varepsilon^2(y_1, y_2) = 0$

$\Leftrightarrow \quad d(x_1, x_2) = 0$

and $\quad e(y_1, y_2) = 0$

$\Leftrightarrow \quad x_1 = x_2$ and $y_1 = y_2$

$\Leftrightarrow \quad (x_1, y_1) = (x_2, y_2).$

Thus $[m1]$ and $[m2]$ are satisfied

$$\text{Now } \delta((x_1, y_1), (x_2, y_2)) = \sqrt{\{d_2(x_1, x_2) + e_2(y_1, y_2))\}}$$
$$= \sqrt{\{d^2(x_2, x_1) + e^2(y_2, y_1)\}}$$
$$= \delta((x_2, y_2), (x_1, y_1)\}.$$

so that $[m_3]$ is also satisfied. Finally we have

$$\delta((x_1, y_1), (x_2, y_2)\} = \sqrt{\{d^2(x_1, x_2) + e^2(y_1, y_2)\}}$$
$$\leq [\{d(x_1, x_3) + d(x_3, x_2)\}^2 + \{e(y_1, y_3) + e(y_3, s_2)\}^2]^{1/2}$$

[using triangle inequality for d and e]

$$\leq [d^2(x_1, x_3) + e^2(y_1, y_3)]^{1/2} + [d^2(x_3, x_2) + e^2(y_3, y_2)]^{1/2}$$

[using Minkowski inequality]

$$= \delta((x_1, y_1), (x_3, y_3)) + \delta((x_3, y_3), (x_2, y_2)).$$

Hence, triangle inequality is also satisfied for δ. Accordingly is a metric for $X \times Y$. We now show that the topology induced by δ is the same as the product topology for $X \times Y$. Let w denote the product topology for $X \times Y$. We must show that every d-open set is w-open and conversely. Let G be any δ open subset of $X \times Y$ and let $(x_0, y_0) \in G$. Then there exists a d-open sphere $S((x_0, y_0), \varepsilon$ with centre at (x_0, y_0) and radius $\varepsilon > 0$ such that $S((x_0, y_0), \varepsilon) \subset G$. Consider

$$A = S(x_0, \varepsilon/\sqrt{2})$$

and $\quad B = S(y_0, \varepsilon/\sqrt{2}).$

Then A is *a* d-open set containing x_0 and B is an e-open set containing y_0. It follows by the definition of product topology that $A \times B$ is *a* w-open set containing (x_0, y_0). It will now be shown that $A \times B \subset G$. Let (x, y) be any arbitrary member of $A \times B$. Then $x \in A$ and $y \in B$ so that $d(x_0, x) < \varepsilon/\sqrt{2}$ and $\varepsilon(y_0, y) < \varepsilon/\sqrt{}_2$. Hence $\delta[(x_0, y_0), (x, y)]$

$$= [d^2(x_0, x) + e^2(y_0, y)]^{1/2} < [(\varepsilon/\sqrt{2})^2 + (\varepsilon/\sqrt{2})^2]^{1/2} = \varepsilon$$

It follows that $(x, y)\ \varepsilon\ S((x_0, y_0), \varepsilon) \subset G$. Thus we have shown that

$$(x, y)\ \varepsilon\ A \times B \Rightarrow (x, y)\ \varepsilon\ G.$$

Thus $(x, y)\ \varepsilon\ A \times B \subset G$ we know that is a w-open subset of $X \times Y$.

Conversely, let G be any w-open subset of $X \times Y$ and let $(x_0, y_0)\ \varepsilon\ G$. By the definition of product topology, there must exist *a* d-open set D and an e-open set E each that

$$(x_0, y_0)\ \varepsilon\ D \times E \subset G.$$

Then $x_0 \in D$ and $y_0 \in E$ and so there exist a d-open sphere S

(x_0, ε_1) such that $S(x_0. \varepsilon_1) \subset D$ and e-open sphere $S(y_0, \varepsilon_0)$ such that $S(y_0, \varepsilon_2) \subset E$. Let $\varepsilon = \min\{\varepsilon_1. \varepsilon_2\}$. We claim that

$$S((x_0. y_0), \subset S(x_0, \varepsilon_1) \times S(y_0, \varepsilon_2).$$

Let (x, y) be any arbitrary element of $S((x_0, y_0). \varepsilon)$.

Then, since $\delta((s_0, y_0), (x, y))$

$$= [d^2(x_0, x) + e^2(y_0, y)]^{1/2}, \text{ we have}$$

$$d(x_0, x) \le \delta((x_0, y_0)) < \varepsilon \le \varepsilon_1$$

and $\quad e(y_0, y) \le \delta((x_0, y_0), (s, s)) < \varepsilon \le \varepsilon_2.$

It follows that

$$(x, y) \in S(x_0, \varepsilon_1) \times S(y_0, \varepsilon_2).$$

Hence, $\quad (x_0, y_0) \in S((x_0, y_0), \varepsilon_1) \times S(y_0. \varepsilon_2).$

But $\quad S(x_0, \varepsilon_1) \times S(y_0, \varepsilon_2) \subset D \times S \subset G.$

Therefore $\quad (x_0, y_0) \in S((x_0, y_0), \varepsilon) \subset G$

Thus, we have shown that every point of G is contained in a d-open sphere which is contained in G and consequently G is an open set with respect to the metric δ.

PRODUCT INVARIANT PROPERTY FOR FINITE PRODUCTS

Theorem 1:

The product spaces $X \times Y$ is connected if and only if X and Y cre connected.

Proof:

Assume the X × Y is connected. Since the projections π_x and π_y are continuous and open mappings, that X and Y are also connected spaces.

Conversely, let X and Y be connected spaces. To show that X × Y is also connected. Let (x_1, y_1) and (x_2, y_2) be any two points of X × Y. The $\{x_1\}$ × Y as homeomorphic to Y and X × $\{y_2\}$ is homeomorphic to X. Hence $\{x_1\}$ × Y and X × $\{y_2\}$. They intersect in the point (x_1, x_2) and hence their union is a connected set by theorem. Since this union contains (x_1y_1) and (x_2, y_2), it follows that X × Y is connected.

Theorem 2:

The product space X × Y is compact if and only if each of the spaces X and Y is compact.

Proof:

Let X × Y be compact. Since the projection maps π_x : X × Y → X and π_y X × Y → Y are continuous and onto, we hence already prove it

Conversely, let X and Y be compact spaces. We want to show that X × Y is compact, it suffices to show that every basic open cover of X – Y has a finite sub-cover. Since a basic open set in X – Y is of the form G × H where G is open in X and H is open in Y, we may denote a basic open cover by c = $\{G_\lambda \times H_\lambda : l \in \Lambda\}$ where G_λ is open in X and H_λ is open in Y. For a given point x ∈ X. the set {x} × Y is homeomorphic to Y is therefore, compact. Since {x} × Y,, being a subset of X × Y is covered by c and {x} × Y is compact, there exist a finite sub-family of c, say

$\{G_\lambda \times H_\lambda : i = 1, 2,..., n\}$, which covers {x} × X. Let $\cap G_\lambda$ = G (x). Then G (x) is open in X and contains x since each G_i contains x. Hence $\{G (x) \times H_\lambda: 1, 2, ..., n\}$ is still a finite open cover of {x} × Y.

Proceeding in this manner of each x ∈ X, we contract the collection {G (x) : x ∈ X} of open sets in X which covers X. By compactness of X, there exists a finite sub-cover $\{G (x_j) : j = 1, 2..., m\}$ of this cover for X. Since each G (x_j) is an intersection of open sets in X which were used to from c.

We may select on open set $G_{\lambda_j} \in$ c such that $G(x)_{\lambda j}) \subset G_{\lambda_j}$ for j = 1, 2,..., m. Therefore $\{Gx_{\lambda_j} : j = 1, 2, ..., ... m\}$ is a finite open cover of X, and for each j, 1 ≤ j ≤ m. $\{G_{\lambda_x} \times H_\lambda : i = 1, 2,...n\}$ covers the subset G (x_j) × Y of X × Y. By its construction the collection $\{G_{\lambda_{xj}} \times H_\lambda : i = 1, 2, ..., n = 1.2, ... m\}$ is then a finite sub-cover of c for X × Y and therefore X × Y is compact.

Theorem 3:

The product space of Hausdorff spaces is Hausdorff.

Proof:

Let X and Y by two Hausdorff spaces. To show that X × Y is also a Hausdorff space, Let (x_1, y_1) and (x_2, y_2) be any two distinct points of X × Y. Then either x_1 ° x_2 or y_1 ° y_2. Take x_1 x_2. Since X is Hausdorff, there exist open sets G_1 and G_2 × X such that $x_1 \in G_1$ and $x_2 \in G_2$ and $G_1 \cup G_2 = \varnothing$. Then G_1 × Y and G_2 × X are open subsets of X × Y such that

$$(x_1, y_1) \in G_1 \times Y, (x_2, y_2) \in G_2 \times Y \text{ and}$$
$$(G_1 \times Y) \cup (G_2 \times Y) = (G_1 \cup G_2) \times Y = \varnothing \times Y = \varnothing$$

It follows that X × Y is Hausdroff.

Theorem 4:

The product space of two second countable spaces is second countable.

Proof:

Let X and Y be two second countable spaces. To show that X × Y is also countable. Let B and C be countable bases for X and Y respectively. Consider the collection

$$D = \{B \times C : B \in B, C \in c\}.$$

Then D is surely a countable collection we already know that D is countable base for X × Y.

PRODUCT OF TWO SPACES

Definition : *Let (X, $\mathfrak{J}$) and (Y,v) be two topological spaces. Then the topology w whose base is E = {G × H : G ∈ $\mathfrak{J}$ and H $\mathfrak{J}$ v} is called the Product topology for X × Y and (X × Y, w) is called the* product space *of X and Y.*

We know that *E* is actually *a* base for sine topology *w* of the above definition.

Theorem 1:

The product topology on a non-empty set X × Y is the weak topology for X × Y determined by the projection maps p_x and p_y from the topology on X and Y

Proof:

The weak topology has for a subbase

$\{C_\lambda : G_\lambda = p_x^{-1}[U_\lambda\}$ or G_λ

$= p_y^{-1}[V_\lambda]$. for some U_λ open in X or V_λ open in $Y\}$.

Then intersection

$$p_x^{-1}[U_l] \cap \ldots \cap p_x^{-1}[U_m] \cap p_y^{-1}[V_l] \cap \ldots \cap p_y^{-1}[V_n]$$

$$= (U_l \times Y) \cap \ldots \cap (U_m \times Y) \cap (X \times V_n) \ldots \cap (X \times V_n) \text{ of a}$$

finite number of such sets has the form

$$(U_l \cap U_2 \cap \ldots \cap U_m) \times (V_l \cap V_2 \cap \ldots \cap Vn)$$

$$= U \times V$$

we have already prove that where U is open in X and V is open in Y. Hence weak topology has the same base as the product topology, and so the two topologies are the same

Note that $\pi_x^{-1}[U_1] = U_1 \times Y$, $\pi_y^{-1}[V_1] = X \times V_1$ etc.].

Theorem 2:

Let $(X. \mathfrak{I})$ and (Y, V) be two topological spaces and B a base for $\mathfrak{I}$ and c a base for v. Then,

$$E = \{B \times C : B \in B \text{ and } C \in c\}$$

is *a* base for the product topology w for $X \times Y$.

Proof:

Let (x, y) be any point of $X \times Y$ and let N be a w-nhd of (x, y). Since $E = \{G \times H : G \in \mathfrak{I}$ and $H \in v)$ is *a* base for w, there exist *a* member $G \subseteq H$ of E such that

$$(x, y) \in G \subseteq H \subset N,$$

Since G is $\mathfrak{I}$-open and B is *a* base for $\mathfrak{I}$, there exists some $B \in B$ such that $x \in B \subset G$. Similarly there exist some $C \subset c$ such that $y \in C \subset H$:. It follows that

$$(x, y) \in B \subseteq C \subset G \times H.$$

Hence, from (1) and (2), we get $(x, y) \in B \times C \subset N$. This implies that E is a•base for w.

Example:

Consider the topology

$\mathfrak{I} = \{\varnothing, \{a\}, X\}$ *for* $X = \{a, b, c\}$ *and the topology*

$v = \{\varnothing, \{p\}, \{q\}, \{p, q\}, \{r, s\}, \{p, r, s\}. \{q. r, s\}, Y\}$

for $Y = \{p, q, r, s\}$ and find *a* base for the product topology of $X \times Y$.

Solution :

We know that

$B = \{\{a\}, X\}$ is *a* base for $\mathfrak{J}$ and

$C = \{\{p\}, \{q\}, \{r, s\}\}$ is *a* base for *v*. Hence *a* base for the product topology is given by

$$E = \{\{a\} \times \{p\}, \{a\} \times \{q\}, \{a\} \times \{r, s\},$$
$$X \times \{p\}, X \times \{q\}, X \times \{r, s\}\},$$
$$\{\{(a, p)\}, \{(a, q)\}, \{a, r\}, (a, s), \{(a, p), (b, p), (c, p)\}\},$$
$$\{(a, p), (b, q), (c, q), (c, q)\}, \{(a, r), (a, s), (b, r), (b, s), (c, r), (c, s)\}\}$$

METRIC SPACES

Classical analysis can be described as that part of mathematics which begins with calculus and, in essentially the same spirit, develops similar subject matter much further in many directions. In the course of its development, classical analysis became so complex and varied that even an expert could find his way around in it only with difficulty. Under these circumstances, some mathematicians became interested in trying to uncover the fundamental principles on which all analysis rests. It is a great nation in the World of Mathematics, with many provinces, a few of which are ordinary and partial differential equations, infinite series (especially Power series and Fourier series), and analytic functions of a complex variable. This movement had associated with it many of the great names in mathematics of the last century: Riemann, Weierstrass, Cantor, Lebesgue, Hilbert, Riesz, and others.

It played a large part in the rise to prominence of topology, modern algebra, and the theory of measure and integration; and when these new ideas began to percolate back through classical analysis, the brew which resulted was modern analysis. As modern analysis developed in the hands of its creators, many a major theorem was given a simpler proof in a more general setting, in an effort to lay bare its inner meaning. Much thought was devoted to analyzing the texture of the real and complex number systems, which are the context of analysis.

It was hoped and these hopes were well founded that analysis could be clarified and simplified, and that stripping away superfluous underbrush would give new emphasis to what really mattered from the point of view of the underlying theory. Analysis is primarily concerned with limit processes and continuity, so it is not surprising that mathematicians thinking along these lines soon found themselves studying (and generalizing) two elementary concepts: that of a convergent sequence of real or complex numbers, and that of a continuous function of a real or complex variable.

We remind the reader of the definitions. First, a sequence

$$\{x_n\} = \{x_1, x_2, \ldots, x_n, \ldots\}$$

of real numbers is said to be *convergent* if there exists a real number x (called the *limit* of the sequence) such that, given $\varepsilon > 0$, a positive integer n_0 can be bound with the property that

$$n \geq n_0 \Rightarrow |x_n - x| < \varepsilon.$$

This condition means that x_n must be "close" to x for all "sufficiently large" n, and it is usually symbolized by

$$x_n \to x \quad \text{or} \quad \lim x_n = x$$

and expressed by saying that x_n *approaches* x or x_n converges to x. Second, a real function f defined on a non-empty subset X of the real line is said to be *continuous* at x_0 in X if for each $\varepsilon > 0$ there exists $\delta > 0$ such that

$$x \text{ in } X \text{ and } |x - x_0| < \delta \Rightarrow |f(x) - f(x_0)| < \varepsilon,$$

and f is said to be *continuous* if it is continuous at each point of X. When X is an interval, this definition gives precise expression to the intuitive requirement that f have a graph without breaks or gaps. The corresponding definitions for sequences of complex numbers and complex functions of a complex variable are word for word the same.

Our purpose in giving these definitions in detail here is a simple one. We wish to point out explicitly that each is dependent for its meaning on the concept of the absolute value of the difference between two real or complex numbers. We wish to observe also that this absolute value is the *distance between the numbers* when they are regarded as points on the real line or in the complex plane.

In many branches of mathematics—in geometry as well as analysis—it has been found extremely convenient to have available a notion of distance which is applicable to the elements of abstract sets. A *metric space* is nothing more than a non-empty set equipped with a concept of distance which is suitable for the treatment of convergent sequences in the set and continuous functions defined on the set. Our purpose in this chapter is to develop in a systematic manner the main elementary facts about metric spaces. These facts are important for their own sake, and also for the sake of the motivation they provide for our later work on topological spaces.

Definition and Examples

Let X be a non-empty set. A *metric* on X is a real function d of ordered pairs of elements of X which satisfies the following three conditions:

(1) d(x, y) ≥ 0, and d(x, y) = 0 ⇔ x = y;

(2) d(x, y) = d(y, x) (*symmetry*);

(3) d(x, y) ≤ d(x, z) + d(z, y) (the *triangle inequality*).

The function d assigns to each pair (x, y) of elements of X a non-negative real number d(x, y), which by symmetry does not depend on the order of the elements; d(x, y) is called the distance between x and y. A *metric space* consists of two objects: a non-empty set X and a metric d on X. The elements of X are called the *points* of the metric space (X, d). Whenever it can be done without causing confusion, we denote the metric space (X, d) by the symbol X which is used for the underlying set of points. One should always keep in mind, however, that a metric space is not merely a non-empty set: it is a non-empty set together with a metric. It often happens that several different metrics can be defined on a single given non-empty set, and in this case distinct metrics make the set into distinct metric spaces.

There are many different kinds of metric spaces, some of which play very significant roles in geometry and analysis. Our first example is rather trivial, but it is often useful in showing that certain statements we might wish to make are not true. It also shows that every non- empty set can be regarded as a metric space.

Example 1:

Let X be an arbitrary non-empty set, and define d by

$$d(x,y) = \begin{cases} 0 & \text{if } x = y, \\ 1 & \text{if } x \neq y. \end{cases}$$

The reader can easily see for himself that this definition yields a metric on X.

Our next two examples are the fundamental number systems of mathematics.

Example 2:

Consider the real line R and the real function | x | defined on R. Three elementary properties of this absolute value function are important for our purposes:

(i) $|x| \geq 0$, *and* $|x| = 0 \Leftrightarrow x = 0;$

(ii) $|-x| = |x|;$

(iii) $|x + y| \leq |x| = |y|.$

We now define a metric on R by

$$d(x, y) = |x - y|.$$

This is called the *usual metric* on R, and the real line, as a metric space, is always understood to have this as its metric. The fact that d actually is a metric follows from the three properties stated above. This is a piece of reasoning which occurs frequently in our work, so we give the details. By

(i) $d(x, y) = |x - y|$ is a non-negative real number which equals $0 \Leftrightarrow x - y = 0 \Leftrightarrow x = y$. By

(ii) $d(x, y) = |x - y| = |-(y - x)| = |y - x| = d(y, x)$. And by

(iii) $d(x, y) = |x - y| = |(x - y) + (z - y)| \leq |x - z| + |z - y| = d(x, z) + d(z, y)$.

Example 3:

The set of points in the preceding example—that is, the set of all bounded continuous real functions defined on the closed unit interval—has another metric which is far more important for our purposes. It is defined by means of

$$||f|| = \sup \{|f(x)| : x \in [0, 1]\},$$

which we usually write more briefly as

$$||f|| = \sup |f(x)|,$$

and $$d(f, g) = ||f - g|| = \sup |f(x) - g(x)|.$$

Properties (i) and (ii) of the norm are obvious, and we ask the reader to prove (iii) in a slightly more general form. This example is typical of a large class of metric spaces which will play a major role in all our work throughout the rest of this book. We denote this space by [0, 1].

So much for the present for specific examples. We now turn to several fundamental principles relating to metric spaces in general.

Let X be a metric space with metric d. Let Y be an arbitrary non- empty subset of X. If the function d is considered to be defined only for points in Y, then (Y, d) is evidently itself a metric space. Y, with d restricted in this way, is called a *subspace* of X. This technique of forming subspaces of a given metric space enables us to obtain an infinity of further examples from the handful described above. For instance, the closed unit interval [0, 1] is a subspace of the real line, as is the set consisting of all the rational points; and the unit circle, the closed unit disc, and the open unit disc are subspaces of the complex plane. Also, the real line itself is a subspace of the complex plane.

It is desirable at this stage to introduce the extended real *number system*, by which we mean the ordinary real number system R with the symbols

$$-\infty \text{ and } +\infty$$

adjoined. An *extended real number* is thus a real number or one of these symbols. We say (by definition) that

$$-\infty < +\infty;$$

also, if x is any real number, then

$$-\infty < x < +\infty.$$

The symbols $-\infty$ and $+\infty$ add nothing to our understanding of the real numbers. They are used mainly as a notational convenience.

Let A be a non-empty set of real numbers which has an upper bound. We defined what is meant by the least upper bound (or supremum) of A: sup A is the smallest upper bound of A, that is, it is the smallest real number y such that $a \le y$ for every a in A. With the stated assumptions about A, sup A always exists and is a real number. If A is a non-empty set of real numbers which has no upper bound, and therefore no least upper bound in R, we express this by writing

$$\sup A = +\infty;$$

and if A is the empty subset of R, we put

$$\sup A = -\infty$$

The greatest lower bound (or infimum) of A is defined similarly: if A is non-empty and has a lower bound, inf A is the largest real number x such that $x \le a$ for every a in A; if A is non-empty and has no lower bound, we put

$$\inf A = -\infty;$$

and if A is empty, we put

$$\inf A = +\infty$$

These remarks illustrate one advantage of the extended real number system: it enables us to speak of sup A and inf A for subsets A of the real line without any restrictions whatever on the nature of A.

Another advantage of having available the symbols $-\infty$ and $+\infty$ is that they make convenient a reasonable extension of our concept of an interval on the real line. The definitions whose scope we are now widening. Let a and b be any two real numbers such that $a \le b$; then the *closed interval* from a to b is the subset of the real line R defined by

$$[a, b] = \{x : a \le x \le b\}.$$

This extends our previous notion in that a closed interval may now consist of a single point (if a = b). If b is a real number and a is an extended real number such that a ≤ b, then the *open-closed interval* from a to & is

$$(a, b) = \{x : a < x \leq b\}.$$

This allows open-closed intervals of the form (−∞, b]. If a is a real number and b is an extended real number such that a < b, then the *closed-open interval* from a to b is

$$[a, b) = \{x : a \leq x < b\}.$$

This permits [a, + ∞) to be considered a closed-open interval. If a and b are extended real numbers such that a < b, then the *open interval* from a to b is

$$(a, b) = \{x : a < x < b\}.$$

This adds to the previously defined open intervals those of the form (−∞, b) where b is real, (a, + ∞) where a is real, and (−∞, + ∞).

Throughout the rest of this book, the term *interval* will always signify one of the four types defined in this paragraph. The extended real numbers a and h are called the *end-points* of these intervals.

We have used the symbols −∞ and +∞ with considerable freedom, and it therefore seems desirable to emphasize that an interval in our present sense is always a non-empty subset of the real number system: it never actually contains either of these symbols.

The very definition of a metric space presents us with the concept of the distance from one point to another. We now define the distance from a point to a set and the diameter of a set.

Let X be a metric space with metric d, and let A be a subset of X. If x is a point of X, then the *distance from* x to A is defined by

$$d(x, A) = \inf [d(x, a) : a \in A\};$$

that is, it is the greatest lower bound of the distances from x to the points of A. The *diameter* of the set A is defined by

$$d(A) = \sup \{d(a_1, a_2) : a_1 \text{ and } a_2 \in A\}.$$

The diameter of A is thus the least upper bound of the distances between pairs of its points. A is said to have *finite diameter or infinite diameter* according as d(A) is a real number or ±∞.

We observe that the empty set has infinite diameter, since $d(\phi) = -\infty$. A *bounded set* is one whose diameter is finite. A mapping of a non-empty set into a metric space is called a *bounded mapping* if its range is a bounded set. Several of the simpler facts about these concepts are brought out in the following problems.

GENERAL PRODUCT SPACES

Recall that if $\{X_\lambda : \lambda \in \Lambda\}$ is an arbitrary collection of sets indexed by L then the ***Cartesian product*** *of this collection is the set of all mapping x defined by $x : \Lambda \to \cap \{X_\lambda : \lambda \in \Lambda\}$ such that $x(\lambda) \in X_\lambda$ for all $\lambda \in \Lambda$ and is denoted by $\pi \{x_\lambda : \in \Lambda\}$ or by $\times \{X_\lambda : \lambda \in \Lambda\}$. The set X_λ is called the λth* ***coordinate*** *set of the product.*

it is customary to use the symbol x_λ for the image $x(\lambda)$ of λ; under the mapping x. Thus if $X = \cap \{X_\lambda : \in \Lambda\}$.

Then $\times \{X_\lambda : \lambda \in \Lambda\}$

$$= \{x : x : \Lambda \to X,$$

where $x(\lambda) \in X_\lambda \forall \lambda \in \Lambda\}$.

If one of the sets is empty then $\times \{X_\lambda : \lambda \in \Lambda\}$ is empty. Also very often we shall denote the mapping x by $\{x_\lambda : \lambda \in \Lambda\}$ where $x_\lambda \in X_\lambda$, x_λ is called the λth coordinate bf x. If $X_\lambda = X$ for all $\lambda \in \Lambda$, then for $\times \{X_\lambda : \lambda \in \Lambda\}$, we write X^Λ Thus X^Λ is the set of all mappings on Λ into X and is called the Λth power of X. For any set X. The set of all mapping on the empty set $\varnothing$ into X consists of $\varnothing$ alone. Thus $X^\varnothing = \{\varnothing\}$.

Definition: *Let $X \to \{X_\lambda : \lambda \in \mathfrak{I}\}$. Then the mapping $x_\lambda : X \to X_\lambda$ defined by $\pi_\lambda(x) = x_\lambda$ for all $x \in X$ is called the λ projection.*

Thus π_λ is the mapping on X which carries each point x of the product onto its λth coordinate x_λ.

Example:

Describe the set $\pi_\lambda^{-1}[A_\lambda]$,

where $A_\lambda \subset X_\lambda$ in words.

Solution:

$\pi_\lambda^{-1}[A_\lambda]$ is the set of all x in $\times [X_\lambda : \lambda \in \Lambda]$ whose λ^{th} coordinates are members of A_λ. Equivalently. $\pi_\lambda^{-1}[A_\lambda]$ is the set of all those members of $\times \{X_\lambda : \lambda \in \Lambda\}$ whose λ^{th} coordinates are restricted to lie in A_λ and all other coordinates are unrestricted in the respective coordinate sets. Symbolically, $\pi_\lambda^{-1}[A_\lambda] = \{x : x \in X \text{ and } x_\lambda = \in A_\lambda\}$

where $X = \times [X_\lambda : \lambda \in \Lambda\}$.

Note that $\pi_\alpha^{-1}[A_\alpha] = \times \{Y_\lambda : \lambda \in \Lambda\}$

where $Y_\alpha = A_\alpha$

and $Y_\lambda = X_\lambda$

whenever $\lambda \neq \alpha$.

LATTICE

Definitions

(i) *A partially ordered set A is said to be a* lattice *iff every pair iff of elements in A has infimum and supremum.* We shall denote the infimum and the supermum of any pair of element x, y of A and x A y and x ∨ y respectively.

(ii) *A lattice A is said to be a complete lattice iff every non-empty subset of A has an infimum and a supermum.*

(iii) *A non-empty subset B of a lattice A is said to be sub-lattice iff* $x \wedge y$ *and* $x \vee y$ *are members of B for every pair of members, y of B.*

Theorem 1:

Let X be any non-empty set and let $\{X_\gamma, X_\lambda : \in L\}$ *be an arbitrary non-empty collection of topological spaces. For each* $l \in L$ *let* f_λ *be mapping of X into* X_λ *and let* $\mathfrak{I}$ *be the week topologies v on X determined by* $\{f_\lambda : \lambda \in \mathfrak{I}\}$. *Than* $\mathfrak{I}$ *is the intersection of all topologies V on X such that* f_λ *is v* $\mathfrak{I}$ *continuous for each* $\lambda \in \mathfrak{I}$, *that is,* $\mathfrak{I}$ *is the coarsest topology on X fir which* f_1 *is continuous for each* $l \in \mathfrak{I}$.

Proof :

We first prove that f_l is $\mathfrak{I}$-$\mathfrak{I}_\lambda$ continuous for each $l \in \mathfrak{I}$. Let *G* be any $\mathfrak{I}_\lambda$-open set. Then by definition of week to be log y we have $f_\lambda^{-1}[G] \in \mathfrak{I}$. Hence f_λ is $\mathfrak{I}$-$\mathfrak{I}_\lambda$ continuous.

Now let v be topology on *X* such that fl is v-$\mathfrak{I}_\lambda$ continuous for each $\lambda \in L$. Then $f_\lambda^{-1}[G] \in v$ for each $G \in \mathfrak{I}_\lambda$ and for each $\lambda \in \Lambda$.

Since *v* is *a* topology on *X*, *v* contains all the unions of finite intersections of member of the collection $\{f_\lambda^{-1}[G] : l \in \Lambda$ and $G \in \mathfrak{I}_\lambda\}$. This means that *v* contains $\mathfrak{I}$, that is $\mathfrak{I}$ is weaker than *v*.

Thus, we have shown that $\mathfrak{I}$ is the coarsest topology on *X* such that f_λ is $\mathfrak{I}$-$\mathfrak{I}_\lambda$ continuous for each $\mathfrak{I} \in \Lambda$. This implies that $\mathfrak{I}$ is the intersection of all topologies *v* on *X* such that f_λ is v-$\mathfrak{I}_\lambda$ continuous for each $\lambda \in \mathfrak{I}$.

Theorem 2:

Let C be the class of all topologies on a non-empty set X. There C is a complete lattice with respect to the relations $\leq$ *defined by setting* $\mathfrak{I}1 \leq \mathfrak{I}2$ *iff* $\mathfrak{I}_1$ *is weaker than* $\mathfrak{I}_2$ *Moreover, c has a least member and a greatest member.*

Proof:

It is easy to see that C is *a* partially ordered set with respect to the relation "is weaker that". If $\Im_1$, $\ddot{A}_2$ are any two members of C, then the $\Im_1 \wedge \Im_1 \cap \Im_2$ and $\Im_1 \vee \Im_2$ is the intersection of all topologies $\Im_1$ such that $\Im_1 \leq \Im_1$ and $\Im_2 \leq \Im 1$ [Note that the intersection of arbitrary collection of topologies on X is a topology on X]. Hence C is lattice. It is also a complete lattice. For if C′ is any subclass of C then inf C′ is the intersection of all the members of C′ is the intersection of all the members of C′ and sup C′ is the intersection of all those members of C which are stronger than every members of C′.

Further, the indiscrete topology I is the weakest topology on X and the discrete topology D is the strongest typology on X. Thus, I is the least and D is the greatest member of C.

Definition : *Let X be any non-empty set and let* $A = \{(X_\lambda, \Im_\lambda) : l \in \Lambda\}$ *be any arbitrary non-empty collection of topological spaces. For each* $\lambda \in \Lambda$ *let* f_λ *be a mapping of X into* X_λ.

Then the topology L on X generated by the collection $B = \{f_\lambda^{-1}[G] : l \in \Lambda \text{ and } G \in \Im_\lambda\}$ *is called the weak topology (or the initial topology) on X determined by the collection all finite intersections of members of A is called the defining subbase of* $\Im$ *and the collection B of all finite intersections of members of A is called the defining base of* $\Im$.

Theorem 3:

Let $(X, \Im)$ and (Y, v) be two topological spaces. Then the collection $E = \{G \times H : G \in \Im \text{ and } H \in v\}$ *is a* base for some topology for $X \times Y$.

Proof:

We shall show that E satisfies the conditions [B*1] and [B*2].

Since $X \times Y \in E$,

we have $X \times X = \cup \{G \times H : G \in \Im \text{ and } H \in v\}$

Thus [B*1] is satisfied. Now let $G_1 \times H_1$ and $G_2 \times H_2$ be any two members of E. We then have

$$(G_1 \times H_2) \cap (G_2 \times H_2) = (G_1 \cap G_2) \times (H_1 \cap H_2). \quad ...(1)$$

Since $\Im$ and v are topologies, we have

$G1 \in \ddot{A}.\ G_2 \in \Im$

$\Rightarrow \quad G_1 \cup G_2 \in \Im$

and $\quad H_1 \in v,\ H_2 \in v$

$\Rightarrow \quad H_1 \cup H_2 \in v.$

Hence, it follows from (1) that $(G_1 \times H_1) \cup (G_2 \times H_2) \in E$. Thus, we have shown that the intersection of any two members of E us again *a* member of E and so [B*2] is also satisfied. Therefore is a base for some topology for $C \times Y$.

Example 1:

Consider the set R of all real numbers. This set is partially ordered by $\leq$ in the usual sense. If x, y be any two real numbers, then $x \wedge y = \min \{x, y\}$ and $x \vee y = \max \{x, y\}$. Hence R is a lattice. But it is not a complete lattice since $\{x \in R : x \geq\}$ is a subset of R which has no supermum. In fact it has no upper bound.

The set N of all natural number and the set Q of all rational numbers are sublattices of R since

	$x \in N, y \in N$
$\Rightarrow$	$x \wedge y \in N$
and	$x \vee y \in N$
and	$x \in Q, y \in Q$
$\Rightarrow$	$x \wedge y \in Q$
and	$x \vee y \in Q$.

Example 2:

Let P (X) be the power set of a non-empty set X so that P (X) is the collection of all subsets of X. Then P (x) is partially ordered with respect to the relation two members of F, then $f \wedge g$ is the real valued function on X defined by $(f \not\leq g) = \min \{f(x), g(x)\}$ and $f \vee g$ is that defined by $(f \vee g)(x) = \max \{f(x), g(x)\}$. Hence F is a lattice here.

Example 3:

Consider the set N of all natural numbers. This set is partially ordered with respect to the relation $\leq$ defined by setting $a \leq$ by iff a divided b. If a, be are any two natural numbers, then $a \wedge b$ is the highest common factor (H.C.F >) of a, and b $a \vee b$ is the least common multiple (L.C.M.) of a and b. Hence $(N, \leq)$ is a lattice. It is however not complete lattice since $\{a \in N : a \geq 1\}$ has no supermum.

OPEN SETS

Let X be a metric space with metric d. If xa is a point of X and r is a positive real number, the *open sphere* $S_r(x_0)$ with *center* x_0 and *radius* r is the subset of X defined by

$$S_r(x_0) = \{x:d\{x, x_0) < r\}.$$

An open sphere is always non-empty, for it contains its center. An open sphere with radius 1 contains only its center. $S_r(x_0)$ is often called the open sphere with radius r centered on x_0; intuitively, it consists of all points in X which are "close" to x_0, with the degree of closeness given by r.

A few concrete examples are in order. It should be easy to visualize the open sphere $S_r(x_0)$ on the real line: it is the bounded open interval $(x_0 - r, x_0 + r)$ with mid-point x_0 and total length 2r. Conversely, it is clear that any bounded open interval on the real line is an open sphere, so the open spheres on the real line are precisely the bounded open intervals. The open sphere $S_r(z_0)$ in the complex plane is the inside of the circle with center z_0 and radius r. The space $\mathscr{C}[0, 1]$: $S_r(f_0)$ consists of all functions f in $\mathscr{C}[0, 1]$ whose graphs lie within the shaded band of vertical width 2r centered on the graph of f_0. A subset G of the metric space X is called an open set if, given any point x in G, there exists a positive real number r such that $S_r(x) \subseteq G$, that is, if each point of G is the center of some open sphere contained in G. Loosely speaking, a set is open if each of its points is "inside" the set, in the sense made precise by the definition. On the real line, a set consisting of a single point is not open, for each bounded open interval centered on the point contains points not in the set.

Similarly, the subset [0, 1) of the real line is not open, because the point 0 in [0, 1) has the property that each bounded open interval centered on it (no matter how small it may be) contains points not in [0, 1), *e.g.*, negative points. If we omit the offending point 0, the resulting bounded open interval (0, 1) is an open set (this is very easy to prove and is a special case of Theorem B below). Further, it is quite clear that any open interval—bounded or not—is an open set, and also that the open intervals are the only intervals which are open sets.

Theorem 1:

In any metric space X, the empty set ϕ and the full space X are open sets.

Proof:

To show that ϕ is open, we must show that each point in ϕ is the center of an open sphere contained in ϕ; but since there are no points in ϕ, this requirement is automatically satisfied. X is clearly open, since every open sphere centered on each of its points is contained in X.

We have seen that [0, 1) is not open as a subset of the real line. However, if we consider [0, 1) as a metric space X in its own right, as a subspace of the real line, then [0, 1) is open as a subset of X, since from this point of view it is the full space. This apparent paradox disappears when we realize that points outside of a given metric space have no relevance to any discussion taking place within the context of that space. A set is open or not open only with respect to a specific metric space containing it, never on its own.

Our next theorem justifies the adjective in the expression "open sphere."

Theorem 2:

Let X be a metric space. A subset G of X is open $\Leftrightarrow$ it is a union of open spheres.

Proof:

We assume first that G is open, and we show that it is a union of open spheres. If G is empty, it is the union of the empty class of open spheres. If G is non-empty, then since it is open, each of its points is the center of an open sphere contained in it, and it is the union of all the open spheres contained in it.

We now assume that G is the union of a class S of open spheres. We must show that G is open. If S is empty, then G is also empty, and by Theorem A, G is open. Suppose that S is non-empty. G is also non-empty. Let a; be a point in G. Since G is the union of the open spheres in S, x belongs to an open sphere $S_r(x_0)$ in S. By Theorem B, x is the center of an open sphere $S_{r_1}(x)$ $\subseteq S_r(x_0)$. Since $S_r(x_0) \subseteq G$, $S_{r_1}(x) \subseteq G$ and we have an open sphere centered on x and contained in G. G is therefore open.

The fundamental properties of the open sets in a metric space are those stated in

Theorem 3:

Let X is a metric space. Then (1) any union of open sets in X is open; and (2) any finite intersection of open sets in X is open.

Proof:

To prove (1), let $(G_1\}$ be an arbitrary class of open sets in X. We must show that $G = \cup_i G_i$ is open. If $\{G_i\}$ is empty, then G is empty, and by Theorem A, G is open. Suppose that $\{G_i\}$ is non-empty. By Theorem C, each G, (being an open set) is a union of open spheres; G is the union of all the open spheres which arise in this way; and by another application of Theorem C, G is open.

To prove (2), let $\{G_i\}$ be a finite class of open sets in X. We must show that $G = \cap_i G_i$ is open. If (G",} is empty, then G = X; and by Theorem A,

G is open. Suppose that $\{G_i\}$ is non-empty and that $\{G_i\} = \{G_1, G_2, \dots G_n\}$ for some positive integer n. If G happens to be empty, then it is open by Theorem A, so we may assume that G is non-empty. Let a; be a point in G. Since x is in each G_i, and each G_i is open, for each i there is a positive real number r_i such that $S_{ri}(x) \subseteq G_i$. Let r be the smallest number in the set $\{r_1, r_2, \dots, R_n\}$. This number r is a positive real number such that $S_r(x) \subseteq S_{ri}(x)$ for each i, so $S_r(x) \subseteq G_i$ for each i, and therefore $S_r(x) \subseteq G$. Since $S_r(x)$ is an open sphere centered on x and contained in G, G is open.

The above theorem says that the class of all open sets in a metric space is closed under the formation of arbitrary unions and finite intersections. The reader should clearly understand that Theorem A is an immediate consequence of this statement, since the empty set is the union of the empty class of open sets and the full space is its intersection. The limitation to *finite* intersections in this theorem is essential. To see this, it suffices to consider the following sequence of open intervals on the real line:

$$(-1,1)\left(-\frac{1}{2},\frac{1}{2}\right),\left(-\frac{1}{3},\frac{1}{3}\right)$$

The intersection of these open sets is the set $\{0\}$ consisting of the single point 0, and this set is not open.

In an arbitrary metric space, the structure of the open sets can be very complicated indeed. Theorem C contains the best information available in the general case: each open set is a union of open spheres. In the case of the real line, however, a description can be given of the open sets which is fairly explicit and reasonably satisfying to the intuition.

Theorem 4:

Every non-empty open set on the real line is the union of a countable disjoint class of open intervals.

Proof:

Let G' be a non-empty open subset of the real line. Let x be a point of G. Since G is open, x is the center of a bounded open interval contained in G. Define I_x to be the union of all the open intervals which contain x and are contained in G. The following three facts are easily proved: I_x is an open interval, which contains x and is contained in G; I_x contains every open interval which contains x and is contained in G; and if y is another point in I_x, then $I_x = I_y$. We next observe that if x and y are any two distinct points of G, then I_x and I_y are either disjoint or identical; for if they have a common point z, then $I_x = I_z$ and $I_y = I_z$, so $I_x = I_y$. Consider the class I of all distinct sets of the form I_x for points x in G. This is a disjoint class of open intervals, and G is obviously its union. It remains to be proved that I is countable. Let G_r be the set of rational points in G. G_r is clearly non-empty. We define a mapping f of G_r onto I as follows: for each r in G_r, let f(r) be that unique interval in I which contains r. G_r is countable, and the fact that I is countable.

A firm grasp of the ideas involved in the theory of metric spaces depends on one's capacity to "see" these spaces with the mind's eye.

The complex plane is perhaps the best metric space to use as a model from which to absorb this necessary intuitive understanding. When we consider an unspecified set A of complex numbers, we usually imagine it as a region bounded by a curve. We think of the point z_1, which is completely surrounded by points of A, as being "inside" the set A, or in its "interior," while z_2 is on the "boundary" of A. More precisely, z_1 is the center of some open sphere contained in A, and each open sphere
centered on z_2 intersects both A and its complement A'. We formulate these ideas for a general metric space in the next paragraph and at the end of the next section.

Let X be an arbitrary metric space, and let A be a subset of X. A point in A is called an *interior point* of A if it is the center of some open sphere contained in A; and the *interior* of A, denoted by Int (A), is the set of all its interior points. Symbolically,

$$\text{Int}(A) = \{x : x \in A \text{ and } S_r(x) \subseteq A \text{ for some } r\}.$$

The basic properties of interiors are the following:

(1) Int (A) is an open subset of A which contains every open subset of A (this is often expressed by saying that the interior of A is the largest open subset of A);

(2) A is open $\Leftrightarrow$ A = Int (A);

(3) Int (A) equals the union of all open subsets of A.

CLOSED SETS

Let X be á metric space with metric d. If A is a subset of X, a point x in X is called a *limit point* of A if each open sphere centered on x contains at least one point of A different from x. The essential idea here is that the points of A different from x get "arbitrarily close" to x, or "pile up" at x.

The subset $\left\{1, \frac{1}{2}, \frac{1}{3}, \ldots\right\}$ of the real line has 0 as a limit point; in fact, 0 is its only limit point. The closed-open interval [0, 1) has 0 as a limit point which is in the set and 1 as a limit point which is not in the set; further, every real number x such that $0 < x < 1$ is also a limit point of this set. The set of all integral points on the real line has no limit points at all, whereas every real number is a limit point of the set of all rationals. Every open sphere of radius less than 1 consists only of its center, so no subset of this space has any limit points.

A subset F of the metric space X is called a *closed set* if it contains each of its limit points. In rough terms, a set is closed if its points do not get arbitrarily close to any point outside of it. Among the subsets of the real line mentioned in the preceding paragraph, only the set of integral points is closed. Every subset is closed.

Theorem 1:

In any metric space X, the empty set ϕ and the full space X are closed sets.

Proof:

The empty set has no limit points, so it contains them all and is therefore closed. Since the full space X contains all points, it automatically contains its own limit points and thus is closed.

The following theorem characterizes closed sets in terms of open sets. We already know a good deal about open sets, so that characterization provides us with a useful tool for establishing properties of closed sets.

Theorem 2:

Let X be a metric space. A subset F of X is closed $\Leftrightarrow$ its complement F' is open.

Proof:

Assume first that F is closed. We show that F' is open. If F' is empty, it is open, so we may suppose that F' is non-empty. Let x be a point in F'. Since F is closed and x is not in F, x is not a limit point of F. Since x is not in F and is not a limit point of F, there exists an open sphere $S_r(x)$ which is disjoint from F. $S_r(x)$ is an open sphere centered on x and contained in F', and since x was taken to be any point of F', F' is open.

We now assume that F' is open and show that F is closed. The only way F can fail to be closed is to have a limit point in F'. This cannot happen, for since F' is open, each of its points is the center of an open sphere disjoint from F, and no such point can be a limit point of F.

If x_0 is a point in our metric space X, and r is a non-negative real number, the closed sphere $S_r[x_0]$ with *center* x_0 and *radius* r is the subset of X defined by

$$S_r[x_0] = \{x : d(x, x_0) \leq r\}.$$

$S_r[x_0]$ contains its center, and when r = 0 it contains only its center. The closed spheres on the real line are precisely the closed intervals. In this connection, we observe that though open spheres on the real line are open intervals, there are open intervals which are not open spheres, *e.g.*, $(-\infty, +\infty)$.

The following theorem justifies the adjective in the phrase "closed sphere."

Theorem 3:

In any metric space X, each closed sphere is a closed set.

Proof:

Let $S_r[x_0]$ be a closed sphere in X. By Theorem B, it suffices to show that its complement $S_r[x_0]'$ is open. $S_r[x_0]'$ is open if it is empty, so we may assume that it is non-empty. Let a; be a point in $S_r[x_0]'$. Since $d(x, x_0) > r$, $r_1 = d(x, x_0) - r$ is a positive real number. We take r_1 as the radius of an open sphere Sr_1 (x) centered on x, and we show that $S_r[x_0]'$ is open by showing that $S_{r_1}(x) \subseteq S_r[x_0]'$. Let y be a point in $S_{r_1}(x)$, so that $d(y, x) < r_1$. On the basis of this and the fact that $d(x_0, x) < d(x + y) + d(y, x)$, we see that $d(y, xa_0) > d(x, x_0) - d(y, x) > d(x, x_0) - n = d(x, x_0) - [d(x, x_0) - r] = r$, so that y is in $S_r[x_0]'$.

The main general facts about closed sets are those given in our next theorem.

Theorem 5:

Let X be a metric space. Then (1) any intersection of closed sets in X is closed; and (2) any finite union of closed sets in X is closed.

Proof:

This theorem is an immediate consequence. We prove (1) as follows. If $\{F_i\}$ is an arbitrary class of closed subsets of X and $F = F \cap_i F_i$, then by Theorem B, F is closed if F' is open; but $F' = \cup_i F_i'$ is open, since by Theorem B each F_i' is open. The second statement is proved similarly.

In Theorem E of the previous section, we gave an explicit characterization of the open sets on the real line. We now consider the structure of its closed sets. Among the simplest closed sets on the real line are the closed intervals (which are the closed spheres) and finite unions of closed intervals. Finite sets are included among these, since a set consisting of a single point is a closed interval with equal end-points. What is the character of the most general closed set on the real line? Since closed sets are the complements of open sets, The most general proper closed subset of the real line is obtained by removing a countable disjoint class of open intervals. This process sounds innocent enough, but in fact it leads to some rather curious and complicated examples. One ,of these examples is of particular importance. It was studied by Cantor and is usually called the *Cantor set*. To construct the Cantor set, we proceed as follows. First, denote the closed unit interval [0, 1] by F_1. Next, delete from F_1 the open interval (1/3, 2/3) which is its middle third, and denote the remaining closed set by F_2. Clearly,

$$F_3 = F_2 = \left[0, \frac{1}{3}\right] \cup \left[\frac{2}{3}, 1\right]$$

Next, delete from F_2 the open intervals $\left(\frac{1}{9}, \frac{2}{9}\right)$ and $\left(\frac{7}{9}, \frac{8}{9}\right)$, which are the middle thirds of its two pieces, and denote the remaining closed set by F_3. It is easy to see that,

$$F_3 = \left[0, \frac{1}{9}\right] \cup \left[\frac{2}{9}, \frac{1}{3}\right] \cup \left[\frac{2}{3}, \frac{7}{9}\right] \cup \left[\frac{8}{9}, 1\right]$$

If we continue this process, at each stage deleting the open middle third of each closed interval remaining from the previous stage, we obtain a sequence of closed sets F_n, each of which contains all its successors. The Cantor set F is defined by

$$F = \bigcap_{n=1}^{\infty} F_n,$$

and it is closed by Theorem D. F consists of those points in the closed unit interval [0, 1] which "ultimately remain" after the removal of all the open intervals $\left(\frac{1}{3}, \frac{2}{3}\right), \left(\frac{1}{9}, \frac{2}{9}\right), \left(\frac{7}{9}, \frac{8}{9}\right), \ldots$ What points, do remain? F clearly contains the end-points of the closed intervals which make up each set F_n:

$$0, 1, \frac{1}{3}, \frac{2}{3}, \frac{1}{9}, \frac{2}{9}, \frac{7}{9}, \frac{8}{9} \ldots$$

Does F contain any other points? We leave it to the reader to verify that 1/4 is in F and is not an end-point. Actually, F contains a multitude of points other than the above end-points, for the set of these end-points is clearly countable, while the cardinal number of F itself is c, the cardinal number of the continuum.

To prove this, it suffices to exhibit a one-to-one mapping/ of [0,1) into F. We construct such a mapping as follows. Let x be a point in [0,1), and let $x = b_1 b_2 b_3 \ldots$ be its binary expansion. Each b_n is either 0 or 1. Let $t_n = 2b_n$, and regard $t_1 t_2 t_3 \ldots$ as the ternary expansion of a real number f(x) in [0, 1). The reader will easily convince himself that f(x) is in the Cantor set F:

since t_1 is 0 or 2, f(x) is not in $\left[\frac{1}{3}, \frac{2}{3}\right)$; since t_2 is 0 or 2, f(x) is not in $\left[\frac{1}{9}, \frac{2}{9}\right)$ or $\left[\frac{7}{9}, \frac{8}{9}\right)$ etc. Also, it is easy to see that the mapping f : [0, 1) → F is one-to-one. According to this, F contains exactly as many points as the entire closed unit interval [0, 1]. It is interesting to compare this conclusion with the fact that the sum of the lengths of all the open intervals removed is precisely 1, since

$$\frac{1}{3}+\frac{2}{9}+\frac{4}{27}+\ldots=1$$

It is also interesting to observe (by doing a little arithmetic) that F_{25} is the union of 16,777,216 disjoint closed intervals of the same length which are rather irregularly distributed along [0, 1]. These facts may suffice to indicate that the Cantor set is a very intricate mathematical object and is just the sort of thing mathematicians delight in. We shall encounter this set again from time to time.

We conclude this section by defining two additional concepts which are often useful.

Let X be an arbitrary metric space, and let A be a subset of –X". The *closure* of A, denoted by $\overline{A}$, is the union of A and the set of all its limit points. Intuitively, $\overline{A}$ is A itself together with all other points in X which are arbitrarily close to A. As an example, if $\overline{A}$ is the open unit disc $\{z:|z| < 1\}$ in the complex plane, then A is the closed unit disc $[z:|s| \le 1\}$. The main facts about closures are the following:

(1) $\overline{A}$ is a closed superset of A which is contained in every closed superset of A (we express this by saying that $\overline{A}$ is the smallest closed superset of A);

(2) A is closed ⇔ A = $\overline{A}$;

(3) A equals the intersection of all closed supersets of A.

Our second concept relates to the discussion of Fig. 2.3 given at the end of the previous section. Again, let X be a metric space and A subset of X. A point in X is called a *boundary point* of A if each open sphere centered on the point intersects both A and A', and the boundary of A is the set of all its boundary points. This concept possesses the following properties:

(1) The boundary of A equals A ∩ A';

(2) The boundary of A is a closed set;

(3) A is closed ⇔ it contains its boundary.

COMPLETENESS, CONVERGENCE AND BAIRE'S THEOREM

As we emphasized in the introduction to this chapter, one of our main aims in considering metric spaces is to study convergent sequences in a context more general than that of classical analysis. The fruits of this study are many, and among them is the added insight gained into ordinary convergence as it is used in analysis.

Let X be a metric space with metric d, and let

$$\{x_n\} = \{x_1, x_2, \dots, x_n, \dots\}$$

be a sequence of points in X. We say that $\{x_n\}$ is convergent if there exists a point x m X such that either

(1) for each $\varepsilon > 0$, there exists a positive integer n_0 such that $n \geq n_0 \Rightarrow d(x_n, x) < \varepsilon$; or equivalently,

(2) for each open sphere $S_\epsilon(x)$ centered on x, there exists a positive integer no such that x_n is in $S_\epsilon(x)$ for all $n \geq n_0$.

The reader should observe that the first condition is a direct generalization of convergence for sequences of numbers as defined in the introduction, and that the second can be thought of as saying that each open sphere centered on x contains all points of the sequence from some place on.

If we rely on our knowledge of what is meant by a convergent sequence of real numbers, the statement that $\{x_n\}$ is convergent can equally well be defined as follows: there exists a point x in X such that $d(x_n,x) \to 0$. We usually symbolize this by writing

$$x_n \to x,$$

and we express it verbally by saying that x_n approaches x, or that x_n *converges to x*. It is easily seen from condition (2) that the point x in this discussion is unique, that is, that $x_n \to y$ with $y \neq x$ is impossible. The point x is called the *limit* of the sequence $\{x_n\}$, and we sometimes write $x_n \to x$ in the form

$$\lim x_n = x.$$

The statements $x_n \to x$ and $\lim x_n = x$

mean exactly the same thing, namely, that $(x_n\}$ is a convergent sequence with limit x.

Every convergent sequence $\{x_n\}$ has the following property: for each $\varepsilon > 0$, there exists a positive integer n_0 such that $m, n \geq n_0 \Rightarrow d(x_m, x_n) < \varepsilon$. For if $x_n \to x$, then there exists a positive integer no such that $n_0 \geq n_0$ $\Rightarrow d(x_n, x) < \varepsilon/2$, and from this we see that

$$m,n \geq n_0 \Rightarrow d(x_m, x_n) \leq d(x_m, x) + d(x, x_n) < \varepsilon/2 + \varepsilon/2 = \epsilon.$$

A sequence with this property is called a *Cauchy sequence*, and we have just shown that every convergent sequence is a Cauchy sequence. Loosely speaking, this amounts to the statement that if the terms of a sequence approach a limit, then they get close to one another. It is of basic importance to understand that the converse of this need not be true, that is, that a Cauchy sequence is not necessarily convergent.

As an example, consider the subspace $X = (0, 1]$ of the real line. The sequence defined by $x_n = 1/n$ is easily seen to be a Cauchy sequence in this space, but it is not convergent, since the point 0 (which it wants to converge to) is not a point of the space. The difficulty which arises in this example stems from the fact that the notion of a convergent sequence is not intrinsic to the sequence itself, but also depends on the structure of the space in which it lies. A convergent sequence is not convergent "on its own"; it must converge to some point in the space. Some writers emphasize the distinction between convergent sequences and Cauchy sequences by calling the latter "intrinsically convergent" sequences.

A *complete metric* space is a metric space in which every Cauchy sequence is convergent. In rough terms, a metric space is complete if every sequence in it which tries to converge is successful, in the sense that it finds a point in the space to converge to. The space (0, 1] mentioned above is not complete, but it evidently can be made so by adjoining the point 0 to it to form the slightly larger space [0, 1]. As a matter of fact, any metric space, if it isn't already complete, can be made so by suitably adjoining additional points.

It is a fundamental fact of elementary analysis that the real line is a complete metric space. The complex plane is also complete, as we see from the following argument. Let $\{z_n\}$, where $z_n = a_n + ib_n$ be a Cauchy sequence of complex numbers. Then $\{a_n\}$ and $\{b_n\}$ are them- selves Cauchy sequences of real numbers, since,

$$|a_m - a_n| \leq |z_m - z_n|$$

and

$$|b_m - b_n| \leq |z_m - z_n|$$

By the completeness of the real line, there exist real numbers a and b such that $a_n \to a$ and $b_n \to b$. If we now put $z = a + ib$, then we see that $z_n \to z$ by means of

$$\begin{aligned} |z_n - z| &= |(a_n + ib_n) - (a + ib)| \\ &= |(a_n - a) + i(b_n - b)| \\ &\leq |a_n - a| + |b_n - b| \end{aligned}$$

and the fact that both final terms on the right approach 0.

The completeness of the complex plane thus depends directly on the completeness of the real line. In this space a Cauchy sequence must be constant (*i.e.*, it must consist of a single point repeated) from some place on, and it converges with that point as its limit.

The problem of completing this space leads to the modern theory of Lebesgue integration, and it would carry us too far afield to pursue this matter to its natural conclusion.

On the other hand, the space $\mathcal{C}[0, \ell]$ defined. The completeness of this space, and of others similar to it, is one of the major focal points of topology and modern analysis.

The terms *limit* and *limit point* are often a source of confusion for people not thoroughly accustomed to them. On the real line, for instance, the constant sequence (1, 1, ... , 1, ...) is convergent with limit 1; but the set of points of this sequence is the set consisting of the single element 1, the point 1 is not a limit point of this set. The essence of the matter is that a sequence of points in a set is not a subset of the set: it is a function defined on the positive integers with values in the set, and is usually specified by listing its values, as in $\{x_n\} = [x_1, x_2, \dots , x_n, \dots\}$, where Xn is the value of the function at the integer n. A sequence may have a limit, but cannot have a limit point; and the set of points of a sequence may have a limit point, but cannot have a limit. The following theorem relates these concepts to one another and is a useful tool for some of our later work.

Theorem 1:

If a convergent sequence in a metric space has infinitely many distinct points, then its limit is a limit point of the set of points of the sequence.

Proof:

Let X be a metric space, and let $\{x_n\}$ be a convergent sequence in X with limit x. We assume that x is not a limit point of the set of points of the sequence, and we show that it follows from this that the sequence has only finitely many distinct points.

Our assumption implies that there exists an open sphere $S_r(x)$ centered on x which contains no point of the sequence different from x. However, since x is the limit of the sequence, all x_n's from some place on must lie in $S_r\{x)$, hence must coincide with x. From this we see that there are only finitely many distinct points in the sequence.

Our next theorem guarantees the completeness of many metric spaces which arise as subspaces of complete metric spaces.

Theorem 2:

Let X be a complete metric space, and let Y be a subspace of X. Then Y is complete ⇔ it is closed.

Proof:

We assume, first that Y is complete as a subspace of X, and we show that it is closed. Let y be a limit point of Y. For each positive integer n, $S_{1/n}(y)$ contains a point y_n in Y. It is clear that $\{y_n\}$ converges to y in X and is a Cauchy sequence in Y, and since Y is complete, y is in Y. Y is therefore closed.

We now assume that Y is closed, and we show that it is complete. Let $\{y_n\}$ be a Cauchy sequence in y. It is also a Cauchy sequence in X, and since X is complete, $\{y_n\}$ converges to a point x in X. We show that x is in Y. If $\{y_n\}$ has only finitely many distinct points, then x is that point infinitely repeated and is thus in Y. On the other hand, if {yn) has infinitely many distinct points, then, by Theorem A, x is a limit point of the set of points of the sequence; it is therefore also a limit point of Y, and since Y is closed, x is in Y.

A sequence $\{A_n\}$ of subsets of a metric space is called a decreasing sequence if

$$A_1 \supseteq A_1 \supseteq A_1 \supseteq ....$$

The following theorem gives conditions under which the intersection of such a sequence is non-empty.

Theorem 3:

(Cantor's Intersection Theorem): *Let X be a complete metric space, and let $\{F_n\}$ be a decreasing sequence of non-empty closed subsets of X such that $d(F_n) \to 0$. Then $F = \bigcap_{n=1}^{\infty} F_n$ Fn contains exactly one point.*

Proof:

It is first of all evident from the assumption $d(F_n) \to 0$ that F cannot contain more than one point, so it suffices to show that F is nonempty. Let x_n be a point in F_n. Since $d(F_n) \to 0$, $\{x_n\}$ is a Cauchy sequence. Since X is complete, $\{x_n\}$ has a limit x.

We show that x is in F, and for this it suffices to show that x is in F_n, for a fixed but arbitrary no. If $\{x_n\}$ has only finitely many distinct points, then x is that point infinitely repeated, and is therefore in F_{n0}. If $\{x_n\}$ has infinitely many distinct points, then a: is a limit point of the set of points of the sequence, it is a limit point of the subset $\{x_n : n \geq n_0\}$ of the set of points of the sequence, it is a limit point of F_{n0} and thus (since F_{n0} is closed) it is in F_{n0}.

A subset A of a metric space is said to be *nowhere dense* if its closure has empty interior. It is easy to see that A is nowhere dense $\Leftrightarrow$ $\overline{A}$ does not contain any non-empty open set $\Leftrightarrow$ each non-empty open set has a non-empty open subset disjoint from $\overline{A}$ $\Leftrightarrow$ each non-empty open set has a non-empty open subset disjoint from A $\Leftrightarrow$ each non-empty open set contains an open sphere disjoint from A. If a nowhere dense set is thought of as a set which doesn't cover very much of the space, then our next theorem says that a complete metric space cannot be covered by any sequence of such sets.

Theorem 4:

If $\{A_n\}$ is a sequence of nowhere dense sets in a complete metric space X, then there exists a point in X which is not in any of the A_n's.

Proof:

For the duration of this proof, we abandon our usual notations for open spheres and closed spheres. Since X is open and A_1 is nowhere dense, there is an open sphere S_1 of radius less than 1 which is disjoint from A_1. Let F_1 be the concentric closed sphere whose radius is one-half that of S_1, and consider its interior. Since A_2 is nowhere dense, Int (F_1) contains an open sphere S_2 of radius less than 1/2 which is disjoint from As.

Let F_2 be the concentric closed sphere whose radius is one-half that of S_2, and consider its interior. Since A_3 is nowhere dense, Int (F_2) contains an open sphere S_3 of radius less than 1/4 which is disjoint from A_3. Let F_3 be the concentric closed sphere whose radius is one-half that of S_3. Continuing in this way, we get a decreasing sequence $\{F_n\}$ of non-empty closed subsets of X such that $d(F_n) \rightarrow 0$. Since X is complete, Theorem C guarantees that there exists a point x in X which is in all the F_n's. This point is clearly in all the S_n's, and therefore (since S_n is disjoint from A_n) it is not in any of the A_n's.

For our purposes, the following equivalent form of Theorem D is often more convenient.

Theorem 5:

If a complete metric space is the union of a sequence of its subsets, then the closure of at least one set in the sequence must have non- empty interior.

Theorems D and E are really one theorem expressed in two different ways. We refer to both (or either) as *Baire's theorem*. This theorem is admittedly rather technical in nature.

He will find, however, that a need for it crops up from time to time, and when this need arises, Baire's theorem is an indispensable tool.

CONTINUOUS MAPPINGS

In the previous section we extended the idea of convergence to the context of a general metric space. We now do the same for continuity.

Let X and Y be metric spaces with metrics d_1 and d_2, and let f be a mapping of X into Y. f is said to be *continuous at a point* x_0 in X if either of the following equivalent conditions is satisfied:

(1) for each $\epsilon > 0$ there exists $\delta > 0$ such that $d_1(x, x_0) < \delta \Rightarrow d_2f(x), f(x_0)) < \epsilon$;

(2) for each open sphere $S_\epsilon(f(x_0))$ centered on $f(x_0)$ there exists an open sphere $S_\delta(x_0)$ centered on a_0 such that $f(S_\delta(x_0)) \subseteq S_\epsilon(f(x_0))$.

The reader will notice that the first condition generalizes the elementary definition given in the introduction to this chapter, and that the second translates the first into the language of open spheres.

Our first theorem expresses continuity at a point in terms of sequences which converge to the point.

Theorem 1:

Let X and Y be metric spaces and f a mapping of X into Y. Then f is continuous if and only if $x_n \to x \Rightarrow f(x_n) \to f(x)$.

This result shows that continuous mappings of one metric space into another are precisely those which send convergent sequences into convergent sequences, or, in other words, which *preserve convergence*. Our next theorem characterizes continuous mappings in terms of open sets.

Theorem 2:

Let X and Y be metric spaces and f a mapping of X into Y. Then f is continuous $\Leftrightarrow f^{-1}(G)$ *is open in X whenever G is open in Y.*

Proof:

We first assume that f is continuous. If G is an open set in Y, we must show that $f^{-1}(G)$ is open in X. $f^{-1}(G)$ is open if it is empty, so we may assume that it is non-empty. Let x be a point in $f^{-1}(G)$.

Then f(x) is in G, and since G is open, there exists an open sphere $S_\epsilon(f(x))$ centered on f(x) and contained in G. By the definition of continuity, there exists an open sphere $S_\delta(x)$ such that $f(S_\delta(x)) \subset S_\epsilon(f(x))$. Since $S_\epsilon(f(x)) \subseteq G$, we also have $f(S_\delta(x)) \subseteq G$, and from this we see that $S_\delta(x) \subseteq f^{-1}(G)$. $S_\delta(x)$

is therefore an open sphere centered on x and contained in $f^{-1}(G)$, so $f^{-1}(G)$ is open.

We now assume that $f^{-1}(G)$ is open whenever G is, and we show that f is continuous. We show that f is continuous at an arbitrary point x in X. Let $S_\epsilon(f(x))$ be an open sphere centered on f(x). This open sphere is an open set, so its inverse image is an open set which contains x. By this, there exists an open sphere $S_\delta(x)$ which is contained in this inverse image. It is clear that $f(S_\delta(x))$ is contained in $S_\epsilon(f(x))$, so f is continuous at x. Finally, since x was taken to be an arbitrary point in X, f is continuous.

The fact just established—that continuous mappings are precisely those which pull open sets back to open sets—will be of great importance for all our work.

We now come to the useful concept of *uniform continuity*. In order to explain what this is, we examine the definition of continuity expressed in condition (1) at the beginning of this section. Let X and Y be metric spaces with metrics d_1 and d_2, and let f be a mapping of X into Y. We assume that f is continuous, that is, that for each point x_0 in X the following is true: given $\epsilon > 0$, a number $\delta > 0$ can be found such that $d_1(x, x_0) < \delta \Rightarrow d_2(f(x), f(x_0)) < \epsilon$. The reader is no doubt familiar with the idea that if xa is held fixed and e is made smaller, then, in general, δ has to be made correspondingly smaller. Thus, in the case of the real function f defined by f(x) = 2x, δ can always be chosen as any positive number $\leq \epsilon/2$, and no larger δ will do. In general, therefore, δ depends on ε. Let us return to our examination of the definition.

It says that for our given ε, a δ can be found which "works" in the above sense at the particular point x_0 under consideration. But if we hold ε fixed and move to another point x_0, then it may happen that this δ no longer works; that is, it may be necessary to take a smaller δ to satisfy the requirement of the definition. We see in this way that δ nay well depend, in general, not only on ε but also on x_0. Uniform continuity is essentially continuity plus the added condition that for each ε we can find a δ which works *uniformly over the entire space* X, in the sense that it does not depend on x_0. The formal definition is as follows.

If X and Y are metric spaces with metrics d_1 and d_2, then a mapping f of X into Y is said to be *uniformly continuous* if for each $\epsilon > 0$ there exists $\delta > 0$ such that $d_1(x, x') < \delta \Rightarrow d_2(f(x), f(x')) < \epsilon$. It is clear that any uniformly continuous mapping is automatically continuous.

The reader will observe that the above real function f defined on the entire real line R by f(x) = 2x is uniformly continuous. On the other hand, the function g defined on R by $g(x) = x^2$ is continuous but not uniformly continuous. Similarly, the continuous function h defined on (0, 1) by $h(x) = 1/x$ is not uniformly continuous.

Uniformly continuous mappings—as opposed to those which are merely continuous—are of particular significance in analysis. The following theorem expresses a property of these mappings which is often useful.

Theorem 3:

Let X be a metric space, let Y be a complete metric space, and let A be a dense subspace of X. If is a uniformly continuous mapping of A into Y, then f can be extended uniquely to a uniformly continuous mapping g of X into Y.

Proof:

Let d_1 and d_2 be the metrics on X and Y. If A = X, the conclusion is obvious. We therefore assume that $A \neq X$. We begin by showing how to define the mapping g. If x is a point in A, we define g(x) to be f(x). Now let x be a point in X – A. Since A is dense, x is the limit of a convergent sequence $\{a_n\}$ in A. Since $\{a_n\}$ is a Cauchy sequence and f is uniformly continuous, $\{f(a_n)\}$ is a Cauchy sequence in Y. Since Y is complete, there exists a point in Y—we call this point g(x)—such that $f(a_n) \to g(x)$. We must make sure that g(x) depends only on x, and not on the sequence $\{a_n\}$. Let $\{b_n\}$ be another sequence in A such that $b_n \to x$. Then $d_1(a_n, b_n) \to 0$, and by the uniform continuity of f, $d_2(f(a_n), f(b_n)) \to 0$. It readily follows from this that $f(b_n) \to g(x)$.

We next show that g is uniformly continuous. Let $\epsilon > 0$ be given, and use the uniform continuity of f to find $\delta > 0$ such that for a and a' in A we have $d_1(a, a') < \delta \Rightarrow d_2(f(a), f(a')) < \epsilon$. Let x and x' be any points in X such that $d_1(x, x') < \delta$. It suffices to show that $d_2(g(x), g(x')) \leq \epsilon$. Let $\{a_n\}$ and $\{a'_n\}$ be sequences in A such that $a_n \to x$ and $a'_n \to x'$.

By the triangle inequality, we see that $d_1(a_n, a'_n) < d_1(a_n, x) + d_1(x, x') + d_1(x', a'_n)$. This inequality, together with the facts that $d_1(a_n, x) \to 0$, $d_1(x, x') < \delta$, and $d_1(x', a'_n) \to 0$, implies that $d_1(a_n, a'_n) < \delta$ for all sufficiently large n. It now follows that $d_2(f(a_n), f(a'_n)) < \epsilon$ for all sufficiently large n.

$$d_2(g(x), g(x')) = \lim d_2f(a_n), f(a'_n)),$$

and from this and the previous statement we see that $d_2(g(x), g(x')) \leq \epsilon$.

All that remains is to show that g is unique, and this is easily proved.

There is an important type of uniformly continuous mapping which often arises in practice. If X and Y are metric spaces with metrics di and dt, a mapping f of X onto Y is called an *isometry* (or an *isometric mapping*) if $d_1(x, x') = d_2(f(x), f(x'))$ for all points x and x' in X; and if such a mapping exists, we say that X is *isometric* to Y. It is clear that an isometry is necessarily one-to-one. If X is isometric to Y, then the points of these spaces can be put into one-to-one correspondence in such a way that the distances between pairs of corresponding points are the same.

The spaces therefore differ only in the nature of their points, and this is often unimportant. We usually consider isometric spaces to be identical with one another. It is often convenient to be able to use this terminology in the case of mappings which are not necessarily onto. If f is a mapping of X into Y which preserves distances in the above sense, then we call f an isometry of X *into* Y, or an isometry of X *onto* the subspace f(X) of Y. In this situation, we often say that Y contains an *isometric image* of X, namely, the subspace f(X).

UNIFORM CONVERGENCE AND POINTWISE CONVERGENCE

Definition : Let $< f(x_n) >$ be a sequence in a metric space (X, d). This sequence is said to converge pointwise to f (x) for every $x \in X$ that is to say,

$$\lim_{n \to \infty} f_n(x) = f(x) \ \forall \ x \in X.$$

By def. of point,

given $\varepsilon > 0$ and $x \in X$, $\exists\ m \in N$ s.t. $n \geq m$

$$\Rightarrow d(f_n(x), f(x) < \varepsilon$$

Here $m = m(\varepsilon, x)$.

If we keep ε fixed and vary x, then we shall get a set of values of m for different $x \in X$. This set of values may or may not have an upper bound. If this set has an upper bound, say n_0, then

$$\forall\ n \geq n_0 \Rightarrow d(f_n(x),$$
$$f(x) < \varepsilon \ \forall \ x \in X.$$

In this case sequence $<f_n(x)>$ is called *uniformly convergent.*

Definition : *A sequence defined on a metric space (X,d) is said to be uniformly convergent if given $\varepsilon > 0$, $\exists\ n_0 \in N$*

s.t. $$n \geq n_0 \Rightarrow d(f_n(x), f(x) < \varepsilon \ \forall x \in X.$$

Remark : It should be noted that ∃ one m ∀ x in case of uniformly convergent where as in case of ordinary convergence or pointwise convergence, one value of x will correspond to one value of m.

SOME IMPORTANT INEQUALITIES

Lemma 1:

If $1 < p < \infty$ *and* $\frac{1}{p}+\frac{1}{q}=1$, *then*

$ab \le \frac{a^p}{p}+\frac{b^q}{q}$ *where a, b are two non-negative reals.*

Proof :

if a = 0 or b = 0 the result is obvious.

Now consider the case in which $a \neq 0, b \neq 0$.

Let $\quad f(t) = 1-\lambda + \lambda t - t^{\lambda}$ for $0 < \lambda < 1$.

Then $\quad f'(t) = \lambda - \lambda t^{\lambda-1}$

$f'(t) = 0 \Rightarrow \lambda(1 - t^{\lambda-1})$,

where $\quad \lambda, \lambda - 1 \neq 0$

$\Rightarrow t = 1$

$f''(t) = -\lambda(\lambda-1)t^{\lambda-2}$.

At $t = 1$, $f''(t) = -\lambda(\lambda - 1) = +$ ive as $0 < \lambda < 1$.

Thus, $\quad \frac{df}{dt} = 0, \frac{d^2f}{dt^2} >$ at $t = 1$.

This proves that f is minimum at $t = 1$.

$\therefore \quad f(1) \le f(t)$

This $\quad \Rightarrow 1 - \lambda + \lambda - 1 \le 1 - \lambda + \lambda t - t^{\lambda}$

or $\quad 1 - \lambda + \lambda t - t^{\lambda} \ge 0$.

or $\quad (1-\lambda) + \lambda t \ge t^{\lambda}$ (Put $\lambda = 1/p$)

or $\quad \left(1-\frac{1}{p}\right)+\frac{t}{p} \ge \frac{t^1}{p}$

But $\quad \frac{1}{p}+\frac{1}{q}=1$

Hence $\quad \frac{1}{q}+\frac{t}{p} \ge \frac{t^1}{p}$. Put $t = \frac{a^p}{b^q}$.

Then $\quad \frac{1}{q} + \frac{a^p}{pb^q} \ge \frac{a}{b^{q/p}}$

or $\quad \frac{b^q}{q}+\frac{a^p}{p} \ge ab^{q-q/p} = ab$.

or $$\frac{a^p}{p}+\frac{b^q}{q}\geq ab$$

as $$\frac{1}{p}+\frac{1}{q}=1 \Rightarrow q-\frac{q}{p}=1$$

CAUCHY SCHWARZ INEQUALITY

If a_i, b_i $(i = 1, 2 \ldots n)$ are real or complex numbers, then

$$\left(\int_a^b fg\,dx \leq \left(\int_a^b f^p\,dx\right)^{1/p} \left(\int_a^b g^q\,dx\right)^{1/q}\right)$$

$$\sum_{i=1}^{n}|a_i b_i| \leq \left[\sum_{i=1}^{n}|a_i|^2\right]^{1/2}\left[\sum_{i=1}^{n}|b_i|^2\right]^{1/2}$$

Proof :

If we replace a_i, b_i , p, q by $|a_i|$, $|b_i|$, 2, 2 respectively, then we get this inequality. An independent proof is given below :

Write $A = (a_1, a_2, a_3, \ldots a_n)$, $b = (b_1 , b_2 , \ldots , b_n)$.

The magnitudes of these vectors are given by

$$\|a\|^2 = \sum_{i=1}^{n}|a_i|^2 \ \|b\|^2\sum_{i=1}^{n}|b_i|^2$$

If a = 0 or b = 0, then the inequality reduces to equality.

So we suppose that $a \neq 0$. $b \neq 0$. We know that

Geometric $\leq$ mean Arithmetic mean.

This $$\Rightarrow \sqrt{(xy)} \leq \frac{x+y}{2}$$

Set $$\sqrt{x} = \frac{|a_i|}{\|a\|}, \sqrt{y} = \frac{|b_i|}{\|b\|}$$

Then $$\frac{2|a_i|.|b_i|}{\|a\|.\|b\|} \leq \frac{|a_i|^2}{\|a\|^2}+\frac{|b_i|^2}{\|b\|^2}$$

or $$2\sum_{i=1}^{n}\frac{|a_i|.|b_i|}{\|a\|.\|b\|} \leq \frac{\sum|a_i|^2}{\|a\|^2}+\frac{\sum|b_i|^2}{\|b\|^2} = 1+1 = 2$$

or $$\sum_i |a_i|.|b_i| \leq \|a\|.\|b\|$$

or $$\sum_{i=1}^{n}|a_i b_i| \leq \left(\sum_{i=1}^{n}|a_i|^2\right)^{1/2}\left(\sum_{i=1}^{n}|b_i|^2\right)^{1/2}$$

Note : If a_i , b_i are real numbers, then

$$\sum_i |a_i b_i| \le \left(\sum_{i=1}^{n} a_i^2\right)^{1/2} \left(\sum_{i=1}^{n} b_i^2\right)^{1/2}$$

MINKOWSKI INEQUALITY

If $p > 1$ and a_i , b_i are non-negative real numbers, then

$$\left[\sum_{i=1}^{n}(a_i + b_i)^p\right]^{1/p} \le \left(\sum_{i=1}^{n} a_i^p\right)^{1/p} + \left(\sum_{i=1}^{n} b_i^p\right)^{1/p}$$

Proof :

By Holder's inequality,

$$\sum_{i=1}^{n} a_i + b_i \le \left(\sum_{i=1}^{n} a_i^p\right)^{1/p} \left(\sum_{i=1}^{n} b_i^q\right)^{1/q} \quad \text{... (1)}$$

with $$\frac{1}{p} + \frac{1}{q} = 1 \quad \text{... (2)}$$

By (2), $q\,(p-1) = p$ ... (3)

$$\sum_{i=1}^{n}(a_i + b_i)^p = \sum_{i=1}^{n}(a_i + b_i)(a_i + b_i)^{p-1}$$

where $\sum_{i=1}^{n} = \sum$

$$= \sum a_i\,(a_i + b_i)^{p-1} + \sum b_i\,(a_i + b_i)^{p-1}$$

Using (1) for each member of R.H.S.,

$$\sum\,(a_i + b_i)^p \le \left(\sum a_i^p\right)^{1/p} \left[\sum (a_i, b_i)^{q(p-1)}\right]^{1/q}$$

$$+ \left(\sum b_i^p\right)^{1/p} \left[\sum (a_i + b_i)^{q(p-1)}\right]^{1/q}$$

Using (3), $\sum\,(a_i + b_i)^p \le \left(\sum a_i^p\right)^{1/p} \left[\sum (a_i + b_i)^p\right]^{1/q}$

$$+ \left(\sum b_i^p\right)^{1/p} \left[\sum (a_i + b_i)^p\right]^{1/q}$$

Dividing by $\left[\sum (a_i + b_i)^p\right]^{1/q}$ we get

$$\left[\sum (a_i + b_i)^p\right]^{1-(1/q)} \le \left(\sum a_i^p\right)^{1/p} + \left(\sum b_i^p\right)^{1/p}$$

or, $$\left[\sum_{i=1}^{m}(a_i+b_i)^p\right]^{1/p} \le \left(\sum a_i^p\right)^{1/p} + \left(\sum b_i^p\right)^{1/p}$$

Note : Minkowski's inequality for integrals

$$\left[\int_a^b (f+g)^p\right] \le \left(\int_a^b f^p\right)^{1/p} + \left(\int_a^b g^p\right)^{1/p}$$

where $p > 1$.

MINKOWSKI'S INEQUALITY OF NORMS

Let $a = (a_1, a_2, \dots, a_n)$, $b = b_1, b_2, \dots, b_n)$ be two n tuples or real or complex numbers, then

$$\|a+b\| \le \|a\| + \|b\|$$

$$\left[\sum_{i=1}^{n}|a_i+b_i|^2\right]^{1/2} \le \left(\sum_{i=1}^{n}|a_i|^2\right)^{1/2} + \left(\sum_{i=1}^{n}|b_i|^2\right)^{1/2}$$

Proof:

We have $\|a+b\|^2 = \sum_i |a_i+b_i|^2$,

or $$\|a\|^2 = \sum_i |a_i|^2$$

$$\|b\|^2 = \sum_i |b_i|^2$$

where $\sum_i$ means $\sum_{i=1}^{n}$

$$\|a+b\|^2 = \sum_i |a_i+b_i|^2 = \sum_i |a_i+b_i|.|a_i+b_i|$$

$$\le \sum_i |a_i+b_i|.(|a_i+b_i|)$$

$$\le \sum_i |a_i+b_i|.|a_i| + \sum_i |a_i+b_i|.|b_i|$$

$$\le \sum_i |(a_i+b_i)a_i| + \sum_i |(a_i+b_i)b_i|$$

Applying Cauchy Schwarz inequality to each sigma or R.H.S., we get

$$\|a+b\|^2 \le \left(\sum |a_i+b_i|^2\right)^{1/2} \left(\sum_i |a_i^2|\right)^{1/2}$$

$$+\left(\sum_i |a_i+b_i|^2\right)^{1/2} \cdot \left(\sum_i |b_i^2|\right)^{1/2}$$

$$= \|a+b\| . \|a\| + \|a+b\| . \|b\|$$

Dividing by $\|a+b\|$, we get

$$\|a+b\| \le \|a\| + \|b\|$$

HOLDER'S INEQUALITY

If a_i, b_i are non-negative real numbers,

then $$\sum_{i=1}^{n} a_i b_i \le \left(\sum_{i=1}^{n} a_i^p\right)^{1/p} \left(\sum_{i=1}^{n} b_i^q\right)^{1/q}$$

where $\frac{1}{p}+\frac{1}{q}=1$ and $p > 1$.

Proof :

Firstly we shall establish a lemma :

Lemma 2:

If $a \ge 0$, $b \ge 0$, $p > 1$, then

$$\frac{a^p}{p}+\frac{b^q}{q} \ge ab$$

where $\frac{1}{p}+\frac{1}{q}=1$.

By the lemma, $$\frac{A^p}{p}+\frac{B^q}{q} \ge AB \qquad \text{... (1)}$$

Put $$A = \frac{a_i}{\left(\sum_i a_i^p\right)^{1/p}}, \quad B = \frac{b_i}{\left(\sum_i b_i^q\right)^{1/q}}$$

$$\therefore \quad \frac{1}{p}\frac{a_i^p}{\left(\sum_i a_i^p\right)} + \frac{1}{q}\frac{b_i^q}{\left(\sum_i b_i^q\right)} \ge \frac{a_i b_i}{\left(\sum_i a_i^p\right)^{1/p}\left(\sum_i b_i^q\right)^{1/q}}$$

Sum the result from

$i = 1$ to n,

$$\frac{1}{p\left(\sum_i a_i^p\right)}\left(\sum_{i=1}^{n} a_i^p\right)+\frac{1}{q\left(\sum_i b_i^q\right)}\left(\sum_{i=1}^{n} b_i^q\right)$$

$$\geq \frac{\sum_i a_i b_i}{\left(\sum_i a_i^p\right)^{1/p}\left(\sum_i b_i^q\right)^{1/q}}$$

or
$$\frac{1}{p}+\frac{1}{q}\geq \frac{\sum_i a_i b_i}{\left(\sum_i a_i^p\right)^{1/p}\left(\sum_i b_i^q\right)^{1/q}}$$

But
$$\frac{1}{p}+\frac{1}{q}=1.$$

Hence
$$\left(\sum_{i=1}^{n} a_i^p\right)^{1/p}\left(\sum_{i=1}^{n} b_i^q\right)^{1/q}\geq \sum_{i=1}^{n} a_i b_i$$

Notes: (1) Holder's inequality for complex numbers :

$$\sum_{i=1}^{n}|a_i b_i|\leq \left(\sum_{i=1}^{n}|a_i|^p\right)^{1/p}\left(\sum_{i=1}^{n}|b_i|^q\right)^{1/q}.$$

(2) Holder's inequality for integrals :

DEFINITION CANTOR'S TENNARY SETS

1. Let 1 = [0, 1]. Divide the closed interval 1 into three equal parts and remov the middle open interval (1/3 , 2/3), this leaves two disjoint closed intervals [0, 1/3] and [2/3 , 1] each of length 1/3. This is first stage or our construction. We now divide the remaining two closed intervals [0, 1/3] and [2/3 , 1] , each in three equal parts. Delete the middle open interval (1/9, 2/9) and (7/9, 8/9). This is the second stage of our construction. Similarly at the nth stage, 2^{n-1} open intervals are removed each or length $1/3^n$. The points which remain constitute Contor's tennary set. We denote this set by E. Let $x = 0.\,a_1\,a_2\,a_3 \ldots$ (in the scale of three) be any point of E. Then

$$x=\frac{a_1}{3}+\frac{a_2}{3^2}+\frac{a_3}{3^3}+\ldots$$

Hence a_n can not take values 3 or greater than 3.

Hence the only possibility is that $a_n = 0, 1, 2 \ \forall \ n$.

At the first step, all the points with $a_2 = 1$ are removed

Hence $a_n = 0$ or $2 \ \forall \ n$.

2. To prove that cardinal number of E is c.,

i.e., is uncountable. Recall that

$$\{0, 2)^N = \{f : N \to \{9, 2\}\}$$

i.e., $\{0, 2\}^N$ is the set of functions from N to 0, 2. In otherwords, $\{0, 2\}^N$ is the set of sequence $< x_n >$ where each $x_n = 0$ or 2.

Evidently, the map $0.\ x_1\ x_2\ x_3 \ldots \to < x_1, x_2 \ldots >$ is one one onto map.

This $\Rightarrow \exists$ one-one onto map between E and $\{0, 2\}^N$.

$$\Rightarrow |E| = \left|\{0,2\}^N\right| = 2^a = C$$

$\Rightarrow$ cardinal number of E is c.

Theorem 1:

In any metric space (X, d) each open sphere is an open set.

Proof:

Let (X, d) be a metric space. Let $S_{r0\,(x0)}$ be an open sphere in X.

To prove that $S_{r0\,(x0)}$ is an open set.

Let $x \in S_{r0\,(x0)}$ be arbitrary, then $d(x, x_0) < r_0$

Write $\quad r = r_0 - d(x, x_0)$, ...(1)

By definition, $\quad S_{r0\,(x0)} = \{y \in X : d(y, x_0) < r_0\}$

$S_{r\,(x)} = \{y \in X : d(y, x) < r\}$.

We claim $\quad S_{r\,(x)} \subset S_{r0\,(x0)}$

Let $\quad y \in S_{r\,(x)}$ be arbitrary.

Then $\quad d(x, y) < r$

$d(y, x_0) \leq d(y, x) + d(x, x_0)$

$< r + d(x, x_0) = r_0$.

$\therefore \quad d(y, x_0) < r_0$ on using (1)

This $\Rightarrow \quad y \in S_{r0\,(x0)}$

and $\quad y \in S_{r\,(x)}$

$\Rightarrow \quad y \in S_{r0\,(x0)}$

This $\Rightarrow$ $S_{r\,(x)} \subset S_{r0\,(x0)}$

Thus we have shown that

given any $x \in S_{r0\,(x0)}$, $\exists\ r > 0$ s.t. $S_{r\,(x)} \subset S_{r0\,(x0)}$.

By definition, this proves that $S_{r0\,(x0)}$ is an open set.

Theorem 2:

In any metric space, any closed sphere is a closed set.

Proof:

Let $Sr_{0\,[x0]}$ denote a closed sphere in a metric space (X, d).

To prove that $Sr_{0\,[x0]}$ is a closed set.

For this we must show that $S'\, r_0\, [x_0]$ is open in X.

Let $x \in S'\, r_{0\,[x0]}$ be arbitrary,

$$x \in S'\, r_{0\,[x0]} \Rightarrow x \notin S'\, r_{0\,[x0]}$$

$$\Rightarrow d\,(x, x_0)\,, <\,, \neq r_0,$$

$$\because Sr_{0\,[x0]} = \{\, y \in X : d\,(y\,,\, x_0) \leq r_0\}$$

$$\Rightarrow d\,(x, x_0) > r_0$$

$$\Rightarrow d\,(x, x_0) - r_0 > 0$$

$$\Rightarrow r > 0, \text{ on taking } r = d\,(x, x_0) - r_0 \qquad \text{...(1)}$$

We claim $S_{r\,(x)} \subset S'\, r_{0\,[x0]}$.

Let $y \in S_{r\,(x)}$ be arbitrary, so that, $d\,(y, x) < r$.

$\therefore$ $d\,(x, x_0) \leq d\,(x, y) + d\,(y, x_0)$.

$\therefore$ $d\,(y, x_0) \geq d\,(x, x_0) - d\,(x, y)$.

$> d\,(x, x_0) - r = r_0$ on using (1)

$\therefore$ $d\,(y, x_0) > r_0$

$\Rightarrow y \notin Sr_{0\,[x0]}$

Thus, any $y \in S_{r\,(x)} \Rightarrow y \in S'r_{0\,[x0]}$

This $\Rightarrow S_{r\,(x)} \subset S'r_{0\,[x0]}$

$\therefore$ Given any $x \in S'r_{0\,[x0]}$, $\exists\ r > 0$ s.t. $S_{r\,(x)} \subset S'r_{0\,[x0]}$

This proves that $S'r_{0\,[x0]}$ is open in x.

Theorem 3:

If a_i, b_i ($i = 1, 2 \dots n$) are non- negative real numbers and $p > 1$, then

$$\sum_{i=1}^{n}(a_i + b_i)^p \leq \sum_{i=1}^{n} a_i^p + \sum_{i=1}^{n} b_i^p$$

Proof :

Let f (x) = 1 + x^p – (1 + x $)^p$.

Then taking derivative ; it can be shown that

$$f(x) \geq 0 \text{ for } x \geq 0.$$

or $1 + x^p - (1 + x)^p \geq 0.$

or $(1 + x)^p \leq 1 + x^p$

put $x = a_i/b_i$, we get

$$\left(1 + \frac{a_i}{b_i}\right)^p \leq 1 + \left(\frac{a_i}{b_i}\right)^p$$

or, $(a_i + b_i)^p \leq (a_i)^p + (b_i)^p$

or $\sum_{i=1}^{n}(a_i + b_i)^p \leq \sum_{i=1}^{n} a_i^p + \sum_{i=1}^{n} b_i^p$

Theorem 4:

Let (x, ρ) be a metric space, Show that ∅ and X are open sets in X.

Proof :

Let (X, ρ) be a metric space.

To prove that ∅ and X are open sets .

To prove that ∅ is open in X, it suffices to show that

any $x \in \Phi \Rightarrow \exists\ \varepsilon > 0$, s.t. $S_{\varepsilon(x)} \subset \varnothing$.

Since ∅ does not contain any element and hence this condition is automatically fulfilled.

Corresponding to every point x ∈ X, ∃ an open sphere with its centre at x which is contained in X, showing there by X is an open .

Example : *Is the subset [0, 3] open in X = [0, 3] under the metric d given by d (x, y) = | x – y?*

Solution:

Since Φ and X both are open sets in any metric space (X, d).

∴ [0, 3] is open in x = [0, 3] .

Theorem 5:

In a metric space (X, ρ), ∅ and X are closed sets.

Proof :

Let (X, ρ) be a metric space.

To prove that $\varnothing$ and X are closed sets.

$\because \quad D(\varnothing) = \varnothing \subset \varnothing$

$\therefore \quad D(\varnothing) \subset \varnothing$

$\because$ This $\Rightarrow \varnothing$ is a closed set.

All the limiting points of X belong to X. For X is the universal set.

i.e., any $x \in D(X) \Rightarrow x \in X$

$\because \quad D(X) \subset X.$

This $\Rightarrow$ X is a closed set.

Theorem 6:

Let (X, d) be a metric space. Show that $F \subset X$, F is closed $\Leftrightarrow$ F' is open.

Proof:

Let (X, d) be a metric space .

Let C be a closed subset of X, so that $D(F) \subset F$.

To prove that F' is open in X.

Let $X \in F'$ be arbitrary. Then $X \notin F$.

$D(F) \subset F.\ x \notin F \Rightarrow x \notin D(F)$

$\Rightarrow (S_{r(x)} - \{x\}) \cap F = \varnothing$ for some $r > 0$.

$\Rightarrow S_{r(x)} \cap F = \varnothing \quad \because x \notin F$

$\Rightarrow S_{r(x)} \subset X - F$

$\Rightarrow S_{r(x)} \subset F'.$

$\therefore$ Given $x \in F'$, $\exists$ any an open sphere $S_{r(x)}$ s.t.

$S_{r(x)} \subset F'.$

By definition, this proves that F' is open.

Conversely suppose that F' is open in X.

To prove that F is closed in X.

Let $x \in F'$ be arbitrary, then $x \notin F$.

$\because$ F' is open, $\exists\ r \subset R^+$ s.t., $S_{r(x)} \subset F'$

This $\Rightarrow S_{r(x)} \cap F = \varnothing$

$\Rightarrow (S_{r(x)} - \{x\}) \cap F = \varnothing$

$\Rightarrow x \notin D(F).$

Thus, any $x \in F' \Rightarrow x \notin D(F)$.

i.e., any $x \in X - F$

$\Rightarrow x \in X - D(F)$

This $\Rightarrow$ $X - F \subset X - D(F)$

or $D(F) \subset F$.

This $\Rightarrow$ F is closed.

Theorem 7:

Let (X, d) be a metric space. Prove that

(i) an arbitrary union of open sets is open in X.

(ii) any finite intersection of open sets is open.

Proof:

Let (X, d) be a metric space.

Let $G_i \subset X$ be open X $\forall$ $i \in N$

Let $G = \bigcap_{i=1}^{\infty} G_i$, $G = \bigcap_{i=1}^{n} G_i$

To prove that

(i) G is open in X.

(ii) G is open in X.

(i) By Theorem 7, each G_i is a union of open spheres and hence G is the union of all open spheres arising in this way.

Hence g is open in X.

(ii) If G $= \Phi$, then G is open in X. For Φ is open in X.

If $G \neq \Phi$, then $\exists$ $x \in \bigcap_{i=1}^{n} G_i$ so that

$x \in G_i$ for $1 \leq i \leq n$.

$x \in G_i$, G_i is open

$\Rightarrow$ $\exists$ $r_i \in R^+$ s.t. $S_{r}(x) \subset G_i$ ...(1)

This is true for each i (= 1, 2, ..., n).

Take $r = \min \{r_1, r_2 ..., r_n\}$

Then $r \leq r_i$ for $1 \leq i \leq n$.

This $\Rightarrow S_r(x) \subset S_{ri}(x)$ for $1 \leq i \leq n$

$\Rightarrow S_r(x) \subset S_{ri}(x) \subset G_i$ by (1)

$\Rightarrow S_r(x) \subset G_i$

$$\Rightarrow S_{r}(x) \subset G = \bigcap_{i=1}^{n} G_i = G \Rightarrow S_{r}(x) \subset G$$

$\therefore$ any $x \in G \Rightarrow \exists\ r \in R^{+}$ s.t. $S_{r}(x) \subset G$

By definition, this $\Rightarrow$ G is open in X.

Remark: The intersection of an infinite number of open sets may or may not be open. If we take an open set, $A_n = (-1/n, 1/n)$, then

$$\bigcap_{n=1}^{\infty} A_n = \{0\} = \text{closed set}]$$

Theorem 8:

In any metric space (X, d), show that

(i) an arbitrary intersection of closed sets is closed.

(ii) any; finite union on closed sets is closed.

Proof:

Let (X, d) be a metric space

Let $F_i \subset X$ be a closed set $\forall\ i \in N$

Let $F = \bigcap_{i=1}^{\infty} F_i,\ F = \bigcup_{i=1}^{n} F_i$

To Prove that F and F are closed sets in X.

By theorem 4, F_i is closed $\Leftrightarrow F_i'$ is open.

By theorem 8, $\bigcup_{i=1}^{\infty} F_i', \bigcap_{i=1}^{n} F_i'$ are open in X. ...(1)

By, De Morgan's law, $\bigcup_{i=1}^{\infty} F_i', = \left(\bigcap_{i=1}^{\infty} F_i\right) = F'$

and $\bigcap_{i=1}^{\infty} F_i' = \left(\bigcup_{i=1}^{\infty} F_i\right) = F'$

From (1), F' , F' are open in X

i.e., F, F are closed in X. [by Theorem 4]

Remark: The union of an infinite number of closed sets may or may not be closed.

Let $F_n = \left[0, \dfrac{n}{n+1}\right]$

Then $\bigcup_{i=1}^{\infty} F_n = [0, 1)$ =semi open set

$\therefore \quad \bigcup_{n=1}^{\infty} F_n$ is not closed, though each F_n is closed.

Theorem 9:

Let (X, d) be a metric space and let $G \subset X$ be an arbitrary set. Show that G is open $\Leftrightarrow$ it is a union of open spheres.

Proof:

Let (X, d) be a metric space. Let $G \subset X$ be an open set.

To prove that G is a union of open spheres.

If $G = \varnothing$, the proof is trivial,

If $G \neq \varnothing$, then by definition of an open set

any $\quad x \in G \Rightarrow \exists\, r \in R^+$ s.t. $S_{r\,(x)} \subset G$ ...(1)

In view of this statement we are justified to write.

$$G = \bigcup_i \{S_{ri(xi)} : x_i \in G\}$$

where each $r_i \in R^+$ is s.t. $S_{ri(xi)}$ satisfies the condition (1).

Conversely suppose that (X, d) is a metric space. Let $G \subset X$ be expressible as a union of open spheres, *i.e.*, $G = \bigcup_i S_{ri(xi)}$.

To prove that G is an open set.

Let $x \in G$ be arbitrary.

Then x is a centre of some open sphere, say.

$$S_{(x,\, r)} \in \{S_{ri(xi)}\}$$

Then $\quad S_{r\,(x)} \subset \bigcup_i S_{ri\,(xi)} = G,$

or $\quad S_{r\,(x)} \subset G,$

$\therefore$ any $\quad x \in G \Rightarrow \exists\, r \in R^+$ s.t. $S_{r\,(x)} \subset G.$

By definition this proves that G is an open set.

Examples:

To find the boundary of set of integers Z and set of rational Q.

$Z^o = \cup \{ G \subset R : G \text{ is open and } G \subset Z \} = \phi$

(1) For every subset of R contains fractions also.

Similarly $(R - Z)^o = \phi$

$$b\,(Z) = R - Z^o \;\cup\; (R - Z)^o = R - \phi \;\cup\; \phi = R$$

or, $b\,(Z) = R$

Similarly $b\,(Q) = R$

(2) Give two examples of limit points.

(i) If $A = \left\{1+\frac{1}{n} : n \in N\right\}$,

$$A = \left\{2, \frac{3}{2}, \frac{4}{3}, \frac{5}{4}, \frac{6}{5}, \ldots\right\}, \text{then}$$

1 is limit point of A for $\lim_{n\to\infty}\left(1+\frac{1}{n}\right) = 1.$

(ii) If $A = \left\{1, \frac{1}{2}, \frac{1}{3}, \frac{1}{4}, \ldots\right\} = \left\{\frac{1}{n} : n \in N\right\}$,

then 0 is the limit point of A.

For $\lim_{n\to\infty}\frac{1}{n} = 0.$

(3) Since N has no limit points and so $D\,(N) = \phi$. *For every open sphere centred at any point of N other than that point.*

$$\overline{N} = N \cup D\,(N) = N \cup \phi = N$$

or $\overline{N} = N$ $\therefore$ *N is closed.*

so that N' is open and so $(N')^o = N'$.

$$ext\,(N) = (N')^o = N'$$

$$b\,(N) = [N^o \cup ext\,(N)]' = [\phi \cup ext\,(N)']$$

$$= (N')' = N$$

Note that N is non-dense set.

For $(\overline{N})^o = N^o = \phi$

***Ex.** : Define non where dense set and give an example.*

(4) Consider the set R. Evidently $Q \subset R$. *Every real number is a limit point so that rational limit points belong to the set Q.*

It means that $D\,(R) = R$, $D\,(Q) = R$

$\because$ $\overline{A} = A \cup D\,(A)$

$\therefore$ $\overline{Q} = Q \cup D\,(Q) = Q \cup R = R$

This proves that the closure of the set of rational numbers w.r.t. usual metric no R is R.

Thus $Q \subset R$, *Q is countable and* $\overline{Q} = R$.

This proves that r with usual metric is separable.

$Q^o = \cup\{G \subset R : G \text{ is open}, G \subset Q = \phi.$

Since every subset of R contains rationals and irrationals. Hence their exists no open set.

$$G \subset R \text{ s.t. } G \subset Q.$$

Thus, $Q^o = \phi.$

Similarly $ext\ (Q) = (R - Q)^o = \phi.$

$int\ (\overline{Q}) = int\ (R) = R \neq \phi.$

$\therefore$ Q is some where dense set in R.

$$b\ (Q) = R - Q^o \cup (R - Q)^o = R - \phi \cup \phi = R$$

(5) If $A = (a, b)$ or $[a,b)$ or $(a, b]$ or $[a, b]$,

then in every case,

$\overline{A} = [a, b],\ A^o = (a, b),\ b(A) = \{a, b\}.$

(6) If $A = [0, 1)$, then $(\overline{A})^o = [0, 1]^o = (0, 1) \neq \phi$. .

and so A is somewhere dense set in R.

(7) To show that N and Z are non-dense sets in R relative to the usual metric.

$$D\ (N) = \phi.$$

For if a is any real number, then any nbd $= (a - \varepsilon, a + \varepsilon) \subset R$ does not contain any point of N other than a

$$Z = \{n : n \in N\} \cup \{0\} \cup \{-n : n \in N\}$$

$$D\ (Z) = \phi \cup \phi \cup \phi = \phi$$

or, $D\ (Z) = \phi \subset Z$

or, $D\ (Z) \subset Z.$

$\therefore$ Z is a closed set. Similarly N is a closed set.

For $D\ (A) \subset A \Leftrightarrow A$ is closed.

$\therefore$ $\overline{Z} = Z$ and $\overline{N} = N$

$Int\ \overline{Z} = Int\ (Z) = \cup \{G \subset R : G \text{ is open and } G \subset Z\} = \phi$

For any open subset $(a_1, a_2) \subset R$ contains fractions also.

Now $Int\ (\overline{Z}) = (\overline{Z})^o = \phi.$

Similarly $(\overline{N})^o = \phi.$

$\therefore$ Z and N are nowhere dense subsets of R.

DEFINITION OF NEIGHBOURHOOD

Let (X, d) be a metric space and $x \in X$, $A \subset X$.

A subset A of X is called a neighbourhood (nbd) of x if $\exists$ open sphere $S_{r(x)}$ s.t. $S_{r(x)} \subset A$.

This means that A is nbd of a point x iff x is an interior point of A.

From the definition of nbd it is clear that :

(1) Every superset of a nbd of a point is also a nbd.

(2) Every open sphere $S_{r(x)}$ is a nbd of x.

(3) Every closed sphere $S_{r[x]}$ is a nbd of x.

(4) Intersection of two nbds of the same point is again a nbd of that point.

(5) A set is open if it contains a nbd of each of its points.

(6) Nbd of a point need not be an open set.

Definition of Boundary of a Point

(i) Boundary or Frontier of a set a is denoted by b (A) or F_r (A) and is defined as b (A) = F_r (A) = $X - A^o \cup (X - A)^o$.

Elements of b (A) are called boundary points of A.

(ii) The exterior of A is defined as the set $(X - A)^o$ and is denoted by ext (A).

Symbolicallyext (A) = $X - A^o$.

(iii) A is said to be dense or everywhere dense in X if $\overline{A} = X$.

(iv) A is said to be somewhere dense if $(\overline{A})^{o}$ ¹ f, *i.e.*, if closure of A contains some open set.

(v) A is said to be non where dense (or nonwhere dense set) if $(\overline{A})^o$ = f.

(vi) A metric space (X, d) is said to be separable if $ A Ì X s.t. A is countable and $\overline{A} = X$.

(vii) A is said to be dense in itself if $A \subset D(A)$.

Theorem 1:

Let A be a subset of a metric space (X, d). Then $x_0 \in X$ is a limit point of A iff every open sphere centred at x_0 contains an infinitely many; points of A, other than x_0.

Proof:

Let a be a subset of a metric space (X, d) and let $x_0 \in X$.

Step I : Given x_0 is a limit point of A.

Aim. $S_{r(x0)}$ contains an infinite number of points of A other than x_0 ...(*)

By def. of limit point,

$$(S_{r(x0)}) - \{x_0\}) \cap A \neq \phi \qquad ...(1)$$

for every open sphere centred at x_0.

Suppose (*) is not true.

Then $S_{r(x0)}$ contains a finite number of points

$$x_1, x_2, \ldots, x_n, \in A \qquad \ldots(2)$$

Let $r = \min \{d(x_0, x_i) : = 1, 2. \ldots, n\}$

and $S_{r(x0)} = \{x \in X : (x, x_0) < r_0\}$

Evidently $(S_{r(x0)} - \{x_0\}) \cap A = \phi$

This $\Rightarrow x_0$ is not a limit point of A.

This is contrary to what is given . This contradiction arises by assuming that $S_{r(x0)}$ contains a finite number of points of A other than x_0. Hence the required result follows.

Step II: Given $S_{r(x0)}$ contains an infinite number of points of A other than x_0.

Aim. x_0 is a limit point of A.

By assumption, $(S_{r(x0)} - \{x_0\}) \cap A \neq \phi$

This $\Rightarrow x_0$ is a limit point of A.

Corollary : *If p is a limit point of A in a metric space, then every nbd N(p) contains infinitely many points of A.*

Or

Every open sphere centred at a limit point of a sat in a metric space contains infinitely many points of the set.

Hint : Open sphere $S_{r(p)}$ can be taken in place of $N_{(p)}$. Here write step I of the above Theorem 13.

Theorem 2:

Let (X, d) be metric space and $A \subset X$. Show that

$$\overline{A} = \{x \in X : d(x, A) = 0\}.$$

Proof:

Let (X, d) be metric space and $A \subset X$.

We claim $\overline{A} = \{x \in X : d(x, A) = 0\}$

Suppose $d(x, A) = \epsilon > 0$

Then (i) $x \notin A$ and

(ii) $(S_{\epsilon/2(x)} - \{x\}) \cap A = \Phi$

Evidently (ii) $\Rightarrow x \notin D(A)$.

Thus, $d(x, A) = \notin \Rightarrow x \notin A, x \notin D(A)$

$$\Rightarrow x \notin A \cup D(A) = \overline{A}$$

$$\Rightarrow x \in \overline{A}$$

Finally, $d(x, A) = \varepsilon \Rightarrow x \notin \overline{A}$

Again $d(x, A) = 0 \Rightarrow$

(iii) $x \in A$,

(iv) $(S_{r(x)} - \{x\}) \cap A \neq \Phi$

$$(iv) \Rightarrow x \in D(A)$$

Finally, $d(x, A) = 0 \Rightarrow x \in A$ or $x \in D(A)$

$$\Rightarrow x \in A \cup D(A) = \overline{A} \quad \Rightarrow x \in \overline{A}$$

Thus, we have shown that

$$d(x, A) = 0 \Rightarrow x \in \overline{A}$$

and $d(x, A) = \varepsilon > 0 \Rightarrow x \notin \overline{A}$

Furthermore, $d(x, A) = \geq 0$.

Consequently the required result follows :

Theorem 3:

A subset in a metric space is open iff it is a nbd of each of its point.

Proof:

Let A be a subset of a metric space (X, d).

Step I : Given A is and of each of its points.

Aim : A is an open set. Recall that a set N is called nbd of a point $x \in X$ if $\exists$ open set $G \subset X$ s.t. $x \in G \subset N$.

X s.t. $p \in G_p \subset A$.

It is true $\forall p \in A$.

Take $A = \cup \{G_p : p \in G_p, G_p$ is open set, $G_p \subset A\}$

= An arbitrary union of open sets

= open set.

$\therefore$ A is an open set.

Step II : Let A be an open subset of X.

Aim : A is a nbd of each of its points. By assumption, we can write

$$p \in A \subset A \ \forall \ p \in A$$

This $\Rightarrow$ A is a nbd of each of its points.

Theorem 4:

A subset A of metric subspace (Y, d) of a metric space (X, d) is open in Y iff $\exists$ an open set G in (X, d) s.t. $A = Y \cap G$.

Proof:

Let $A \cap Y$ and let (Y, d) be a subspace of a metric space (X, d).

Then $\quad A \subset Y \subset X$ ...(1)

Step I : Let A be open in Y. ...(2)

Aim : $A = Y \cap G$ for some open set G in X. By def. of open set,

$\forall\ y \in A \Rightarrow \exists$ open sphere $S_{1\,(Y,\,r)}$ in s.t.

$$S_{1\,(Y,\,r)} \subset A \qquad ...(3)$$

Then A is expressible as

$$A = \cup\ \{S_{1\,(Y,\,r)} : y \in A\} \qquad ...(4)$$

Suppose S(Y, r) is an open sphere in X.

Then $\quad S_{1\,(Y,\,r)} = Y \cap S(Y, r)$

Now (4) is expressible as

$$A = \cup\ \{S_{(Y,r)} \cap Y : y \in A\}$$

$$= Y \cup [S_{(Y,r)} : y \in A\}$$

or, $\quad A = Y \cap G,$

where $\quad G = \cup\ \{S_{(Y,r)} : y \in A\}$

= arbitrary union of open spheres in X

= open set in X.

$\therefore\ A = Y \cap G$, where G is open in X.

For each open sphere is an open set.

Step II : Let $\quad A = Y \cap G$ ...(5)

where G is an open set in X.

Aim : A is open in Y.

By def. of open set in X,

$\forall\ x \in G \subset \Rightarrow \exists$ open sphere $S_{(x,\epsilon)}$ in X. s.t.

$$S_{(x,\epsilon)} \subset G \qquad ...(6)$$

Write $\quad S_{1(x,I)} = Y \cap S_{(x,I)}$ ...(7)

Then $S_{1(x,\epsilon)}$ is an open sphere in Y.

By (6), $S_{(x,\epsilon)} \cap Y \subset G \cap Y$

Using (7), we get $S_{1(x,\epsilon)} \subset G \cap Y = A$, by (5)

or, $x \in S_{1(x,\epsilon)} \subset A$

This $\Rightarrow$ A is open in Y, be def. of open set.

Corollary : *The necessary and sufficient condition that a subset a metric subspace (Y, d) of a metric space (X, d) be closed in (Y, d) is ∃ a set F closed in (X, d) s.t.*

$$A = Y \cap F.$$

Proof:

Firstly we shall prove a lemma :

A is open in Y $\Leftrightarrow$ A Y $\cup$ G, G is open in X.

Now we come to the proof of the actual theorem.

F is closed in X $\Rightarrow$ X – F is open in X.

A is closed in Y $\Leftrightarrow$ Y – A is open in Y.

Be the lemma,

$\therefore$ $Y - A = Y \cap (X - F)$

or, $Y - Y \cap (X - F) = A$

or, $Y - (Y \cap X - Y \cap F) = A$

or, $Y - (Y - Y \cap F) = A$

or, $Y \cap F = A$

Theorem 5:

If A_1 is an open set in (X_1, d) and A_2 is an open set in (X_2, d_2), then $A_1 \times A_2$ is open in the product metric space (X,d), where $X = X \times X_2$

Proof:

Let $x = (x_1, x_2) \in A_1 \times A_2$ be arbitrary, then $x_1 \in A_1$, $x_2 \in A_2$. By def. of open set,

∃ open spheres $S(x_1, r_1)$, $S(x_2, r_2)$ s.t.

$$\left.\begin{array}{l} S(x_1, r_1) \subset A_1 \\ S(x_2, r_2) \subset A_2 \end{array}\right\} \quad ...(1)$$

Let $r = \min. \{r_1, r_2\}$, then $r \leq r_1$, $r \leq r_2$.

$\therefore$ $S(x_1, r) \subset S(x_1, r_1) \subset A_1$

$S(x_2, r) \subset S(x_2, r_2) \subset A_2$

This $\Rightarrow S(x_1, r) \times S(x_2, r_1) \subset A_1 \times A_2$

$\Rightarrow x = (x_1, x_2) \in S(x, r)$

$\Rightarrow S(x_1, r) \times S(x_2, r_1) \subset A_1 \times A_2$

$\Rightarrow x \in S(x, r) \subset A_1 \times A_2$

It is true $\forall x \in A_1 \times A_2$

$\therefore$ $A_1 \times A_2$ is open subset of (X, d).

Here S (x, r) is an open sphere in (X, d).

Theorem 6:

Let (X, d) be a metric space. A subset A of X is closed if given any $x \in X - A$, $d(x, A) \neq 0$.

Proof :

Let (X, d) be a metric space and $A \subset X$ be an arbitrary closed set.

To prove that

given any $x \in X - A$, $d(x, A) \neq 0$

A is closed $\Rightarrow$ X − A is open.

By definition of open set,

any $x \in X - A \Rightarrow \exists\, r \in R^+$ s.t. $S_{r(x)} \subset X - A \Rightarrow S_{r(x)} \cap A = \Phi$

$\Rightarrow d(x, A) \geq r \Rightarrow d(x, A) \neq 0$.

Conversely let A be any subset of a metric space (X, d). Let

any $x \in X - A \Rightarrow d(x, A) \neq 0$

To prove that A is closed.

Let $x \in X - A$ be arbitrary so that, by assumption,

$d(x, A) = r \neq 0 \Rightarrow S_{r(x)} \cap A = \phi$

$\Rightarrow$ $S_{r(x)} \subset X - A$

$\therefore$ $x \in X - A \Rightarrow \exists\, r \in R^+$ s.t. $S_{r(x)} \in X - A$.

By definition, this $\Rightarrow$ X − A is open

$\Rightarrow$ A is closed.

Theorem 7:

Let (X, ρ) be a metric space.

Define $d(x, y) = \dfrac{\rho(x, y)}{1 + \rho(x, y)} \;\forall\, x, y, \in X.$

Then (X, y) is metric space.

Proof:

Let (X, ρ) be a metric space

Let $d(x, y) = \dfrac{\rho(x, y)}{1+\rho(x, y)} \; \forall \; x, y, \in X.$

To prove that (X, d) is a metric space.

Let x, y, z, ∈ X be arbitrary.

∵ (X, ρ) is a metric space.

∴ (i) $\rho(x, y) \geq 0$

(ii) $\rho(x, y) = \rho(y, x)$

(iii) $\rho(x, y) = 0$ iff $x = y$

(iv) $\rho(x, y) + \rho(y, z) \geq \rho(x, z)$.

Now (i)' $d(x, y) \geq 0$ in accordance with (i).

(ii)' $d(x, y) = d(y, x)$.

For $d(y, x) = \dfrac{\rho(y, x)}{1+\rho(y, x)}$

$= \dfrac{\rho(x, y)}{1+\rho(x, y)} = d(x, y)$ in accordance with (ii)

(iii)' $d(x, y) = 0$, iff $x = y$

$d(x, y) = 0 \Leftrightarrow \dfrac{\rho(x, y)}{1+\rho(x, y)} = 0$

$\Leftrightarrow \rho(x, y) = 0 \Leftrightarrow x = y$, acoording to (iii).

(iv)' $d(x, y) + d(y, z) \geq d(x, z)$.

For $d(x, y) + d(y, z) = \dfrac{\rho(x, y)}{1+\rho(x, y)} = \dfrac{\rho(y, z)}{1+\rho(y, z)}$

$\geq \dfrac{\rho(x, y)}{1+\rho(x, y)+\rho(y, z)} + \dfrac{\rho(y, z)}{1+\rho(y, z)+\rho(x, y)}$

$\geq \dfrac{\rho(x, y)+\rho(y, z)}{1+\rho(x, y)+\rho(y, z)}$

$\geq \dfrac{\rho(x, z)}{1+\rho(x, z)}$

$= d(x, z)$

i.e., d (x, y) + d (y, z) ≥ d (x, z).

The results (i)', (ii)', (iii)', (iv)' taken together prove the required result.

Theorem 8:

Let (X, y) be a metric space and let m a positive real number. Show that there exists a metric d' for X such that the metric space (X, y) is bounded with d(X) ≤ m.

Proof:

Let (X, y) be a metric space and m > 0.

we define d (x, y) = $\dfrac{md\left(x,y\right)}{1+d\left(x,y\right)}\ \forall\ x,y\in X.$

To prove that (X) (X, d) is a metric space and m > 0.

Let x, y, x ∈ X be arbitrary.

d is metric ⇒ d (x, y) ∈ R

Also m ∈ R

⇒ d' (x, y) $\dfrac{md\left(x,y\right)}{1+d\left(x,y\right)}\in R.$

(X, d) is metric space ⇒

(i) d (x, y) ≥ 0

(ii) d (x, y) = d (y, x)

(iii) d (x, y) = 0 iff x = y

(iv) d (x, y) + d (y, z) ≥ d (x, y).

(1) Evidently d' (x, y) ≥ 0, according to (i) as m > 0.

(2) d' (x, y) = d' (y, x)

For d' (y, x) $\dfrac{md\left(y,x\right)}{1+d\left(y,x\right)}$

$= \dfrac{md\left(x\,y\right)}{1+d\left(x,y\right)}$, according to (ii)

= d' (x, y)

(3) d' (x, y) = 0 iff x = y.

For d' (x, y) = 0 ⇔ $\dfrac{md\left(x\,y\right)}{1+d\left(x,y\right)} = 0$

⇔ md (x ,y) = 0

$\Leftrightarrow$ d (x, y) = 0

$\Leftrightarrow$ x = y, according to (iii).

(4) d' (x, y) + d' (y, z) $\geq$ d' (x, z).

For $$d'(x, y) + d'(y, z) = \frac{md(x,y)}{1+d(x,y)} + \frac{md(y,z)}{1+d(y,z)}$$

$$= m\left[\frac{d(x,y)}{1+d(x,y)} + \frac{d(y,z)}{1+d(y,z)}\right]$$

$$\geq m\left[\frac{d(x,y)}{1+d(x,y)+d(y,z)} + \frac{d(y,z)}{1+d(x,y)+d(y,z)}\right]$$

$$= m\left[\frac{d(x,y)+d(y,z)}{1+d(x,y)+d(y,z)}\right] \geq \frac{md(x,z)}{1+d(x,z)}$$

= d' (x, z)

The results (1), (2), (3) and (4) taken together prove that (X, d') is metric space. Remains to prove that d' is bounded.

$$d'(x, y) = \frac{md(x,y)}{1+d(x,y)} = \frac{m}{1+\dfrac{1}{d(x,y)}}$$

d' (x, y) $\leq$ m $\forall$ x, y, $\in$ X.

This $\Rightarrow$ (x, d') is a bounded metric space.

Remark: If $d'(x, y) = \dfrac{d(x,y)}{1+d(x,y)}$

Then d' (x, y) $\leq$ 1 so that (X, d') is bounded.

Theorem 9:

Let (X, d) be a metric space and let

$$d'(x, y) = \min\{1, d(x, y)\}.$$

Then d' is a metric for X.

Proof:

Let (X, d) be a metric space so that

(i) d (x, y) $\geq$ 0

(ii) d (x, y) = d (y, x)

(iii) d (x, y) = 0 iff x = y

(iv) d (x, y) + d (y, x) ≥ d (x, z), where x, y, z ∈ X.

Let d' (x, y) = min, {1, d (x, y)}.

Then $d'(x,y) = \begin{cases} 1 \\ or\ d(x,y) \end{cases}$ ...(1)

(2) d' (x, y) ≥ 0, according to (1) and (i),

(3) d' (x, y) = d' (y, x).

For if d' (x, y) = 1, then d' (y, x) = 1 so that d' (x, y) = d' (y, x).

Again if d' (x, y) = d (x, y), then d' (y, x) = d (y, x) using (ii) and combining the last two, we get

d' (x, y) = d' (y, x)

∴ In either case, d' (x, y) = d' (y, x)

(4) d' (x, y) = 0 iff x = y.

For d' (x, y) = 0 ⇔ d (x, y) = 0 according to (1)

⇔ x = y according to (iii).

(5) To prove d' (x, y) + d' (y, z) ≥ d' (x, z)

If d' (x, y) = 1, then d' (x, z) = 1, d' (y, z) = 1 and

so d' (x, y) + d' (y, z) > d' (x, z)

If we take d' (x, y) = d (x, y), then

d' (y, z) = d (y, z),

d' (x, z) = d (x, z)

and so d' (x, y) + d' (y, z) = d (x, y) + d (y, z)

≥ d (x, z) = d', (x, z) d by (iv)

or, d' (x, y) + d' (y, z) ≥ d' (x, z)

∴ (5) is true in both cases.

From what has been it follows that d' is a metric on X.

Similar Problem : *Show that if d is a pseudo metric on X and e(x,y) = min {1, d (x, y)}, then e is also a pseudo metric.*

Theorem 10:

In any metric space, prove that difference between the two sides of a triangle is less than the third side.

Or

Let (X, d) be a metric space and let x, y, z be any three points of X. Then |d (x, z) – d (y, z)| ≤ d (x, y)

Proof :

Let x, y, z denote the co-ordinates of vertices of a triangle.

If we show that $|d(x, z) - d(y, z)| \leq d(x, y)$, the result will follow.

We know that

$$d(x, y) + d(y, z) \geq d(x, z)$$

Similarly d $(y, x) + d(x, z) \geq d(y, z)$

From Which $d(x, y) \geq d(x, z) - d(y, z)$ ...(1)

and $d(y, x) \geq -[d(x, z) - d(y, z)]$

The last gives $-d(x, y) \leq d(x, z) - d(y, z)$ ...(2)

For $d(x, y) = d(y, x)$.

Combining (1) and (2),

$$-d(x, y) \leq d(x, z) - d(y, z) \leq d(x, y)$$

This $\Rightarrow |d(x, z) - d(y, z)| \geq d(x, y)$.

Theorem 11:

Let (X, d) be a metric space, then $\forall$ *x, y, x', y', in X,* $|d(x, y) - d'(x' y')| \leq d(x, x') + d(y, y')$.

Proof:

For proving the required result, we make repeated use of "triangle inequality."

$$d(x, y) \leq d(x, x') + d(x', y)$$

$$\leq d(x, + d(x', y') + d(y',$$

or, $d(x, y) - d(x', y') \leq d(x, + d(y',$ ...(1)

as $d(y, = d(y',$

Again $d(x', y') \leq d(x', + d(x, y')$

$$\leq d(x', + d(x, y) + d(y,$$

Using the fact that

$d(a, b) = d(b, a)$,

we get $d(x', y') - d(x, y) \leq d(x, + d(y,$

or, $d(x, y) - d(x', y') \geq -[d(x, + d(y,$ ...(2)

Combining (1) and (2), we get,

$-[d(x, + d(y, \leq d(x, y) - d(x', y') \leq d(x, + d(y,$

or, $|d(x, y) - d(x', y')| \leq d(x, + d(y,$

Theorem 12:

Let (X, d) be a metric space. If $\langle x_n \rangle$ and $\langle y_n \rangle$ are sequences in X, such that $x_n \to x$, $y_n \to y$, then show that

$$d(x_n, y_n) \to d(x, y).$$

Proof :

Suppose $\langle x_n \rangle$ *and* $\langle y_n \rangle$ are sequences in a metric space (X, d).

Let $x_n \to x$, $y_n \to y$.

To prove that

$$d(x_n, y_n) \to d(x, y).$$

Let x, y, z ∈ X be arbitrary.

Firstly we shall prove that

$$|\, d(x, z) - (y, z)| \le d(x, y) \qquad ...(3)$$

Now write the proof of Theorem 20.

$\because \quad x_n \to x,\ y_n \to y.$

$\therefore \quad \varepsilon > 0,\ \exists\ n_0, m_0, \in N$ s.t.

$\forall\ n \ge n_0 \Rightarrow d(x_n, x) < \varepsilon/2$

$\forall\ n \ge m_0 \Rightarrow d(y_n, y) < \varepsilon/2$

Take $\quad k_0 = \max.\{m_0, n_0\}$

Then $\quad d(x_n, x) < \varepsilon/2,\ d(y_n, y) < \varepsilon/2\ \forall\ n \ge k_0$

Now

$$\begin{aligned} & |d(x_n, y_n) - d(x, y)| \\ &= |d(x_n, y_n) - d(x_n, y) + d(x_n, y) - d(x, y)| \\ &\le |d(x_n, y_n) - d(x_n, y)| + |d(x_n, y) - d(x, y)| \\ &= |d(y_n, x_n) - d(y, x_n)| + |d(x_n, y) - d(x, y)| \\ &\le d(y_n, y) + d(x_n, x), \text{ according to (3)} \\ &< \varepsilon/2 + < \varepsilon/2 = \varepsilon \end{aligned}$$

$\therefore |d(x_n, y_n) - d(x, y)| < \varepsilon/2\ \forall\ n \ge k_0$

This $\quad \Rightarrow d(x_n, y_n) \to d(x, y)$

Theorem 13:

Let (X, d) be a metric space. Let $\langle x_n \rangle$ be a Cauchy sequence in X and $\langle y_n \rangle$ be a sequence in X s.t.

$$d(x_n, y_n) < \frac{1}{n}\ \forall\ n \in N.$$

Then show that

(i) $\langle y_n \rangle$ *is a Cauchy sequence,*

(ii) $\langle y_n \rangle$ *converges to x iff* $\langle x_n \rangle$ *converges to x.*

Proof :

Let $\langle x_n \rangle$ be a Cauchy sequence. Then given $\varepsilon > 0$.

$$\exists\, n_0 \in N \text{ s.t. } m, n \geq n_0$$

$$\Rightarrow \quad d(x_n, x_m) < \varepsilon/3 \qquad ...(1)$$

Choose $m_0 \in N$ s.t. $\frac{1}{m_0} < \frac{\varepsilon}{3}$.

Then $\qquad n \geq m_0 \Rightarrow \frac{1}{n} \leq \frac{1}{m_0} < \frac{\varepsilon}{3}$.

or, $\qquad d(x_n, y_n) < \frac{1}{n} \leq \frac{1}{m_0} < \frac{\varepsilon}{3}$.

or, $\qquad d(x_n, y_n) \leq \frac{\varepsilon}{3} \;\; \forall\, n \geq m_0 \qquad ...(2)$

Choose $k_0 = \max\{n_0, m_0\}$.

Then $\forall\, n, m \geq k_0$, we have

$$d(y_n, y_m) \leq d(y_n, x_n) + d(x_n, x_m) + d(x_m, y_m)$$

$$< \frac{\varepsilon}{3} + \frac{\varepsilon}{3} + \frac{\varepsilon}{3} \text{ by (1) \& (2)}$$

or, $\qquad d(y_n, y_m) < \varepsilon$,

Showing that $\langle y_n \rangle$ is a Cauchy sequence.

(ii) Let $\lim x_n = x$, then $\exists\, m_1 \in N$ s.t.

$$d(x_n, x) < \varepsilon/3 \;\; \forall\, n \geq m_1 \qquad ...(3)$$

But $\qquad d(y_n, x) \leq d(y_n, x_n) + d(x_n, x) \qquad ...(4)$

Choose $\qquad m_2 = \max\{m_1, m_0\}$

Using (2) and (3), in (4).

$$d(y_n, x) < \frac{\varepsilon}{3} + \frac{\varepsilon}{3} = \frac{2\varepsilon}{3} < \varepsilon \;\; \forall\, n \geq m_2$$

This $\Rightarrow \lim y_n = x$

Similarly, $y_n \to x \Rightarrow x_n \to x$.

Theorem 14:

Any sequence in a metric space converges to the unique point.

Or, A sequence in a metric space can converge to almost one point.

Or, If $\langle a_n \rangle$ is a sequence in a metric space (X, ρ) s.t. $a_n \to a_0$, $a_n \to a_0'$, then $a_0 = a_0'$.

Proof :

Let $\langle a_n : n \in N \rangle$ be a sequence in metric space (X, ρ) s.t.

$a_n \to a_0$, $a_n \to a_0'$,

To prove that $a_0 = a_0'$.

Suppose the contrary.

Then $a_0 \neq a_0'$.

Let $\rho(a_0, a_0') = 2r$.

Then $S_r(a_0) \not\in S_r(a_0') = \phi$.

$\because$ $a_n \to a_0$, $a_n \to a_0'$

$\therefore$ Given any $r > 0$, $\exists\ n_0, n_0' \in N$ s.t.

$\forall\ n \geq n_0 \Rightarrow a_n \in S_r(a_0)$

$\forall\ n \geq n_0' \Rightarrow a_n \in S_r(a_0')$

Taking $k_0 \Rightarrow \max\{n_0, n_0'\}$, we obtain

$\forall\ n \geq k_0 \Rightarrow a_n \in S_r(a_0)\ a_n \in S_r(a_0')$

$\Rightarrow a_n \in S_r(a_0) \cap S_r(a_0')$

$\Rightarrow S_r(a_0) \cap S_r(a_0') \neq \phi$. A contradiction.

Hence our initial assumption $a_0 \neq a_0'$ is wrong. Consequently

$a_0 = a_0'$.

Theorem 15:

Let <f_n (x)> be a sequence of continuous functions defined on a metric space (X,d). Let this sequence converge uniformly to f on X. Then f(x) is continuous on X.

Or

Uniform limit of a sequence of continuous function is continuous.

Proof:

Since <f_n (x)> converges uniformly to f on (X, d). Hence given $\varepsilon > 0$, $\exists\ n_0 \in N$ independent of $x \in X$ s.t. $n \geq n_0$.

$$\Rightarrow d(f_n(x), f(x) < \varepsilon/3 \qquad ...(1)$$

Let $a \in X$ be arbitrary. To prove that f is continuous on X, we have to prove that f is continuous at $x = a$. For this we have to show that given $\varepsilon > 0$, $\exists\ \delta > 0$, 0 s.t. $d(x, a) < \delta$.

$$\Rightarrow d(f(x), f(a) < \varepsilon \qquad ...(2)$$

Continuity of f_n at $a \in X$

$$\Rightarrow d(f_n(x), f(x) < \varepsilon/3 \text{ for } d(x, a), \delta \qquad ...(3)$$

By (1), $d(f_n(a), f(a)) < \varepsilon/3\ \forall\ n \geq n_0$ ...(4)

If $d(x, a) < \delta$, then

$$d(f(x), f(a)) \leq d[f(x), f_n(x)] + d[f_n(x), f_n(a) + d[f_n(a), f(a)]$$

$$< \frac{\varepsilon}{3} + \frac{3}{3} = \frac{\varepsilon}{3} o = \varepsilon \text{ by (1), (3), \& (4)}$$

or, $d(f(x), f(a)) < \varepsilon$ for $d(x, a) < \delta$. Hence the result (2).

Theorem 16:

Every non empty open set $G \subset R$ is the union of a countable collection of pairwise disjoint open intervals.

Proof:

Let G be a non-empty open subset of R and $x \in G$ be arbitrary. Then $\exists$ an interval I_x with centre x s.t. $x \in G$.

Let C_x denote the union of all such open intervals then $C_x = \cup \{I_x : x \in I_x \subset G\}$. Evidently C_x is the largest open interval containing x s.t. $x \in C_x \subset G$.

Moreover this also suggests that $G = \cup C_x$.

$\forall\ x, y \in C_y$ we have either $C_x \cap C_y = \phi$ or $C_x \cap C_y \neq \phi$

Consider the case $C_x \cap C_y \neq \phi$.

Evidently $C_x \cap C_y$ is an open interval containing both x and y

s.t. $\quad C_x \subset C_x \cup C_y \subset G,$

$\quad C_y \subset C_x \cup C_y \subset G.$

By def. of C_x, and C_y, $C_x \cup C_y \subset C_x$,

$\quad C_x \cup C_y \subset C_y$

Hence $\quad C_x = C_x \cup C_y = C_y$

or $\quad C_x = C_y$

$\therefore\ \forall\ x, y \in G$

we have either $C_x \cap C_y = \phi$ or $C_x = C_y$

s.t. $G = \cup C_x$.

IInd Part: It remains to show that the intervals C_x form a countable collection. Let $G = \{C_x\}$ be a collection of open intervals s.t. $C_n \neq \Phi \ \forall n$ and $C_n \cap C_m = \phi$ for $n \neq m$.

To prove that G is countable.

We know that each C_n contains an infinite number of rationals. Let K be the set of all rationals belonging to $\cup C_n$. Define a map $f : K \to G$ s.t.

$f(x) = C_x$ where $C_x \in G$ s.t. $x \in C_x$.

For each $x \in K$, there is unique open interval C_x containing x, the rational number. Hence f is a well defined map. For each $C_a \in G$, has at least one rational $a \in K$ s.t. $f(a) = C_a$, showing thereby f is onto.

Also $f(x) = f(y) \Rightarrow C_x = C_y$

where $x \in C_x, y \in C_y \Rightarrow x = y$

Now f is one-one onto. Hence $K \sim G$. But K is countable. Hence G is countable.

For $K \subsetneq Q$, Q is countable $\Rightarrow$ K is countable.

Theorem 17:

Frechet Space: *Let F be the set of infinite sequences of real numbers. $x, y, z \in F$, then*

$$x = \langle x_n \rangle = \langle x_1, x_2 \ldots \rangle,$$

$$y = \langle y_n \rangle, \ z = \langle z_n \rangle$$

where $x_n, y_n, z_n \in R$

Proof:

We define a map

$$d : F \times F : \to R \text{ s.t.}$$

$$d(x,y) = \sum_{n=1}^{\infty} \frac{1}{2^n} \frac{(x_n - y_n)}{[1 + |x_n - y_n|]}$$

To show that d is metric on F.

(i) $d(x, y) \geq 0$ For $|x_n - y_n| \geq 0 \ \forall_n$

(ii) $d(x, y) = 0 \Leftrightarrow x = y$

For $$d(x,y) = 0 \Leftrightarrow \sum_{n=1}^{\infty} \frac{1}{2^n} \frac{(x_n - y_n)}{[1 + |x_n - y_n|]} = 0$$

$$\Leftrightarrow \frac{1}{2^n} \frac{(x_n - y_n)}{\left[1 + |x_n - y_n|\right]} = 0 \ \forall \ n$$

$$\Leftrightarrow |x_n + y_n| = 0 \ \forall \ n$$

$$\Leftrightarrow |x_n + y_n| = \forall \ n$$

$$\Leftrightarrow x = y$$

(iii) $d(x, y) = d(y, x)$

For $|x_n + y_n| = |y_n + x_n|$

(iv) $d(x, y) \geq d(x, z) + d(z, y)$

Here we use the fact that

$$\frac{|\alpha + \beta|}{1 + |\alpha + \beta|} \leq \frac{|\alpha|}{1 + |\alpha|} + \frac{|\beta|}{1 + |\beta|}$$

In view of this we have

$$d(x, y) = \sum_{n=1}^{\infty} \frac{|x_n - y_n|}{1 + |x_n - y_n|} \cdot \frac{1}{2^n}$$

$$= \sum_{n=1}^{\infty} \frac{1}{2^n} \cdot \frac{|(x_n - z_n) + (z_n - y_n)|}{1 + |(x_n - z_n) + (z_n - y_n)|}$$

$$\geq \sum_{n=1}^{\infty} \frac{1}{2^n} \cdot \frac{|x_n - z_n|}{1 + |x_n - z_n|} + \sum_{n=1}^{\infty} \frac{1}{2^n} \cdot \frac{|z_n - y_n|}{1 + |z_n - y_n|}$$

$= d(x, z) + d(z, y)$.

Thus d is metric on F. The pair (F, d) is a metric space and this metric space is called *Frechet space.*

Theorem 18:

In a metric space (X, d), prove that

F is closed $\Leftrightarrow D(F) \subset F$*.*

Or

Prove that a subset F of a metric space X contains all its limit points iff X – F is open.

Proof :

Let (X, d) be a metric space and $F \subset X$.

We know that F is closed $\Leftrightarrow$ X – F is open.

Step I : Let X – F be open so that F is closed.

Aim. D (F) $\subset$ F.

Let $x \in X - F$ be arbitrary. The $X - F$ is an open set containing x s.t. $(X - F) \cap F = \phi$.

This $\Rightarrow$ is not a limit point of F

$$\Rightarrow x \notin D(F) \Rightarrow x \in X - D(F)$$

Thus, " $x \in X - F \Rightarrow x \in X - D(F)$

$\therefore$ $X - F \subset X' - D(F)$

or, $D(F) \subset F$

Step II : Given $D(F) \subset F$ ...(1)

To prove F is closed.

Let $y \in X - F$, then $y \notin F$

$$y \notin F,\ D(F) \subset F \Rightarrow y \notin D(F)$$

$\Rightarrow \exists$ open set G with $y \in G$ s.t.

$(G - \{y\}) \cap F = \phi$

$\Rightarrow G \cap F = \phi$ as $y \notin F$

$\Rightarrow G \subset X - F$

Thus we have shown that

any $y \in X - F \Rightarrow \exists$ open set G with $y \in G$ s.t. $G \subset X - F$

This $\Rightarrow X - F$ is open $\Rightarrow$ is closed.

Theorem 19:

Let $\langle x_n \rangle$ be a sequence in a metric space (X, d) and the sets A_n be defined by

$$A_1 = \{x_1, x_2, x_3, ...\},$$
$$A_2 = \{x_2, x_3, ...\},$$
$$A_n = \{x_n, x_{n+1}, ...\},\ \forall n \in N.$$

Then the sequence $< x_n >$ is a Cauchy sequence iff $d(A_n) = 0$ as $n \to \infty$

Proof :

Let $\langle x_n \rangle$ be a sequence in a metric space (X, d) and let

$A_n = \{x_n, x_{n+1}, ...\}$, $\forall n \in N$ so that

and $$\left.\begin{array}{r} n < m \Rightarrow A_m \subset A_n \\ r \geq n \Rightarrow x_r \in A_n \end{array}\right\}$$ ...(1)

Step (i) : Let $\langle x_n \rangle$ be a Cauchy sequence so that given any

$\varepsilon > 0$, $\exists\ n_0 \in N$ s.t.

$n, m \geq n_0 \Rightarrow d(x_n, x_m) < \varepsilon$ ...(2)

To prove that $d(A_n) = 0$ as $n \to \infty$.

Let $n \geq n_0$, then

$$d(A_n) = \sup \{d(x_r, x_s) : x_r, x_s \in A_n\}$$
$$= \sup \{d(x_r, x_s) : r, s \geq n\} < \varepsilon, \text{ by (1), (2)}$$

$\therefore$ $d(A_n) < \varepsilon\ \forall\ n \geq n_0$

This $\Rightarrow \lim_{n \to \infty} d(A_n) = 0$

Step (ii) : Let us suppose $\lim_{n \to \infty} d(A_n) = 0$, so that given any

$\varepsilon > 0$, $\exists\ n_0 \in N$ s.t.

$n \geq n_0 \Rightarrow d(A_n) < \varepsilon$

$\Rightarrow$ $d(A_{n0}) < \varepsilon$.

To prove that $< x_n >$ is a Cauchy sequence, it is enough to prove that

$d(x_r, x_s) < \varepsilon\ \forall\ r, s \geq n_0$.

Now, $r, s \geq n_0 \Rightarrow x_r, x_s \in A_{n0}$, by (1)

$\Rightarrow d(x_r, x_s) < \varepsilon$

$\because$ $d(A_{n0}) < \varepsilon$

Theorem 20:

Let A be a subset of a metric space (X, d).

Then A is non-dense in X $\Leftrightarrow (\overline{A}) = X - \overline{A}$ is dense in X.

Or

A closed set is nowhere dense $\Leftrightarrow$ its complement is everywhere dense.

Proof :

Recall that a is dense in X if $\overline{A} = x$ and A is nowhere dense in X if int $(\overline{A}) = \phi$.

A is non-dense (or no where dense) in X

$\Leftrightarrow$ int $(\overline{A}) = \phi \Leftrightarrow \exists$ no open nbd N of any point of $\overline{A}$ s.t. $N \subset \overline{A}$

$\Leftrightarrow \overline{A}$ contains no nbd. $\Leftrightarrow S_r(x) \not\subset \overline{A}\ \forall\ r > 0\ \&\ \forall\ x \in X$

$\Leftrightarrow S_r(x) \cap (\overline{A})' \neq \phi$

$\Leftrightarrow$ x is adherent point of $(\overline{A})'$ $\forall$ x $\in$ X

$\Leftrightarrow$ $[(\overline{A})'] = X = \Leftrightarrow (\overline{A})'$ is dense in X.

Theorem 21:

Let A and B be disjoint closed sets in a metric space (X, d). Then $\exists$ disjoint open sets G, H s.t. $A \subset G$, $B \subset H$.

Proof:

If either A or B is empty, say $A = \phi$, then $\exists$ disjoint open sets ϕ, X s.t. $A \subset \phi$, $B \subset X$.

Hence we may assume $A \neq \phi$, $B \neq \phi$.

Let $a \in A$, $b \in B$ arbitrary.

$$A \cap B = \phi,$$

$\Rightarrow$ $a \notin B$, $b \notin A$

$\Rightarrow$ $d(a, B) > 0$, $d(b, A) > 0$.

Take $d(a, B) = \delta_a > 0$ and $d(b, B)\ \delta_b > 0$.

Set $S_a = S(a, 1/3\ \delta_a)$, $s_b = (b, 1/3\ \delta_b)$.

Evidently $a \in S_a$, $b \in S_b$.

Write $G = \cap \{S_a : a \in A\}$, $H = \cap \{S_b : a \in B\}$

Being arbitrary union of open spheres, G and H are open sets.

$a \in S_a \Rightarrow A \subset G$ and $b \in S_b \Rightarrow B \subset H$.

Now we claim $G \cap H = \phi$.

If possible let $p \in G \cap H$, then $p \in G$, $p \in H$.

Consequently $\exists\ a_0 \in A$, $b_0 \in B$ s.t.

$p \in S(a_0, 1/3\ \delta_{a0})$, $p \in S(b_0, 1/3\ \delta_{b0})$. Let $d(a_0, b_0) = \varepsilon > 0$.

Then $d(a_0, B) = \delta_{a0} \leq \varepsilon$, $d(b_0, A) = \delta_{b0} \leq \varepsilon$

$\varepsilon = d(a_0, b_0) \leq d(a_0, p) + d(p, b_0)$

$$< \frac{1}{3}\delta_{a0} + \frac{1}{3}\delta_{b0} < \frac{\varepsilon}{3} + \frac{3}{3} = \frac{2\varepsilon}{3} \text{ or } \varepsilon < \frac{2\varepsilon}{3}.$$

which is not possible. Hence $G \cap H = \phi$.

INTERIOR

Let (X, d) be a metric space and $A \subset X$.

A point $X \in A$ is called an interior point of A if $\exists\ r \in R^+$ s.t. $S_{r(x)} \subset A$.

The set of all interior points of A is called the interior of A and is denoted by A", or by Int (A).

Thus,

$$A^{\circ} = \text{int. }(A) = \{x \in A : S_{r}(x) \subset A \text{ for some } r\}.$$

Alternatively, we define

$$A^{\circ} = \cup (S_{r}(x) : S_{r}(x) \subset A)$$

Evidently,

(i) A^{o} is an open set,

For an arbitrary union of open sets is open.

(ii) A^{o} is the largest open subset of A.

CLOSURE

Let (X, d) be a metric space and $A \subset X$.

The closure of A, denoted by $\overline{A}$ is defined as the intersection of all closed sets that contain A. Symbolically,

$$\overline{A} = \cap \{ F \subset X : F \text{ is closed, } F \subset A\}. \qquad ...(1)$$

Evidently,

(i) A^{o} is closed set.

For an arbitrary intersection of closed sets is closed.

(ii) $\overline{A} \subset A$.

(iii) $\overline{A}$ is the smallest closed set which contains A.

Alternatively we define

$$\overline{A} = A \cup D(A). \qquad ...(2)$$

A point $x \in \overline{A}$ is called a point of closure of A.

Alternatively, a point $x \in X$ is called a point of closure of A iff $x \in A$ or $x \in D(A)$.

Later on we shall show that both the definitions (1) and (2) are equivalent.

Theorem 28:

Let (X, d) be a metric space. Suppose A is a closed subset of X and x is a point of X s.t. $x \notin A$. Then $\exists$ disjoint open subsets U and V of X s.t. $x \in U$, $A \subset V$.

Proof:

$$x \notin A \Rightarrow \exists\ \delta > 0 \text{ s.t. } S_{\delta}(x) \cap A = \phi.$$

Choose ε, η with $0 < \varepsilon < \eta < \delta$.

write $U = S_\varepsilon(x)$, $V = X - S_\eta[x]$.

Since each open sphere is an open set and each closed sphere is a closed set. Also complement of closed set is open set. These prove that U and V are open sets s.t. $U \cap V = \phi$.

For $\quad U = S_\varepsilon(x) \subset S_\eta(x) \subset S_\delta(x)$.

Also $A \subset V$, by construction of V.

Theorem 18:

If A and B are subsets of a metric space (X, d), then

(i) $(A \cap B)^o = A^o \cap B^o$.

(ii) $A^o \cup B^o \subset (A \subset B)^o$.

Also show that $A^o \cup B^o \neq (A \cup B)^o$ *by means of an example.*

Proof:

(i) Any $x \in (A \cap B)^o$

$$\Rightarrow x \text{ is an interior point of } A \cap B$$
$$\Rightarrow S_r(x) \subset A \cap B$$
$$\Rightarrow S_r(x) \subset A \cap B \subset A,\ S_r(x) \subset A \cap B \subset B$$
$$\Rightarrow S_r(x) \subset A,\ S_r(x) \subset B$$
$$\Rightarrow x \in A^o,\ x \in B^o$$
$$\Rightarrow x \in A^o \cap B^o$$
$$(A \cap B^o) \subset A^o \cap B^o.$$

Conversely, any $x \Rightarrow A^o \cap B^o$. ...(1)

$$\Rightarrow x \in A^o \text{ and } x \in B^o$$
$$\Rightarrow \exists \text{ open spheres } S_{r1}(x),\ S_{r2}(x), \text{ s.t.}$$
$$S_{r1}(x) \subset A,\ S_{r2}(x) \subset B$$
$$\text{Choose } r = \min.(r_1, r_2)$$
$$\Rightarrow S_r(x) \subset S_{r1}(x) \subset A,\ S_r(x) \subset S_{r2}(x) \subset B$$
$$\Rightarrow S_r(x) \subset A,\ S_r(x) \subset B$$
$$\Rightarrow S_r(x) \subset A \cap B$$
$$\Rightarrow x \in (A \cap B)^o$$

$\therefore \quad A^o \cap B^o \subset (A \cap B)^o$ ... (2)

Combining (1) and (2), we get the result (i)

(ii) Any $x \in A^o \cup B^o$

$$\Rightarrow x \in A^o \text{ or } x \in B^o.$$
$$\Rightarrow S_{r1}(x) \subset A \text{ or } S_{r2}(x) \subset B$$

Choose the two spheres s.t. one is contained in the other so that $r_2 \geq r_1$ and $S_{r1}(x) \cup S_{r2}(x) = S_{r2}(x)$.

$$\Rightarrow S_{r2}(x) = S_{r1}(x) \cup S_{r2}(x) \subset A \cup B$$
$$\Rightarrow S_{r2}(x) \subset A \cup B$$
$$\Rightarrow x \in (A \cup B)^o$$

$\therefore$ $A^o \cup B^o \subset (A \cup B)^o$.

Example :

Let $A = [0, 1], B = [1, 2]$

Then $A^o = (0, 1), B^o = (1, 2).$

$A^o \cup B^o = (0, 1), \cup (1, 2)$

$= \{x \in R : 0 < x < 1 \text{ or } 1 < x < 2\}$

$\because$ $(A \cup B) = [0, 2]$

$\therefore$ $(A \cup B)^o = (0, 2).$

Evidently $(A \cup B)^o \neq A^o \cup B^o$.hilbert space

Definition (or l_2-space) : Let H be the set of infinite real sequences $< x_n >$ s.t. $\sum_{i=1}^{n} x_i^2 < \infty$.

Let x , y $\in$ H s. t $x = < x_n >, y = < y_n >$. Define a map

$$d : H \times H \to R \text{ s.t. } d(x, y) = \left[\sum_{i=1}^{\infty}(x_i - y_i)^2\right]^{1/2}$$

Aim: d is a metric in H.

(1) To prove that d is well defined map

$$d(x, y) = [\Sigma \{x_i + (-y_i)\}^2]^{1/2}$$
$$\leq (\Sigma x_i^2)^{1/2} + (\Sigma(-y_i)^2)^{1/2}$$

by Minkowski inequality

$$= \Sigma x_i^2 + \Sigma y_i^2 < \infty \text{, by given condition,}$$

or, $d(x, y) < \infty$,

meaning thereby d s a finite real number. Hence d is well defined map.

(2) $d(x, y) \geq 0$

For x_i, y_i, are real

$\Rightarrow \quad (x_i - y_i)^2 \geq 0$

$\Rightarrow \quad d(x, y) \geq 0$

(3) $d(x, y) = d(y, x)$

For $(x_1 - y_i)^2 = (y_i - x_i)^2$

$\Rightarrow \quad \{\Sigma (x_1 - y_i)^2\}^{1/2} = \{\Sigma (y_i - x_i)^2\}^{1/2}$

$\Rightarrow \quad d(x, y) = d(y, x)$

(4) $\quad d(x, z) \leq d(x, y) + d(y, z)$.

where $z = < z_n > \in H$.

$$d(x,z) = \left[\sum_{i=1}^{\infty}(x_i - z_i)^2\right]^{1/2}$$

$= [\Sigma \{(x_i - y_i) + (y_i - z_i)\}^2]^{1/2}$

$\leq [\Sigma (x_i - y_i)^2]^{1/2} + [\Sigma (y_i - z_i)^2]^{1/2}$,

by Minkowski inequality

$= d(x, y) + d(y, z)$.

The above facts prove that d is a metric on H. The space (H, d) is called Hilbert space or l_2-space.

SPACES OF CONTINUOUS FUNCTIONS

We gave a brief description of the metric space $\mathcal{C}[0, 1]$. The reader will recall that the points of this space are the bounded continuous real functions defined on the closed unit interval [0, 1] and that its metric is defined by $d(f, g) = \sup |f(x) - g(x)f$. We have two aims in this section: to generalize this very important example by considering functions defined on an arbitrary metric space, and to place all function spaces of this type in their proper context by giving the details of the structural pattern, which they all have in common with one another. We begin with the second, and define the algebraic systems which are relevant to our present interests.

Let L be a non-empty set, and assume that each pair of elements x and y in L can be combined by a process called *addition* to yield an element z in L denoted by $z = x + y$. Assume also that this operation of addition satisfies the following conditions:

(1) $x + y = y + x$;

(2) $x + (y + z) = (x + y) + z$;

(3) there exists in L a unique element, denoted by 0 and called the *zero element*, or the *origin*, such that $x + 0 = x$ for every x',

(4) to each element x in L there corresponds a unique element in L, denoted by $-x$ and called the *negative* of x, such that $x + (-x) = 0$.

We adopt the device of referring to the system of real numbers or to the system of complex numbers as the scalars. We now assume that each scalar a and each element x in L can be combined by a process called *scalar multiplication* to yield an element y in L denoted by $y = ax$ in such a way that

(5) $\alpha(x + y) = ax + \alpha y$;

(6) $(\alpha + \beta)x = \alpha x + \beta x$;

(7) $(\alpha\beta) = a(\alpha x)$;

(8) $1.x = x$.

The algebraic system L defined by these operations and axioms is called a *linear space*. Depending on the numbers admitted as scalars (only the real numbers, or all the complex numbers), we distinguish when necessary between *real linear spaces and complex linear spaces*. A linear space is often called a *vector space*, and its elements are spoken of as *vectors*.

We are not concerned here with developing the algebraic theory of linear spaces. Our only interest is in making available some pertinent concepts and terminology which are useful as a background against which to view the metric spaces we wish to study. With this in mind, we mention a few simple facts which are quite easy to prove from the axioms for a linear space: $0 + x = x$ for every x; $x + z = y + z \Rightarrow x = y$ (hint: add $-z$ to both sides on the right); $\alpha . 0 = 0$ (hint: $\alpha . 0 + \alpha x = \alpha(0 + x) = \alpha x = 0 + \alpha x$); $0 . x = 0$ (hint: $0 . X + \alpha x = (0 + \alpha)x = \alpha x = 0 + ax$); and $(-l)x = -x$ (hint: $x + (-l)x = 1 . x + (-l)x = (1 + (-l))x = 0 . x = 0$). The reader will notice that in the relation $0 . x = 0$ we have used the symbol 0 with two different meanings: as a scalar on the left and as a vector on the right.

Several other meanings will be given to this single symbol, but fortunately it is always possible to avoid confusion by attending closely to the context in which it occurs. It is convenient to introduce the operation of *subtraction* by using the symbol $x - y$ as an abbreviation for $x + (-y)$; $x - y$ is called the *difference* between x and y.

A non-empty subset At of a linear space L is called a *linear subspace* of L if $x + y$ is in M whenever x and y are and if ax is in M (for any scalar α) whenever x is. Since M is non-empty, 0 . x = 0 shows that 0 is in M. Since −x = (−l)x, −x is in At whenever x is. It will be seen at once that a linear subspace of a linear space is itself a linear space with respect to the same operations.

A *normal linear space* is a linear space on which tnere is defined a *norm*, *i.e.*, a function which assigns to each element x in the space a real number $\| x \|$ in such a manner that

(1) $\| x \| \geq 0$, and $\| x \| = 0 \Leftrightarrow x = 0$;

(2) $\| x + y \| \leq \| x \| + \| y \|$;

(3) $\| \alpha x \| = | \alpha | \, \| x \|$.

In general terms, a normed linear space is simply a linear space in which there is available a satisfactory notion of the distance from an arbitrary element to the origin. From (3) and the fact that −x = (−l)x, we obtain $\| -x \| = \| x \|$. A normed linear space is a metric space with respect to the induced metric defined by

$$d(x, y) = \| x - y \|.$$

A *Banach space* is a normed linear space which is complete as a metric space. Any closed linear subspace of a Banach space is itself a Banach space with respect to the same algebraic operations and the same norm.

So much for the technical framework. We now turn to the metric spaces which really concern us. They are all *function spaces*, in the sense that they are linear spaces whose elements are functions defined on some non-empty set X with addition and scalar multiplication defined pointwise, *i.e.*, by $(f + g)(x) = f(x) + g(x)$ and $(\alpha f)\{x) = \alpha f(x)$. We note that the zero element in such a linear space is the constant function 0 whose only value is the scalar 0 and that $(-f)(x) = -f(.x)$.

Suppose, then, that X is an arbitrary non-empty set, and consider the set L of all real functions defined on X. It is clear that L is a real linear space with respect to the operations described above. We now restrict ourselves to the subset B consisting of the bounded functions in L. B is obviously a linear subspace of L, so it is a linear space in its own right. Even more, if we define a norm on B by $\|f\| = \sup |f(x)|$, then B is a Banach space.

We next assume that the underlying set X is a metric space. This enables us to consider the possible continuity of functions defined on X. We define $\mathcal{C}(X, R)$ to be that subset of B which consists of continuous functions. $\mathcal{C}(X, R\}$ is thus the set of all bounded continuous real functions defined on the metric space X.

Theorem 1:

The set $\mathcal{C}(\mathcal{X}, \mathcal{R})$ of all bounded continuous real functions defined on a metric space X is a real Banach space with respect to pointwise addition and scalar multiplication and the norm defined by $\|f\| = \sup |f(x)|$.

It is desirable at this stage to make a clear distinction between two types of convergence for sequences of functions. Let X be a metric space, and let $\{f_n\}$ be a sequence of real functions defined on X. If for each x in X it happens that $\{f_n(x)\}$ is a Cauchy sequence of real numbers, then by the completeness of the real number system we can define a limit function f by $f(x) = \lim f_n(x)$. We then say that f_n converges pointwise to f, or that f is the pointwise limit of f_n. It is often important to know what properties of the functions f_n carry over to the limit function f, but unless we strengthen the mode of convergence, very little can be said along these lines. The stronger type of convergence normally needed to conclude anything of interest is called *uniform convergence.* In order to explain what this is, we inspect a little more closely what is involved in pointwise convergence. To say that f_n converges pointwise to f is to say the following: for each point x in X, if $\epsilon > 0$ is given, then a positive integer n_0 can be found such that $|fn(x) - f(x)| < \epsilon$ for all $n \geq no$. In general, the integer no may depend on x as well as ϵ. If, however, for each given ϵ an integer n_0 can be found which serves for all points x, then we say that f_n *converges uniformly* to f, or that f is the uniform limit of f_n. The reader will observe that these concepts are quite independent of the assumption that X is a metric space and that they are meaningful for functions defined on an arbitrary non-empty set.

It will be seen at once that convergence in the function space $\mathcal{C}$(X, R) is precisely uniform convergence as we have just defined it. The fact that $\mathcal{C}$(X, R) is complete can be restated as follows in the language of uniform convergence: if a bounded real function f defined on X is the uniform limit of a sequence $\{f_n\}$ of bounded *continuous* real functions defined on X, then f is also continuous. In other words, in the presence of uniform convergence, continuity carries over from the f_n's to the limit function f.

A moment's thought will convince the reader that the entire discussion given above, beginning with our definition of the linear space L, could perfectly well have been based on complex functions. Without going again through all the details, we state the following theorem and consider it proved.

Theorem 2:

The set $\mathcal{C}$(X, C) of all bounded continuous complex functions defined on a metric space X is a complex Banach space with respect to pointwise addition and scalar multiplication and the norm defined by

$$\|f\| = \sup |f(x)|.$$

In summary, we associate with each metric space X two linear spaces of continuous functions defined on X. The first—$\mathscr{C}(X, R)$—contains only real functions, and the second—$\mathscr{C}(X, C)$—consists of complex functions.

Further, all functions considered are assumed to be bounded, so that the norm defined by $\|f\| = \sup |f(x)|$ is always a real number. In the special case in which X is a closed interval [a, b] on the real line, we write $\mathscr{C}(X, R)$ in the simpler form $\mathscr{C}[a, b]$.

EMBEDDING

Definition. *By an embedding of a topological space X into another space Y, we mean a map $f : X \to Y$ which defines a homeomorphism of X onto $f[X]$.*

Example:

If X is a subspace of a space Y, then the inclusion map $i : X \to Y$ defined by $i(x) = f$, for every $x \in X$ is an embedding.

EMBEDDING THEOREMS AND METRIAZABILITY

Lemma 1:

For each λ in a set $\mathfrak{I}$ let f_λ be a mapping of a topological space X into a topological space Y_λ. Then the evaluation map e of X into the product space $\times \{Y_\lambda : \lambda \in \Lambda\}$ is continuous iff f_λ is continuous for each l in $\mathfrak{I}$.

Proof:

Let f_λ be continuous for each $\lambda \in L$ and let $\pi_\lambda^{-1}[G]$ be any subbasic open subset of the product space $\times \{Y_\lambda : \lambda \in \Lambda\}$ where G is open in Y.

$$\pi_\lambda \circ o = f_\lambda \text{ for all } \lambda \in \Lambda.$$

Hence $\quad e^{-1}[\pi_\lambda^{-1}(G)] = (e^{-1} 0\, \pi_\lambda^{-1})[G] = \{\pi_\lambda \circ e\}^{-1}[G]\, f_\lambda^{-1}[G]$ ...(1)

Since f_λ is continuous $f_\lambda^{-1}[G]$ is open in X. It follows from (1) that $e^{-1}[\pi_\lambda^{-1}(G)]$ is open in X.

Thus we have shown that the inverse image under e^{-1} of every subasic open set of the product space $\times \{Y_\lambda : \lambda \in \Lambda\}$ is open in X. It follows that e is continuous.

Thus arrives the 'if' part of theorem. Conversely, let the evaluation map e be continuous. By definition of the product space each projection π_λ of the product space $\times \{Y_\lambda : \lambda \in L\}$ onto the coordinate space Y_λ is continuous. It follows that the composition map $\pi_\lambda \circ e$ is continuous. But $\pi_\lambda \circ e = f_\lambda$. Accordingly f_λ is continuous for every $\lambda \in \Lambda$.

Lemma 2:

For each $\lambda \in \Lambda$, let f_λ be a mapping of a topological space X into a topological space Y_λ. Then the evaluation map e of X into the product space $\times \{Y_\lambda : \lambda \in L\}$ is one-one iff the family $F = \{f_\lambda : \lambda \in \Im\}$ distinguishes points of X.

Proof:

Let e be one-one and let x and y be distinct points of X. Then

$x \neq y \Rightarrow e\ (y)$ [$\because$ e is one-one]

$\Rightarrow \quad \{f_\lambda\ (x)\ L : \in L\} \neq \{f_\lambda\ (y) : \lambda \in \Lambda\}$

$\Rightarrow \quad f_{\lambda_0}\ (x) \neq f_{\lambda_0}\ (y)$ for some $\lambda_0 \in \Lambda\}$.

It follows that the family $\{f_\lambda : \lambda \in L\}$ distinguishes points of X.

Conversely let $\{f_\lambda : \lambda \in \Im\}$ distinguish points of X, Then for each pair of distinct points x and y of X, there exists a $\lambda_0 \in \Im$ such that f_{λ_0}

$(x) \neq f_\lambda\ (y)$.

Thus $\quad x \neq y \Rightarrow f_{\lambda_0}$

$(x) \neq f_{\lambda_0}\ (y)$ for some $\lambda_0 \in \Lambda$

$\Rightarrow \quad \{f_\lambda\ (x) : \lambda \in \Lambda\} \neq \{f_\lambda\ (y) : \lambda \in \Lambda\}$

$\Rightarrow \quad e\ (x) \neq e\ (y)$.

This shows that e is one-one.

Lemma 3:

For each $\lambda \in \Lambda$, let f_λ be a mapping of a topological space X into a topological space Y_λ. Then the evaluation map e of X into the product space $Y = \times \{Y_\lambda \times \in \Lambda\}$ is an open map, of X onto the subspace e [X] of Y if the family $F = \{f_\lambda : \lambda \in \Lambda\}$ distinguishes points and closed sets.

Proof:

Let G be an open subset of X. Let $y \in e\ [G]$. Then there is some $x \in G$ such that e (x) = y. Let G = X – G. Then H is a closed subset of X such that $x \in X - H = G$. Since F distinguishes points and closed sets, there exists $f_\lambda \in F$ such that $f_\lambda\ (x) \notin \overline{(f_\lambda\ [H])}$ in Y_λ.

Let $V = Y_\lambda - (f_\lambda\ [H])$. Since $\overline{(f_\lambda\ [H])}$ is a closed subset of Y_λ, it follows that V is an open subset of Y_λ. Hence $\pi_1^{-1}\ [V]$ is open in Y being a subbasic member of Y. Let $W = e\ [X] \cap \pi_\lambda^{-1}\ [V]$. We clam that W is open set in the subspace e [X] such that $e\ (x) = y \in W$ and $W \subset e\ [G]$. For we have, $x \in X \Rightarrow e\ \ (x) \in e\ (X)$

and $f_\lambda(x) \notin \overline{(f_\lambda[H])} \Rightarrow f_\lambda(x) \in Y_\lambda - \overline{f_\lambda[H]} = V$

$\Rightarrow x \in f_\lambda^{-1}[V] = (\pi_\lambda \, oe)^{-1}[V] = e^{-1}\pi_\lambda^{-2}[V]]$

$\Rightarrow e(x) \in \pi_\lambda^{-1}[V].$

Thus $e(x) \in e[X] \cap \pi_\lambda^{-1}[V]$

Further $V = Y = f_\lambda[H] \Rightarrow V \subset Y_\lambda - f_\lambda[H]$

$\Rightarrow \pi_\lambda^{-1}[V] \subset \pi_\lambda^{-1}[Y_\lambda] - \pi_\lambda^{-1}[f_\lambda[H]] = Y - \pi_\lambda^{-1}[\pi_\lambda \, oe)[H]]$

$\Rightarrow e[X] \cap \pi_\lambda^{-1}[V] \subset e[X] \cap [Y - e[H]] = e[X] - e[H]$

$[\because e[H] \subset e[X] \subset Y]$

$\Rightarrow W \subset e[X \ \ H] = e[G]$ $(\because e[X] - e[H] \subset e[X - H]).$

It follows the e [G] is an open subset of e [X]. Hence e is an open map of X onto e [X].

Theorem 1:

For each $\lambda \in L$, let f_λ be a continuous mapping of a topological space X into another space Y_λ. If the family $F = \{F_\lambda : \lambda \in \Lambda\}$ distinguishes points as well as points and closed sets, then the evaluation map e of X into the product space $Y = \times\{Y_\lambda : \lambda \in \Lambda\}$ is a homeomorphism of X into the subspace [X] of Y. That is, the evaluation map e is embedding.

Proof:

Corollary : *If in addition, X is T_1-space then the result of the above theorem holds if F distinguishes points and closed sets.*

This follows form the fact that in a T_1-space singleton sets are closed.

Theorem 2:

(Tychonoff Embedding Theorem): *Every Tychonoff space X can be embedded as a subspace of a cube.*

Proof:

Recall that a Tychonoff space is a completely regular T_1-space. Let F denote the family of all continuous real functions $f_\lambda : X \to Q$ on the Tychonoff space X into the closed unit interval Q = [0, 1]. By using complete regularity of X, it can be easily proved that F distinguishes points and closed subsets of X. Again since X is also T_1, every singleton subset of X is closed and consequently F also distinguishes points.

It follows by the preceding theorem that the evaluation map $e : X \to Q$ is an embedding. Hence every Tychonoff space can be embedding as a subspace of a cube.

Theorem 3:

A space X is Tychonoff iff X is homeomorphic to a subspace of a cube.

Proof:

For the 'only if' part of the theorem. To prove the 'if part' we observe that since each unit interval Q is compact. Hausdorff space, the cube Q^F is also a compact Hausdorff space by Tychonoff theorem.

It follows that Q^F is a Tychonoff space (where $\Im$ is an arbitrary index set) and so is every subspace of Q^L,. Hence if X is homeomorphic to subspace of a cube, that X is 3 Tychonoff space.

Theorem 4:

For each $n \in N$, let (X_n, d_n) be a metric space such that $\delta_n (X) \leq 1$ and let $X = \times \{X_n : n \in N\}$, For $x, y \in X$

$$\textit{define} \quad = d(x, y) \sum_{n=1}^{\infty} \{2^{-n} d_n (x_n, y_n)\}.$$

Then (i) d is metric for X, and (ii) the d-metric topology $\Im$ is the same as the product topology F for $X = \times \{X_n : n \in N\}$ where X_n has the d-metric topology.

Proof:

We first note the series for d (x, y) is convergent by comparison test for we have $2^{-n} d_n (x_n, y_n) \leq 2_n$ and the geometric series

$$\sum_{n=1}^{\infty} 2^{-n} \text{ is convergent.}$$

(i) We now prove that d is a metric for X. The axioms $[m_1]$, $[m_2]$ and $[m_3]$ immediately follow from the corresponding axioms for d_n.

To prove $[m_4]$ (the triangle inequality), let x, y, z be arbitrary members of X. Then since each d_n is a metric, we have.

$$\sum_{n=1}^{n=k} 2^{-n} d_n (x_n, y_n) \leq \sum_{n=1}^{n=k} 2^{-n} [d_n (x_n, a_n) + d_n (z_n, y_n)]$$

$$= \sum_{n=1}^{n=k} 2^{-n} d_n (x_n, z_n) + \sum_{n=1}^{n=k} 2^{-n} [d_n (z_n, y_n). \qquad ...(1)$$

proceeding to the limits when $k \to \infty$ we get from (1)

$$d (x, y) \leq d (z, x) + d (z, y).$$

Hence d is a metric for X.

(ii) We first prove that $F \in \Im$. Let U be any member of the defining subbase for F. Then U consists of all points $x = \{x_n : n \in B]$ such that $x_k \in G_k$ for some open set G_k in X_k and $x_n \in X_n$ whenever $n = k$. Let $y \in U$.

Then $y_k \in G_k$. Since G_k is open in X_k, there exists $r > 0$ such that

$$D (y_k, r) \subset G_k$$

Let $V = S (y, \eta)$ where $n = 2^{-k} r$ and let $z \in V$. Then

$$2^{-k} d_k (z_k, y_k) \le d (z, y) \text{ by def. of } d < 2^{-k} r.$$

Therefore $d_k (z_k, y_k) < r$.

Then from (2), it follows that $z_k \in G_k$ and consequently $z \in U$ by definition of U. This we have shown that $z \in V \Rightarrow z \in U$. Hence $V \subset U$. Thus to each $y \in U$, there exists an open sphere centred at y such that $V \subset U$.

It follows that $U \in \Im$, that is, U is open with respect to the d-metric topology for X. Now since each subbase member of F belongs to $\Im$, It is easy to see that

$$F \subset \Im.$$

To prove $\Im \subset F$. let $z \in G$ where $G \in \Im$. Then there exists some $m \in N$ such that

$$S (z, 2^{-m}) \subset G.$$

Let $W = \{x : d_n (x_n, z_n) > 2^{-m-n-2}$ for each $n \in N_{m+2}\}$.

[Note that for convenience we write N_{m+2} for the set

$$\{1, 2, 3, m + 2\}.$$

Then $W \subset G$ as shown below. Let $x \in W$. Then.

$$d (x, z) = \sum_{n=1}^{\infty} [2^{-n} d_n (x_n, z_n)].$$

$$= \sum_{n=1}^{n=m+2} 2^{-n} d_n (x_n, z_n) + \sum_{n=m+3}^{n=\infty} [2^{-n} d_n (x_n, z_n)]$$

$$= < \sum_{n=1}^{n=m+2} d_n (x_n, z_n) + \sum_{n=m+3}^{n=\infty} 2^{-n}.$$

$$[\because 2^{-n} < 1 \text{ and } d_n (x_n, z_n) \le 1]$$

$$< \sum_{n=1}^{m+2} 2^{-m-n-2} + \sum_{n=m+3}^{\infty} 2^{-n}$$

$$\frac{2^{-m-2} [1-(\frac{1}{2})^{m+2}]}{1-\frac{1}{-}} + \frac{2^{-m-3}}{1-\frac{.1}{-}}$$

$$= 2^{-m-2}[1-(\frac{1}{2})^{m+2}]+2^{-m-2}$$

$$< 2^{-m-1} + 2^{-m-1} = 2^{-m}.$$

Hence $x \in S(z, 2^{-m}) \subset G$.

Thus it is shown that

$$x \in W \Rightarrow x \in G.$$

Hence $\quad W \subset G$.

But evidently $z \in W$. Also it is easy to see that W is a base member of the product topology F. Thus, we have shown that to each $z \in G$, there exists a base member W for the product topology F such that

$$z \in W \subset G.$$

It follows that G is F-open.

Hence $\quad \mathfrak{I} \subset F$.

From (3) and (4) we have $\mathfrak{I} = F$.

Definition : *A topology space $(X, \mathfrak{I})$ is said to be Metrizable iff there exists a metric d for X which induces the same topology as the topology $\mathfrak{I}$.*

It immediately follows from this definition that a topological space is merrizable it is homeomorphic with a metric space.

Theorem 1:

Every subspace Y of a metrizable space X is metrizable, that is, metrizability is a herediatary property.

Proof:

Let $d : X \to R$ be a metric for X which induces the topology of X. Consider the restriction e of d to the subspace Y, that is, let, e=d/Y. Then e is evidently a metric for X which induces the relative topology of Y.

Theorem 2:

The space (R. U) is metrizable.

Proof :

Do yourself.

Theorem 3:

The topological product of a countable family of metrizable spaces is metrizable.

Proof:

We first prove that the topological product $X_1 \times X_2 \times ... \times X_n$ of a finite number of metrizable spaces $X_1, X_2,... X_n$ is metrizable. Let $d_i : X_i \times X_i \to R$ $(1 \leq i \leq n)$ be a metric for X_i which induces the topology of X_i. Consider the mapping.

$$d : X \times Y \to R$$

defined as follows. Let $x = (x_1,..., x_n)$ and $y = (y_1,...y_n)$ be any two points in X. Then d (x, y) is defined by

$$[d(x, y)] = \sum_{i=1}^{n} [d_i(x_i, y_i)]^2 .$$

It is easy to see that d is a metric for X which induces the product topology.

Now consider a denumerable collection $\{X_n : n \in N\}$ of metrizable spaces. Let d_i be a metric for x_i, i = 1, 2, 3,... Then there exists a metric for the cartesian product $X = \times \{X_n : n \in N\}$ which induces the product topology and consequently X is metrizable.

***Corollary** : The Euclidean n-spaces R_n, the unit n-cube Q_n and the Hilbert cube Q_n are metrizable since R and Q = [0, 1] are metrizable.*

Theorem 4:

(Urysohn Metrization Theorem): *Every regular T_1-space X satisfying the second axiom of countability is metrizable.*

Proof:

If suffices to establish the existence of a countable family of closed function from X into Q = [0, 1], which distinguishes points and closed sets. For then it will follow that X can be embedded as subspace of the Hilbert, cube Q^N which is metrizable and so X will be metrizable.

Let B be a countable base for the topology on X. This is possible since X is second countable. Consider the subcollection c of the cartesian product B × B consisting of pairs (U, V) such that $U \subset V$. Surely, c is countable. Let (U, V) be any member of c. Since X is regular $U \subset V$, we have $U \cap V = \emptyset$. Also $\bar{U}$ and B are closed in X, Since X is regular and second countable, it is normal. Hence by Urysohn's lemma there exists a continuous function f from X into Q such that f (x) = 0 if $x \in \bar{U}$ and f (x) = 1 if $x \in V$. Let F be the family of function so obtained. Then F is countable and it remains to show that F distinguishes points and closed sets. So let F be any closed subset of X and x any point in F′. Since F′ is an open neighbourhood of x, there exists a member V of B such that $x \in V \subset F'$. Again since X is regular, there is an open subset G of X such that $x \in G$ and $G \subset V$, Then by definition of base, we can choose. U in B such that $x \in U \subset G$ so that $x \in U \subset G$. Thus $x \in \bar{U} \subset V \subset F'$ and so $(U. V) \in c$. If f be the corresponding member of F1 then

$$f(x) = 0 \varnothing \{1\} = f[V] \supset f[F] \quad ...(1)$$

$$[\because V \subset F' \Rightarrow V' \supset F \Rightarrow f[V] \supset f[F]]$$

Since {1} is closed in Q_n, we have

$$\{1\} = \{1\} = f(V) \subset \overline{f[F]}.$$

It follows from (1) and (2) that $f(x) \varnothing f[F]$. Hence F distinguishes points and closed sets and so by our remark in the beginning of the proof, X is metrizable.

Theorem 5:

Let X be a T_1-space. Then the following statements are equivalent:

(i) *X is regular and second countable.*

(ii) *X is homeomorphic to a subspace of the Hilbert cube Q^N*

(iii) *X is metrizable and separable.*

Proof:

We shall show: (i) ⇒ (iii) ⇒ (i).

Now (i) ⇒ (ii) follows from the previous theorem. To prove (ii) ⇒ (iii), we first observe that Q^N is metrizable and second countable.

Since metrizability and second coutability, are hereditary properties, it follows that each subspace of Q^N is metrizable and second countable and is therefore separable.

Since by hypothesis X is homeomorphic to a subspace of Q^N, it follows that X is metrizable. Thus (ii) ⇒ (iii).

Finally assume that X is metrizable and separable. Then it X is regular and second countable and so (iii) ⇒ (i).

Theorem 6:

Let f be a mapping of a topological space (X, $\mathfrak{I}$) onto a set Y. Then the collection v of all subsets G of Y such f^{-1} [G] is open in X is the largest topology for Y such that f is $\mathfrak{I}$-v- continuous. Further a subset F of Y is v-closed if and only if f^{-1} [F] is $\mathfrak{I}$-closed.

Proof:

We first show that v is a topology for Y.

[T1] : Since f^{-1} [Ø] = Ø is open in X, we have Ø v. Also since f^{-1} [Y] = X is open in X, we have T ∈ 'v.

[T2] : Let G_1, G_2 ∈ v. Then f^{-1} [G_1] and f^{-1} are open in X by hypothesis and so f^{-1} [G_1] ∩ f^{-1} [G_1 ∩ G_2] is also open in X. Therefore G_1 ∩ G_2 ∈ v.

[T3] : Let G_λ ∈ v for every λ ∈ L where L is an arbitrary index set. Then each f^{-1} [G_λ] is open in X by hypothesis and so their union ∪ {f^{-1} [G_λ] : λ ∈ L} = f^{-1} [∪ {G_λ : λ ∈ L} is also open in X and therefore ∪ {G_λ : λ ∈ Λ} ∈ v. Hence v is a topology for Y. It then follows that f is ℑ-v continuous.

Now let v be another topology for Y such that f is ℑ v′ continuous. Then f^{-1} [H] ∈ v by hypothesis. Hence v′ ⊂ v. In other words, v is the largest topology for Y such that f is ℑ-v continuous.

Finally, let F be subset of Y. Then F is V-closed ⇔ f^{-1} [Y – F] = X – F^{-1} [F] is ℑ-open {Theorem (1.8), ch. 4] ⇔ f^{-1} [F] is ℑ-closed.

QUOTIENT TOPOLOGY SPACE

Definition : *Let (X, ℑ) be a topology space, Y a set, and f a mapping of X onto. Then the largest topology v for Y such that f is ℑ-v continuous is called the quotient topology for Y (relative to f and ℑ) and shall be denoted by ℑf and the map f is called the quotient map.*

Theorem 1:

A subset G of Y is open in the quotient topology (relative to f : X → Y) if and only if f^{-1} [G] is an open subset of X.

Proof:

First let f be continuous and open. We want to show that 3t = bv where $ℑ_f$ denotes the quotient topology for Y. $ℑ_f$ is the largest topology for Y making f continuous. It follows that v ⊂ ℑ.

Conversely, let G ∈ ℑ γ. Then f^{-1} [G] is-open *i.e.*, G ∈ V and therefore ℑf *i.e.*, V is the quotient topology for Y.

Now let f be continuous and closed. Then V ⊂ ℑ as before. Conversely, if G ∈ ℑf then f^{-1} [G] is ℑ-open and so X = f^{-1} [G] is ℑ-closed. But

$$X - f^{-1}[G] = f^{-1}[Y] - f^{-1}[G] = f^{-1}[Y - G].$$

Since f is a closed and onto map. f [f^{-1}[Y – G]] = Y – G is V-closed and so G ∈ V. Therefore $ℑ_f$ ⊂ V. Thus, $ℑ_f$ = V in this case also.

Theorem 2:

Let f a continuous map of a space X onto a space R and let Y have the quotient topology (relative to f). Then a mapping g on Y onto a topology space Z is continuous if and only if the composition map gof f is continuous.

Proof:

Suppose g o f is continuous and let G be open set in Z. Then $(gof)^{-1}$ [G] $= f^{-1}$ $[g^{-1}[G]]$ is open. X and so g^{-1} [G] is open in Y. Hence g is a continuous map.

Conversely, let g : Y → Z be continuous. Since f : X → Y is given to be continuous, it follows that gof f is continuous map.

Remark : It can be shown that the quotient topology and the properties of open and closed maps have little to do with the range space. In fact if f is continuous map of a space X onto a space Y with quotient topology and the family of all sets of the from f^{-1} (y) with y in Y.

The continuous can be described as follows. Let D be the collection of all subsets X of the from f^{-1}. | f (x) | for all x ∈ X. Define g : Y → D by g (y) = f^{-1} (y). Then it is easy to see that g is a one-one map of Y onto D. Also go f = F g = 1 o F.

If now we assign the quotient topology to D (relative to F), then shows that g and g^{-1} are continuous (since g o f = F and g^{-1} of = f). Hence g is a homeomorphism.

The range space Y is essentially extraneous to our discussion, and so the remaining theorems of this section will be formulated so as to display this fact.

The Decomposition (or partition) of a non-empty set X is the collection of non-empty disjoint subset of X whose union is X.

Theorem 3:

If f is a continuous map of a compact space (X, ℑ) onto a Hausdorff space (Y, V), then V must be quotient topology for Y relative to a Hausdorff space (Y, V), then V must be quotient topology for Y relative to f.

Proof:

Let F be any ℑ-closed subset of X. Then F is compact and so its continuous image f [F] is also a compact subset of Y.

Since Y is Hausdorff, f [F] is a closed subset of Y and therefore f is a closed mapping. Hence V must be the quotient topology for Y.

TYCHONOFF TOPOLOGICAL SPACE

Definition : *For each λ in an arbitrary index set Λ, let $(X_\lambda, \Im_l)$ be a topological space and let $X = \times \{X_\lambda : \lambda \in \Lambda\}$. Then topology $\Im$ for X which has subbase the collection.*

$$B = \{\pi_\lambda^{-1} [G_\lambda] : \lambda \in \Im, G_\lambda \in \Im_\lambda\}$$

is called the product topology (or the Tychonoff Topology) for X. and $(X, \Im)$ is called the product space of the given spaces.

Note that here π_λ denotes as usual the lth projection. The collection B, is called the defining subbase for $\Im$.

The collection B of all finite intersections of elements of B, would then from a base for $\Im$..

In particular, if we have a countable collection of topological spaces $\{X_1, X_2, X_3, ...\}$ the product space $X = \times \{X_n : n \in N\}$ consists of all sequences $x = \{x_1, x_2, x_3, ...\}$ where $x_n \in X_n$

Moreover, $\pi_a^{-1} [G_a] = X_1 \times X_2 \times ... X_{a-1} \times X_a \times X_a + 1 \times ...$

Theorem 1:

Let $\{(X_\lambda, \Im_\lambda) : \lambda \in L\}$ be an arbitrary collection of topological space and let $X = \times \{X_\lambda : \lambda \in \Lambda\}$. Let $\Im$ be a topology for X. Then the following statements are equivalent:

(a) $\Im$ *is the product topology for X.*

(b) $\Im$ *is the smallest topology for which the projections are continuous.*

Proof:

(a) ⇒ (b) : Let π_λ be the λth projection maps and let G_λ be any $\Im_\lambda$-open subset of Xl. Then by (a), $\pi\lambda^{-1} [G_\lambda]$ is a member of the subbase for $\Im$ and hence $\pi_\lambda^{-1} [G_\lambda]$ must be $\Im$-open. It follows that $\pi\lambda$ is $\Im$-$\Im$ continuous. Now let v be any topology on X such that $\pi\lambda$ is v-$\Im_\lambda$ continuous for each $\lambda \in \Lambda$. Then $\pi^{-1} [G_\lambda]$ is open for every $G_\lambda \in \Im_\lambda$. Since V is a topology for X, V contains all the unions of finite intersections of members of the collection

$$\{\pi_\lambda^{-1} [G_\lambda] : \lambda \in \Im \text{ and } G_\lambda \in \Im_\lambda\}.$$

This implies that V contains $\Im$, that is, $\Im$ is coarser than v. It follows that $\Im$ is the smallest topology for X such that π_λ is $\Im - \Im_\lambda$ continuous for each $\lambda \in \Im$.

(b) ⇒ (a) : Let B. Be the collection of all sets of the form $\pi^{-1} [G_\lambda]$ where G_λ is an open subset of X_λ for $\lambda \in L$. The topology v for X will make all the projections π_λ continuous iff $B, \subset v$.

We know that the smallest topology for X which makes all the projections continuous is the topology determined by B, as a subbase. That is it is the product topology for X.

Note : Throughout this chapter, whenever we refer to the product topology, we shall mean the topology generated by the collection of all sets of the from $\pi_\lambda^{-1}[G_\lambda]$ where $\lambda \in \Im$ and $G_\lambda \in \Im_\lambda$.

Note that $\pi_a^{-1}[G_a]$ consists of all points $x = \{x_\lambda : \lambda \in \Lambda\}$ in X such that $\pi_a(x) \in G_a$. In other words $\pi_a^{-1}[G_a] = \times\{Y_\lambda : \lambda \in \Lambda\}$ where $Y_a = G_a$ and $Y_\lambda = X_\lambda$ whenever $\lambda \neq \alpha$. Λ base member for this topology is a set B of the from $\cap\{\pi_\lambda^{-1}[G_\lambda] : \lambda \in \Im\}$ where Λ' is a finite subset of Λ and $G_\lambda \in \Im_\lambda$ for some $\lambda \in \Lambda$. Observe that $B = \times\{Y_\lambda : \lambda \in \Lambda\}$ where for each $\lambda \in \Lambda$, $Y_\lambda \in \Im_\lambda$ and $Y_\lambda = X_\lambda$ for all but a finite number of λ' s.

It follows that every open set in the product space X must contain a set (a base member) of in Λ. Thus all but a finite number of coordinates of points of a non-empty open subset of the product space are unrestruct d in the respective coordinate spaces. Hence if L is infinite then a set H of the form $\times\{G_\lambda : \lambda \in \Lambda\}$ is surely not open in the product space if G_λ is an open product subset of X_λ for every $\lambda \in L$ since all the points H are restricted to lie in the respective G_λ' s.

It is easy to see that the collection c of all sets of the from $\times\{G_\lambda : \lambda \in \Lambda\}$, where $G_\lambda \subset \Im_\lambda$, from each $\lambda \in \Lambda$, is a base for some topology $\Im^*$ for the product set $\times\{X_\lambda : \lambda \in \Lambda\}$, but as seen above this topology $\Im^*$ is not in general, the same as the product topology $\Im$. The reader may ask. "why is the definition adopted for the cartersian product of two spaces, is not extended to the product of an arbitrary collection of spaces". The reason simply is that the topology $\Im$ as mentioned above is not so useful to the mathematical as the topology $\Im$ as defined, since $\Im$ does not yield desired theorems and results.

Theorem 2:

Let $\{(X_\lambda, \Im_\lambda) : \lambda \in \Lambda\}$ be an arbitrary collection of topological spaces and let $X = \times\{X_\lambda : \lambda \in \Lambda\}$. Show that the collection c of all sets of the form $\times\{G_\lambda : \lambda \in \Lambda\}$ where $G_\lambda \in \Im_\lambda$ for each $\lambda \in \Lambda$, is a base for some topology for X

Proof:

We know that c satisfies the conditions [B*1] and [B*2]

[B*1] : Let $x \in X$ so that $x = \{x_\lambda : \lambda \in \Lambda\}$ where $x_\lambda \in X_\lambda$. Them the there exists a $\Im_\lambda$ open set G_λ (which may be X_λ) such that $x_\lambda \in G_\lambda$.

Hence, x is an element of a set of the form $\times \{G_\lambda : \lambda \in \Lambda\} = G$ say. Thus to each $x \in X$, there exists a member G of x such that $x \in G$. It follows that

$$X = \cup \{G : G \in c\}$$

[B*2] Let $G \in c$ and $G' \in c$. Then $G = \times \{G_\lambda : \lambda \in \Lambda\}$

and $\quad G' = \times \{G_\lambda : \lambda \in \Lambda\}$

where $\quad G_\lambda \in \Im_\lambda$ and $G_\lambda' \in \Im_\lambda$ for every $\lambda \in \Lambda$. Now

$$G \cap G' = (\times \{G_\lambda : \lambda \in \Im\}) \cap (\times \{G_\lambda' : \lambda \in \Lambda\}) \quad ...(1)$$

$$= \times \{G_\lambda \cup G_\lambda' : \lambda \in \Lambda\}$$

Since $\Im_\lambda$; is a topology for X_λ, we have

$$G_\lambda \in \Im_\lambda' \text{ and } G_\lambda' \in \Im_\lambda \Rightarrow G_\lambda \cap G_\lambda' \in \Im_\lambda.$$

It follows from (1) that $G \cap G' \in c$.

Thus [B*2] is also satisfied.

Example:

Let B be a non-empty member of the base B for the product topology. Show that π_λ [B] $= X_\lambda$ for all except a finite number of λ's. Deduce that if G is a non-empty open subset of the product space X, then π_λ [G] $= X_\lambda$ for all except a finite number of λ' s.

Solution:

Now $B = \times \{Y_\lambda : \lambda \in \Lambda\}$ where $Y_\lambda \in \Im_\lambda$ and $Y_\lambda = X_\lambda$ for all but to finite number of λ's. Hence by the definition of the projection map π_λ, we have π_λ [B] $= Y_1$ where $Y_\lambda = X_\lambda$ for all except a finite contains a non empty open subset G of the product space functions a non-empty member B of the base B, we also have π_λ [G] $= X_\lambda$ for all except finite number of λ' s.

Theorem 3:

Each projection map π_λ is an open map.

Proof:

Let $\alpha \in \Lambda$ and let B be any member of the base B for the $Y_\lambda \in \Im_\lambda$ and $Y_\lambda = X_\lambda$ and $Y_\lambda = X_\lambda$ for all but a finite number of λ' s. So that each Y_λ is an open subset of X_λ. Now by definition of π_α, we have pa $\{B\} = Y_\alpha$ which is open in X_α. Thus the image under π_α of every basic open subset of X is open. Now let G be any open subset of X. Then G is the union of collection λ of base members of the form B. Hence.

$$\pi_\alpha [G] = \pi_\alpha[\cup \{A : A \in \Lambda\}] = \cup \{\pi_\alpha [A] : A \in \Lambda\}.$$

As shown above π_α [A] is an open subset of X_α for each A $\in \Lambda$ and hence their union π_α [G] must also be open in X_α. Thus we have shown that image under π_α of every $\Im$-open set G is $\Im_\alpha$-open. It follows that π_α is an open mapping.

Theorem 4:

Let f be a mapping of a space Y into product space $X = \times \{Xl : l \in \Im\}$. Then f is continuous iff the composition pl o f : $Y \to X_\lambda$ is continuous.

Proof:

Let f be continuous. Since all projections are continuous. We know that π_λ of is also continuous.

Conversely, let each composition π_l of be continuous and ant U be any member of the defining subbase B, of the product space X. Then $U=\pi_\lambda^{-1}$ [G] for some $\lambda \in \Im$ and some G $\in \Im_l$.

Also $\quad f^{-1}[U] = f^{-1} \{\pi_\lambda^{-1}[G]] = (\pi_\lambda$ of $-1[G]$.

Since π_λ of is continuous, it follows that (π_λ) f^{-1} {G} = f^{-1} {U} is open in Y. Thus, we have shown that the inverse image under f of every sub-basic open set in the product space X is open in Y. It follows that f is continuous.

Theorem 5:

Let X be the non-empty product space $X [X_\lambda : \lambda \in \Lambda\}$. Then a non-empty product subset $F = \{F_\lambda : \times \{F_\lambda : \lambda \in \Lambda\}$ is closed in X if and only if each F_λ is closed is X_λ.

Proof:

Let F_λ be closed in X_λ for every $\lambda \in \Im$. Since the projection p_λ is continuous for each $\lambda \in \Im$. p_λ^{-1} $[F_\lambda]$ is closed in X. It is easy to see that

$$F = \cup \{p_\lambda^{-1} [F_\lambda] : \lambda \in \Lambda\}.$$

It follows that F is closed in X, being an intersection of closed sets.

Conversely, let F = $\times$ {F $\in \Lambda$} be closed in X. To show that each F_λ is closed in X_λ. Let $\mu \in \Lambda$ be arbitrary. We shall show that F_μ is closed. Let z_μ be any limit point of F_μ in X_μ. Consider the point z where π_μ (z) = z_μ and p_λ (z) is any element of F_λ for $\lambda \neq \mu$. Let G be any basic open set for the product topology containing z.

Then π_μ [G] is open and contains z_μ. Since z_μ is a limit point of F_μ, F_λ [G] must contain a point x_μ of F_μ different from z_μ. Therefore, G contains the point x where π_λ (x) = π_λ (z) for $\lambda \neq \mu$ and π_μ (x) = x. Evidently x $\in$ F. Also since and z differ in μ^{th} coordinate, we have x $\neq$ z. Thus we have shown that every basic open set containing z contains a point of F different from z. Hence, z is a limit point of F. Since F is closed in X, z ∈ F which implies that π_μ (z) ∈ π_μ [F]. Since of F. π_μ (z) = z and π_μ [F] = F we have z_μ ∈ F. Thus F contains all its limit points and so Fm is closed. Since μ was arbitrary, we see that F_λ is closed for every λ ∈ Λ.

Theorem 6:

Let $\times (X_\lambda : \lambda \in L\}$ *be a product space and* $A_\lambda \subset X_\lambda$ *for every* $l \in L$.

Then $\overline{\times\{A_\lambda : \lambda \in \Lambda\}} = \times\{\overline{A_\lambda} : \lambda \in \Lambda\}$

Proof:

By theorem (above), $\times \{A_\lambda : \lambda \in A\}$ is closed since each $\overline{A_\lambda}$ is closed so that $\overline{\times\{A_\lambda : \lambda \in \Lambda\}} = \times\{\overline{A_\lambda} : \lambda \in \Lambda\}$

Now $\times \{A_\lambda : \lambda \in \Im\} \subset \times \{\overline{A_\lambda} : \lambda \in \Lambda\}$.

$$\Rightarrow \times \{\overline{\times\{A_\lambda : \lambda \in \Lambda\}} \subset \overline{\times\{A_\lambda : \lambda \in \Lambda\}} = \times \{\overline{A_\lambda} : \lambda \in \Lambda\} \quad ...(1)$$

If for some λ, $A_\lambda = \varnothing$,

then $\times \overline{\{A_\lambda : \lambda \in \Lambda\}} = \times\{\overline{A_\lambda} : \lambda \in \Lambda\} = \varnothing$.

Now assume that $G_\lambda \neq \varnothing$, for all λ and let $x = \{x_\lambda : \lambda \in \Im\}$ be any element of $\times \{\overline{A} : \lambda \in \Lambda\}$ so that $x_\lambda \in \overline{A_\lambda}$ for every $\lambda \in A$ and therefore x_λ is an adherent point of A_λ.

Let G be any basic open set for the product topology containing x. Then $G = \times \{G_\lambda : \lambda \in L\}$ where G_λ is open in X_λ and $G_\lambda = X_\lambda$ for all our a finite number of λ's.

Let $G_\lambda = X_\lambda$ for $\lambda \neq \lambda_1, \lambda_2 ..., \lambda_3$ where n is finite. Since $x_\lambda \in \overline{A}$ for every $\lambda \in \Im$, $x_{\lambda_3} \in G_{\lambda_3}$ implies G_{λ_2} contains a point of AG_{λ_2}, for k = 1, 2, ..., n, by the definition of adherent point. Also for $\lambda \neq \lambda_1$ ..., ln, $G_\lambda = X_\lambda$ so that $A_\lambda \subset G_\lambda$ in this case.

It follows that G contains a point of $\times \{A_\lambda : \lambda \in \Im\}$. Thus we have shown that every basic open set containing x contains a point of $\times \{A_\lambda : \lambda \in \Im\}$ and therefore

$$x \in \times \{A_\lambda : \lambda \in \Im\}$$

It follows that,

$$\times\{\overline{A} : \lambda \in \Lambda\} \subset \overline{\times\{A_\lambda : \lambda \in \Lambda\}} \quad ...(2)$$

From (1) and (2),

$$\times\{\overline{A_\lambda : \lambda \in \Lambda}\} = \times\{\overline{A_\lambda} : \lambda \in \Lambda\}$$

Theorem 7:

Let B, be a sub-base for a topology on a space Y and let s be sequence in Y. Also let $x \in Y$. If s is eventually in each member of B. Containing **x,** *then s converges to x.*

Proof:

Let N be a nhd of x. Then there exists a basic nhd $M_1 \cap M_2 \cap \ldots M_k$ of x, such that

$$x \in M_1 \cap M_2 \cap \ldots \cap M_k \subset N.$$

where $M_i \in B$, for $i = 1, 2, \ldots k$.

Since s is eventually in each M_i for each $i = 1, 2, \ldots k$, there exists a positive integer m_i such that $s(n) \in M_i$ for $n > m$.

If we set $m = \max \{m_1 \cdot m_2 \ldots m_1\}$.

We have $s(n) \in M_1 \cap M_2 \cap \ldots \cap M_k \subset N$ for every $n \geq m$.

Thus we have shown that S is eventually in each nhd of x. It follows that s coverages to x.

Theorem 8:

Let s be a sequence in the product space $X = X\{X_\lambda : \lambda \in A\}$. Then s converges to a point x in X iff π_λ as converges to $\pi_\lambda(x)$ (= x) for each $\lambda \in A$.

Proof:

Let the sequence s converge to $x \in X$. Since each projection π_λ is continuous. It sequence π_λ os converges to $\pi_\lambda(x) = x_\lambda$.

Conversely, let the sequence π_λ os) *i.e.* the sequence $\pi_\lambda(s_\lambda)$. $\pi_\lambda((s_2)$. (s_3)... converge to $\pi_\lambda(x)$ (= x_λ) for every $\lambda \in \Lambda$, nd let $\pi_\lambda^{-1}[G]$ be any member of the sub-base B, of the product space X such that $x \in \pi_\alpha^{-1}[G]$, where $G \in \mathfrak{I}_\alpha$. Then $\pi_\alpha(x) \in G$. That is. $x_\alpha \in G$. Also by hypothesis. π_λ os converges to $\pi_\lambda(\xi)$ (= x_α). Hence π_α os is eventually in every nhd of x_α. In particular pa os is eventually in G. So there exists a positive integer m such that

$$(\pi_\alpha \, os)(n) = \pi_\alpha(s(n)) \in G \Rightarrow s(n) \in \pi_\alpha^{-1}[G]$$

for every $n \geq m$. Thus s is eventually in $\pi_\alpha^{-1}[G]$. Since $\pi_\alpha^{-1}[G]$ was chosen to be an arbitrary member of B, containing x, it converges to x.

PRODUCT INVARIANT PROPERTIES

Theorem 1:

The product space $X = \times\{X_\lambda : \lambda \in \Lambda\}$ is T_1 iff each coordinate space X_λ is T_1.

Proof:

The 'iff' part : Let each Coordinate, space X_λ be T_1 and let x = {x_λ : $\lambda \in \Lambda$} be any element of X. The $x_\lambda \in X_\lambda$ for every $\lambda \in \Lambda$. Since X_λ is T_1, (x_λ) is closed in X_λ for every $\lambda \in \Lambda$. Since projection map π_λ is continuous, it follows that π_λ^{-1} [{x_λ{] is closed in X for every $\lambda \in \Lambda$. Hence their intersection $\cap$ (π_λ^{-1} [{x_λ}] is also closed in X. But evidently $\cap\ \pi_\lambda^{-1}$ [{x_μ}] = {x}

{For $z \in \cap$ {x_λ^{-1} [{x_λ{] $\Rightarrow z \in \pi_\lambda^{-1}$ [{x_λ}] $\forall\ \lambda \in \Lambda$

$\Rightarrow \quad z_\lambda = x_\lambda \ \forall\ \lambda \in \Lambda \Rightarrow z = x$]

it follows that {x} is closed in X. Thus we have shown that every singleton subset of X is closed. Hence X is a T_1-space.

The 'only if' part : Let the product space X be T_1. Let $m \in \Lambda$ be arbitrary. We shall show that X_μ is T_1. Let a_μ and b_μ be any two distinct points of X_μ. Choose x and y such that x and y differ only in the μth coordinate and such that $x_\mu = a_\mu$ and $y_\mu = b_\mu$. Since X is T_1, there exists open sets G and H in X such that $x \in G$ but $y \notin G$ and $y \in H$ but $x \notin H$. Hence, by definition of base, there exists a basic open set U = × {U_λ : $\lambda \in \Lambda$} such that $x \in U \subset G$ but $y \notin U$. Similarly there exists a basic open set V = × (V_λ : $\lambda \in \Lambda$} which contains y but not x. It follows that U_μ is open in X_μ and contains $x_\mu = a_\mu$ but not b_μ. Similarly V_μ is open in X_μ and contains b_μ but not a_μ. Hence X_μ is T_1. [Note that $U_\lambda = X_\lambda$ and $V_\lambda = X_\lambda$ for all but a finite number of λ' s but this is immaterial here].

Alternative proof of the 'only if' part : Let the product space X be T_1. We show that each coordinate space is homeomorphic to a certain subspace of X. Let a = {a_λ : $\lambda \in \Lambda$} be a fixed element in X and for each μ consider the subset Am of X consisting of all points {x_λ : $\lambda \in \Lambda$} such that $x_\lambda = a_\lambda$ if $\lambda \neq \mu$ while x_μ may be any point of X_λ. Let g_μ be the restriction of x_μ to A_μ. Then g_μ is a continuous mapping of A_μ onto X_μ. Also it is evident by the construction of A_μ that g_μ is one-one. We now show g_μ is open. Let B_μ be any basic open set B in X so that B = × {G_λ : $\lambda \in \Lambda$} where $G_\mu \in \Im_\lambda$ [Note that G_λ = X for all but a finite number of λ's but this is immaterical here]. But $B_\mu = A_\mu \cap B$ is either empty or consists of those points of A_μ for which $x_\mu \in G_\mu$. Then image of B_μ under G_μ under Gm is either empty or G_μ and so open in either case. Now let U_μ be any open set in A_μ. Then U_μ is the union of a collection of base members. Hence (v), g_μ (U_μ) is the union of images under g_μ of members of that collection and since such image is shown to be open, it follows that gm (U_μ) is an open set and consequently g_μ is an open mapping. Thus g_μ is homeomorphism of g_μ onto X_μ. But Am onto X_μ. But A_μ is T_1 .Hence its homeomorphic image X_μ is also T_1.

Theorem 2:

The product space $X = \times \{X_\lambda : \times \lambda \in L\}$ and $y\ \{y_\lambda : \lambda \in \mathfrak{I}\}$ be two distinct points of the product space X. Then $x_\mu \neq y_\mu$ for some $\mu \in \mathfrak{I}$ where $x_\mu \in X_\mu$ and y_μ. Since X_μ is Hausdorff,

Proof:

The 'if part' : Let each coordinate space X_λ,, be T_2 and let $x = \{x_\lambda : \lambda \in \Lambda\}$ and $y = \{y_\lambda : \lambda \in \Lambda\}$ be two distinct points of the product space X. Then $x, \neq y_\mu$ for some $\mu \in \Lambda$ where $x_\mu \in X_\mu$ and $y_\mu \in X_\mu$. Since X_μ is Hausdorff, there exist open sets G_μ and G_μ in X_μ such that

$$x_\mu \in G_\mu,\ y_\mu\ H_\mu \text{ and } G_\mu \cap H_\mu = \varnothing. \quad ...(1)$$

Since $\pi_\mu(x) = x_\mu$ and $\pi_\mu(y) = y_\mu$, the relations (1) imply that $x \in \pi_\mu^{-1}[G_\mu]$, $y \in \pi_\mu^{-1}[H_\mu]$

and $\quad \pi_\mu^{-1}[G_\mu \cap H_\mu] = \pi_\mu^{-1}[\varnothing] = \varnothing$

or $\quad \pi_\mu^{-1}[G_\mu] \cap \pi_\mu^{-1}[H] = \varnothing$.

But $\pi_\mu^{-1}[G_\mu]$ and $\pi_\mu^{-1}[H_\mu]$ are open in X, being sub basic members of the product topology. Thus we have shown that to each pair x, y of distinct points of X, there exist two disjoint open sets one containing x and the other containing y. It follows that the product space X is Hausdorff.

The 'only if' part : Let the product space X be Hausdroff. We shall show that the coordinate space X_μ is Hausdorff for arbitrary $\mu \in L$. Let a_μ and b_μ be any two distinct points of X_μ. Choose x and y in X such that x and y differ only in the μth coordinate and such that $x_\mu = a_\mu$ and $y_\mu = b_\mu$. Since X is Hausdorff, there exist open sets G and H in X such that $x \in G$ and $y \in H$ and $G \cap H = \varnothing$. There exist basic open sets $U = \times \{U_\lambda : \lambda \in L\}$ and $V = \times \{V_\lambda : \lambda \in \mathfrak{I}\}$ such that $x \in U \subset H$ and $U \cap V = \varnothing$. It follows that U_μ and V_μ are open sets in X that $x_\mu = a_\mu \in U_\mu$ and $y_\mu = b_\mu \in V_\mu$ and $U_\mu \cap V_\mu = \varnothing$. Hence X_μ is Hausdorff.

Theorem 3:

Let $(X_\lambda : \mathfrak{I}l) : \lambda \in \Lambda\}$ be an arbitrary collection of topological sauces and let $(X, \mathfrak{I})$ be their product space where $X = \times \{X_\lambda : \lambda \in \Lambda\}$. Then $(X.\ \mathfrak{I})$ is first countable iff each of the coordinate spaces is first countable and all but a countable number of the coordinate spaces are indiscrete.

Proof:

The 'if' part. Let (X_λ, A_λ) be first countable for every $\lambda \in L$ and let all but a countable number of spaces (X_λ, A_λ) be indiscrete. Let $\Lambda' = \{\lambda : \lambda \in \Lambda$ and X_λ is not indiscrete$\}$. Then by hypothesis, Λ' is countable subset of Λ.

Let x ∈ X. Then $x = \{x_\lambda : \lambda \in \Im\}$ where $x_\lambda \in X_\lambda$. Since each coordinate space X_λ is first countable, there exists a countable local base in x_λ for every ∈ Λ. Let X_λ be countable local base at x_λ. Since for every $\lambda \in \Lambda - \Lambda'$. X_λ is indiscrete. We have $B_\lambda = (X_\lambda)$ for each such λ. For every fixed λ ∈ N, let c_λ be the countable collection

$$\{\pi_\lambda^{-1}[G] : G \in B_\lambda\}. \text{ Let}$$

$$c = \{H : H \in c_\lambda \text{ for } \lambda \in \Im\} \cup \{c_\lambda : \lambda \in \Lambda\}$$

Then c is a countable collection, being a union of a countable collection of countable collections, Hence there are only a countable number of the finite subcollections of c. Then D is countable. It is easy to see that D is a local base at x. Hence X is first countable.

The 'only if part. Let (X, ℑ) be first countable and let $x_\lambda \in X_\lambda$. Also let $x = \{x_\lambda : \lambda \in \Lambda\}$ where $x_\lambda \in X_\lambda$.

Then x ∈ X. Since is first countable, there exists a countable local base at x. say $E = \{E_n : n \in N\}$ where $E_n \in \Im$.

Theorem 4:

The product space × $\{X_\lambda : l \in L\}$ is a second axiom space iff each coordinate space X_λ is second axiom and all but a countable number are indiscrete.

Proof:

Do yourself.

Theorem 5:

Let $\{X_\lambda : \lambda \in \Lambda\}$ be a non-empty collection of topological spaces. Then the non-empty product space $X = \times \{X_\lambda : \lambda \in L\}$ is regular if and only if each coordinate space X_λ is regular.

Proof:

Let each X_λ be regular. Let $x = \{x_\lambda : \lambda \in \sigma \Lambda\}$ be any point of the product space X and G any open set in X containing x. Then there exists a basic open set V in X such that $x \in V \subset G$. Now $V = \times \{V_\lambda : \lambda \in \Lambda\}$ where V_λ is open in X_λ. [Note that $V_\lambda = X_\lambda$ for all but a finite number of λ's but this is immaterial here]. Since each X_λ is regular and V_λ is an open set in X_λ containing x_λ, there exists an open set U_λ in X_λ. Such that $x_\lambda \in U_\lambda$ and $\overline{U_\lambda} \subset V_\lambda$. Let W $= \times \{U_\lambda : {}_\lambda \in L\}$. Then W is open in X and contains x. Also $\overline{W} = \times \overline{\{U_\lambda : \lambda \in \Lambda\}} = \times \{\overline{U_\lambda} : \lambda \in \Lambda\}$. Further, since $U_\mu \subset V_\mu$ for every λ, we have $\overline{W} = \times \{\overline{U_\lambda} : \lambda \in \Lambda\} \subset \times \{V_\lambda : \lambda \in V\} = V$. Thus, we have shown that for every point $x \in X$ and every open set G containing x there exists an open set W in X such that $x \in W$ and $W \subset V \subset G$. It follows that X is regular.

Conversely, let X be a non-empty regular product space $\times \{X_\lambda : \lambda \in \Lambda\}$. Let $\mu \in \Im$ be arbitrary. We shall show that Xm is regular. Let am be any point of X_μ and G_μ any open set in X_μ containing a_μ.

Since X is non-empty, there is an element $x = \{x_\lambda : \lambda \in \Lambda\}$ in X such that x_μ. Since X is non-empty, there is an element $x = \{x_\lambda : \lambda \in \Lambda\}$ in X such that $x_\mu = a_\mu$. Let $G=\pi_\mu^{-1}[G]$. Then $x \in G$ and G is open in X (since π_μ is continuous).

Since X is a regular space, there exists a basic open set $V = \times \{V_\lambda : \lambda \in \Lambda\}$ such that $x \in V$ and $\overline{V} \subset G$. It follows that $x_\lambda = a_\lambda \in V_\mu$. It $\times \{V_\mu : \lambda \in \Im\} = \times \overline{\{V_\lambda : \lambda \in \Lambda\}}$. Hence $\overline{V} = \times \{\overline{V_\lambda} : \lambda \in \Lambda\} \subset G$ which implies that $\overline{V_\mu} \subset G_\mu$. Thus we have shown that to each am $\in X_\mu$ and exists an open set V_μ in X_μ each open set G_μ in X_μ containing a_μ, there exists an open set V_μ in X_μ such that $a_\mu \in V_\mu$ and $\overline{V}_\mu \subset G_\mu$. Hence X_μ is regular.

Corollary : *A non-empty product space $\times \{X_\lambda : \lambda \in \Im\}$ is a T_3 space if each coordinate space X_λ is T_3.*

Proof:

Before proving the main theorem we prove the following lemma.

Lemma : *Let B, be a subspace of topological space $(X, \Im)$. If for each U in B. and each $p \in U$ there is a continuous map of X into [0, 1] such that $f(p) = 0$ and $f[X - U] \subset \{1\}$, then X is completely regular.*

Proof of the Lemma : Let $p \in X$ and let G be a $\Im$-open set containing p. Then (unless $G = X$ which is a trivial case), there exist members $U_1, U_2, \ldots,$ U_n of B, such that

$$p \in U_1 \cap U_2 \cap \ldots \cap U_n \subset G.$$

[$\because$ every nhd of p must contain a basic nhd of p].

Then by hypothesis for each $i = 1, 2, \ldots, n$, there is a continuous map.

$f_i : X \to [0, 1]$ such that $f_i(p) = 0$ and $f_i[X - U_3] = \{1\}$.

Now we define a map.

$f : X \to [0. 1]$

such that hypothesis for each $i = 1, 2, \ldots, n\}$. Evidently $f(p) = 0$

and $f[X - G] = \{1\}$.

We now show that f is continuous as follows, For $n = 2$, we have

$$f(x) = f(x)\frac{|f_1(x)+f_2(x)|}{2}+\frac{|f_1(x)-f_2(x)|}{2} \qquad ...(1)$$

It can be easily prove that the continuity of f_1, f_2 implies the continuity of $f_1 + f_2$ defined by $(f_1 + f_2)(x) = f_1(x) + f_2(x)$ and that the continuity of any function g implies the continuity of | g | defined by $|g|(x) = |g(x)$. It then follows from (1) that f is a continuous map in case n = 2. Hence by induction f is continuous. It follows that X is completely regular.

Proof of the Main Theorem : The 'if' part. Let each coordinate space X_λ be completely regular. To prove that the product space is also completely regular. To prove this we shall use the above lemma. Let U be any member of the usual subbase for the product topology and let $x = \{x_\lambda : \lambda \in \Im\} \in U$. Then $U = \pi_\lambda^{-1}[G]$ here G is an open subset of X_λ containing x_λ for some $\lambda \in \Im$. Since X_λ is completely regular, there exists a continuous mapping

$$f : X_\lambda \to [0, 1]$$

such that $f(x_\lambda) = 0$ and $f(z) = 1$ if $z \in X_\lambda - G$. Since the projections are continuous of X that $f \circ \pi_\lambda$ is continuous mapping of X into [0, 1] such that

$$(f \circ \pi_\lambda)(x) = f(\pi_\lambda(x)) = \begin{cases} 0 \text{ if } x \in U \\ 1 \text{ if } x \in X - U \end{cases}$$

$$[\because x \in U = \pi_\lambda^{-1}[G] \Rightarrow \pi_\lambda(x) \in \pi_\lambda(\pi_\lambda^{-1}[G]) = G$$

$$\Rightarrow f(\pi_\lambda(x)) = 0 \text{ by def. off}$$

$$\text{and } \; x \in X \;\; U \Rightarrow x \in \pi_\lambda^{-1}[X_\lambda] - \pi_\lambda^{-1}[G]$$

$$\Rightarrow x \Rightarrow \pi_\lambda^{-1}[X_\lambda - G]$$

$$\Rightarrow \pi_\lambda(x) \in X_\lambda - G$$

$$\Rightarrow f(\pi_\lambda(x)) = 1 \text{ [by def. off]}$$

It then follows from the above lemma that the product space X is completely regular.

The 'only if part' : Since complete regularity is both a topological and a hereditary property, we advise the reader to write his own proof just as in the alternative proof of the 'only if', functions (f, g) we mean the function h defined by

$$h(x) = \begin{cases} f(x) & \text{if } f(x) \geq g(x) \\ -f(x) & \text{if } f(x) < g(x). \end{cases}$$

and min (f, g) defined likewise. The following identities are then evident.

$$\max \quad (f, g) = \frac{|f+g|}{2}+\frac{|f-g|}{2}$$

and $\min(f, g) = \frac{|f+g|}{2} - \frac{|f-g|}{2}$.

This result can evidently be extended to any finite set of functions.

Theorem 6:

The product space $X = \times \{X_\lambda : \lambda \in \mathfrak{I}\}$ is a Tychonoff space iff each coordinate space X_λ is a Tychonoff space.

Proof:

Since a Tychonoff space is a completely regular T_1 space the proof.

Theorem 7:

The product space $X = \times \{X_\lambda : \lambda \in L\}$ is connected iff each coordinate space X_λ is connected.

Proof :

The 'if part' : Let each coordinate space X_λ be connected and $a = \{a_\lambda : \lambda \in \Lambda\}$ be a fixed point of X. Let C be the component of X to which a belongs. We shall prove that every point of X belongs to C. Let $\times \{G_\lambda : \lambda \in L\}$ be an arbitrary basic open set containing a point $x = \{x_\lambda : \lambda \in \Lambda\}$ of X where G_λ is open in X_λ and $G_\lambda = X_\lambda$ if $\lambda \neq \mu_1, \mu_2, \ldots, \mu_n$. The $A_{\mu_1} \times A_{\mu_2} \times \ldots \times A_{\mu_n}$, (that is, the set of all points $\{z_\lambda : \lambda \in \mathfrak{I}\}$ such that $z_\lambda = a_\lambda$ if $\lambda \neq \mu_1\ \mu_2, \ldots, \mu_n$) is homeomorphic to $X_{\mu_1} \times X_{\mu_2} \times \ldots \times X_{\mu_n}$. But $X_{\mu_1} \times X_{\mu_2} \times \ldots \times X_{\mu_n}$ is connected by theorem.

Hence, its homeomorphic image $A_{\mu_1} \times A_{\mu_2} \times \ldots \times A_{\mu_n} = A$, say, is connected. Since by definition C is maximal connected subset of X, it follows that $A \subset C$. But the set A contains the point $\{z_\lambda : \lambda \in \mathfrak{I}\}$ for wish $z_\lambda = a_\lambda$ if $\lambda \neq \mu_1, \mu_2, \ldots \mu_n$ and $z_\lambda = x_\lambda$ for $\lambda = \mu_1, \mu_2, \ldots, \mu_n$. This point $\{z_\lambda : \lambda \in \Lambda\}$, however, lies in $\times \{G_\lambda : \lambda \in \Lambda\}$, which was an arbitrary basic open set containing $\{x_\lambda : \lambda \in \Lambda\}$. It follows that the point $x = \{x_\lambda : \lambda \in \Lambda\}$ is in the closure of C. Since we have shown that each basic nhd of x contains a point of A and hence a points of C. But C is closed so that $\bar{C} = C$.

Hence $\{X_1\ \lambda \in \Lambda\} \in C$. Thus every point of X belongs to C so that $X \subset C$. But surely $C \subset X$. Hence $X = C$. But by definition C is a connected set. It follows that X is connected.

Theorem 8:

The product space $X = \times \{X_\lambda : \lambda \in \mathfrak{I}\}$ is locally connected iff each coordinate space X_λ is locally connected and all but a finite number are connected.

Proof:

The 'if' part : Let the coordinate space X_λ be locally connected for every $\lambda \in L$ and connected for $\lambda \neq \mu_1, \mu_2, \ldots \mu_n$. To prove that the product space is locally connected. Let $x = \{x_\lambda : \lambda \in \Lambda\}$ be an arbitrary point of X and let x be contained in an arbitrary basic open set $G = \times \{G_\lambda : \lambda \in \Lambda\}$ where G_λ is open in X_λ and $G_\lambda = X_\lambda$ for $\lambda \neq \mu_1^*, \mu_2^*, \ldots, \mu_m^*$. Since each X_λ is locally connected and $x_\lambda \in G_\lambda$, there must exist a connected open set X_λ in X_λ such that $x_\lambda \in C_\lambda \subset X_\lambda$. Now consider the subset $H = \times \{H_\lambda : \lambda \in \Lambda\}$ of G where $H_\lambda = C_\lambda$ if $\lambda \neq \mu_1, \mu_2, \ldots \mu_n\ \mu_1^*, \mu_2^* \ldots, \mu_m$ and $H_\lambda = X_\lambda$ otherwise so that each H_λ is connected. It follows by the preceding theorem that H is connected. Also H is evidently a basic open set containing x and contained in G. Thus we have shown that to each basic open set G containing an arbitrary point x of X, there exists a basic open connected set H such that $x \in H \subset G$. It follows that the product space X is connected.

The 'only if' Part : Let the product space X be locally connected and let x_μ be an arbitrary point of X_μ contained in some open set G_λ in X_λ.

Consider the point $y = \{y_\lambda : \lambda \in \Lambda\}$ with $y_\mu = x_\mu$. Then evidently $y \in$ [illegible]G] which is an open subset of X (since the projections are continuous). By local connectedness of X, there must exist a connected open set C such that $y \in C \subset \pi^{-1}$ [G]. Then

$$\pi_\mu (y) \in \pi_\mu [C] \subset \pi\, [\pi_\mu^{-1} [G_\mu]].$$

That is, $x_\mu = y_\mu \in \pi_\mu\ [C] \subset G_\mu$.

Since the projections are open mapping, π_μ [G] is an open subset of X_μ. Again since the projections are continuous π_μ [G] is connected. Thus we have shown that every open set G_μ in X_μ containing an arbitrary point x_μ contains a connected open set π_μ [C] containing. It follows that Xm is locally connected.

Now let a be an arbitrary point of the product space X. By local connectedness of X. There exists a connected open set C such that $y \in C$. Then there must exist basic open set

$$G = \times\ \{G_\lambda : \lambda \in \Lambda\}$$

where $G_\lambda \in \mathfrak{I}_\lambda$ for every $\lambda \in \Lambda$ and $G_\lambda = X_\lambda$ for all but a finite number of values of λ such that $y \in G \subset C$. Then π_λ [G] $= X_\lambda$ for all but a finite number of values of and consequently π_λ [C] $= X_\lambda$ for all but a finite number of l's. But π_λ [C] is connected, being a continuous image of a connected set. Thus all but a since number of coordinate spaces are connected.

TYCHONOFF THEOREM

We now prove the most celebrated theorem of general topology known as. Tychonoff theorem. This theorem on the product of compact spaces is undoubtedly the most useful theorem on compactness. While looking forward for the proof, we first prove some preliminary theorems.

Definition : *Let X be a topological space. We say that a collection c of subsets of X is short iff c does not cover X. and we say that c is finitely short iff no finite subfamily of c covers X.*

Lemma 1:

Let X be a topological space and let M be a maximal finitely short family of open sets. If some member of zm contains $G_1 \cap G_2 \cap ... \cap G_n$ where each G_i is open then $G_m \in M$ for some M where $1 \leq m \leq n$.

Proof:

We shall prove the contrapositive of the lemma for n = 2. Suppose $G_1 \notin M$ and $G_2 \notin M$ where G_1, G_2 are open subsets of X. Then by the maximality of M, there exists members E_1, E_2 ... E_r of M such that

$$E_1 \cup E_2 \cup E_3 \cup ... \cup E_3 = X. \quad ....(1)$$

Hence no open set containing G_1 belongs to M. Similarly, there are members F_1, F_2,..., F_s of M such that

$$F_1 \cup F_2 \cup F_3 \cup ... \cup F_s = X \quad ...(2)$$

From (1) and (2), it is easy to see that

$$(G_1 \cap G_2) \subset E_1 \cup E_2 \cup \; ... \cup E_r \cup \; F_1 \cup F_2 \cup ... \cup F_s = X \; ...(3)$$

Since M is finitely short, it follows from (3) that no member of M can contain $G_1 \cup G_2$. Thus, we have proved the contrapositive of the lemma for n = 2. Hence by finite induction, lemma is true for any positive integer n.

Lemma 2:

Let (X, $\Im$) be a topological space and let F be a finitely short subfamily of $\Im$. Then there exists a maximal finitely short subfamily. M of $\Im$ such that $F \subset M$.

Proof:

Let C be the class of all finitely short subfamilies of $\Im$, and let c be ordered by the inclusion relation $\subset$. Then $F \in c$. Now the singleton {F} is a trivial simply ordered set. Hence by.

Hausdorff Maximal Principle there exists a maximal simply ordered subclass D of c such that $\{F\} \subset D$, that is $F \in F$. Let $M = \cup D$. Then M is a collection of open sets since each member of D is such a collection. It will now be shown that

(i) M is finitely short and

(ii) M is a maximal finitely short subfamily of $\Im$.

To prove (i), let $A_1, A_2, ..., A_n \in M$. Then, for every i in N_k, there exists some c_j in D such that $A_i \in c_i$. Since D is simply ordered, one of these c_i s, say c_j, contains each of the other c_i' s, and so $A_i \in c_j$ for each $i \in N_k$. Since c_j is finitely short, it follows that $A_1 \cup A_2 \cup ... \cup A_k \neq X$. Hence M is finitely short.

We now prove (ii). Suppose that (ii) does not hold, that is, suppose that there is some open set $G \notin M$ such that $M \cup \{G\}$ is still finitely short. Since M contains each member of D, $D \cup \{M \cup \{G\}\}$ would be simply ordered. Also since $G \notin M = \cup D$, $D \cup \{M \cup \{\{G\}\}$ will properly contain D. But this contradicts the maximality of D.

Hence (ii) must hold. Also $F \subset M$ since $f \in D$ and $M = \cup D$. Thus M is the required maximal finitely short subfamily of $\Im$ such that $F \subset N$.

Theorem 1:

(Alexander, Subbase Lemma): *A topological space (X, $\Im$) is compact iff every subbasic open cover for X has a finite subcover.*

Proof:

The 'only if' part follows immediately from the definition of compactness. For, if the space is compact, every open cover for X has a finite subcover and so in particular every sub-basic open cover must have a finite subcover.

We now prove the contrapositive of the 'if part', that is, we shall show that each finitely short subfamily of $\Im$ is short. Let F be any finitely short subfamily M of $\Im$-such that $F \subset M$. It will now be shown that M is short. Let B. Be a subbase for $\Im$. Since M is finitely short, the subfamily $M \cap B.$ of M is also finitely short. Hence, by hypothesis, $M \cap B,$ is short. This implies that $\cup (M \cap B,) \neq X$.

Now suppose $x \in \cup M$. Then $x \in$ for some $A \in M$. Since A is open and B, is a subbase

$$x \in B_1 \cap B_2 \cap ... \cap B_k \subset A.$$

for some B_i's in E, By lemma (5.13), $B_i \in M$ for some $j \in \{1, 2, ... k)$. Hence $x \in B_j$. But $B_j \in B$, and $B_j \in M$ implies that $B_j \in M \cap B$. It follows that $x \in \cap M \Rightarrow x \in \cup (M \cap B.)$.

Therefore $\cup M \subset \cap (M \cap B.)$. It then follows that M is short which certainly implies that F is short. Hence X is compact.

Another Statement : The above lemma can be stated contrapositively as follows : X is compact if easy finitely short subfamily of subbasic members is short.

Theorem 2:

(Tychonoff): *Let $\{(X_\lambda, \Im_\lambda) : \lambda \in \Im\}$ be a collection of topological spaces and let $X = \times \{X_\lambda : \lambda \in \Im\}$. Then X is compact relative to the product topology iff each space X_λ is compact.*

Proof:

Let X be compact. Since each projection π_λ. is a continuous mapping of X into X_λ. It follows that X_λ is a continuous image of the compact space X and hence each X_λ is compact.

To prove that converse, we shall make use of Alexander Subbase Lemma. Let B, the defining subbase for the product topology on X. that is, let

$$B = \{\pi_\lambda^{-1} [G_y] : \lambda \in \Im \text{ and } G_\lambda \in \Im_\lambda\}.$$

Then by Alexander subbase lemma, the space X will be compact if each finitely short subfamily A of B, is short, that is, if each subfamily A of B, such that no finite sub-family of A covers X fails to cover X. For each index $1 \in \Im$, let B_λ be the family of open sets G_λ in X_λ such that $\{\pi^{-1} [G_\lambda], \in A\}$. Let $B_\lambda = \{G_\lambda : G_\lambda \in \Im_\lambda, \pi_\lambda^{-1} [G] \in A\}$. Then B_λ is finitely short in X_λ otherwise A would not be finitely short in X. A Since X_λ is compact B_λ does not cover X_λ. This implies that there exist a point $x_\lambda \in X_\lambda$ such that $x_\lambda \notin G_\lambda$ for any $G_\lambda \in B_\lambda$. It follows that the point x whose λ^{th} coordinate is x_λ, belong to no member of A and hence A is not a cover of X, that is, A is short. Consequently, X is compact.

Theorem 3:

(Generalized Heine Borel Theorem): *A subset of Euclidean n space is compact iff it is closed and bounded.*

Proof:

Let A be a compact subset of Euclidean space E_n. Since E_n is a Hausdorff space (being a metric space), A is a closed subset of E_n. Since A is compact, it can be covered by a finite collection of follows that A is bounded.

Conversely, let A be a closed and bounded subset of E_n. Then $\pi_i [A]$ is bounded in R for each i = 1, 2, ... n, there exists a closed interval li in R such that $\pi_i [A] \subset \lambda_i$. Now each I_i is compact. Hence by the preceding theorem, $\times \{I_i : i = 1, 2, ... n)$ is a compact subset of En (with due consideration of relative topologies and the relation between compact subset and compact subspaces). We now show that A is a closed subset of $\times \{1_i : i = 1, 2, ...n\}$. Let $A = A_1 \times A_2 \times ... \times ... A_n$ where each A_i is a bounded subset of R.

Then $\quad \pi_i [A] = A_i \subset 1_i \qquad [\because \pi_i [A] \subset I_i]$.

Therefore $A = A_1 \times A_2 \times ... \times A_n \subset I_1 \times I_2 \times ... I_n$.

Thus we have show that A is a closed subset of a compact space $\times \{I_i : i \in N_n\}$ and consequently A is compact.

TYCHONOFF CUBES

Theorem 1:

For every λ in an index set L, let X_λ, denote the closed unit interval

$$Q = \{x : x \in R, 0 \leq x \leq 1\} = [0, 1].$$

with the relativized usual topology. Then the product space $\times \{X_\lambda : \lambda \in \mathfrak{I}\}$ is called a ***Tychonoff cube,*** *and may be denoted by Q^N,. If L = N, then Q^N is called the* ***Hilbert cube.***

Theorem 2:

Every Tychonoff cube is normal.

Proof:

Let $X = \times \{X_\lambda : \lambda \in \Lambda\}$ where each X_λ is the closed unit interval Q = [0, 1] with the relativized usual topology so that X is a Tychonoff cube. Follows that each X_λ is Hausdroff and consequently the product space X is also Hausdorff each X_λ is compact and hence by Tychonoff theorem the product space X is also compact. Thus, the Tychonoff cube is a compact. Hausdorff space and consequently, X is normal.

Theorem 3:

A non empty product space $X = \times \{X_\lambda : \lambda \in \mathfrak{I}\}$ where each Xl is a Hausdorff space is locally compact if and only if :

(i) each Xl is locally compact, and

(ii) all but a finitely many X_λ are compact.

Proof:

Let X be locally compact. Since each X_λ is a Hausdorff space and each projection π_λ is continuous and open such that $\pi_\lambda [X] = X_\lambda$. It we know that each X_λ is locally compact. This proves (i). We now prove (ii). Since X is locally compact, for $x \in X$ there exists a nhd N of x in X such that $G \subset N \subset N$ and hence $\pi_\lambda [G] \subset \pi_\lambda [N]$. But $\pi_\lambda [G] = X_\lambda$ for $\lambda \neq \lambda_1, \lambda_2 ... \lambda_n$. Thus $X_\lambda \subset \pi_\lambda [\bar{N}]$. But $\pi_\lambda [\bar{N}] \subset X_\lambda$ always. Hence $\pi_\lambda [N] = X_\lambda$. Since N is compact and π_λ is continuous, it we know that $\pi_\lambda [\bar{N}] = X_\lambda$ is compact for $\lambda \neq \lambda_1, \lambda_2, ... \lambda_n$,. Thus (ii) is also proved.

Conversely, let (i) and (ii) hold. We shall prove that X is locally compact. Let x be any point of X. By (ii). X_λ is compact for $\lambda \neq \lambda_1, \lambda_1, \lambda_2 ... \lambda_n$ say.

By (i), $\pi_\lambda(x) \in X_\lambda$ has an open nhd U_{λ_i} in X_λ such that $\overline{U_\lambda}$ is compact for i = 1, 2, ... n. Also X_λ is given to be compact for $\lambda \neq \lambda_1, \lambda_2 \dots, \lambda_n$. Now U = × $\{U_\lambda : \lambda \in \Lambda\}$ where $U_\lambda = X_\lambda$ for $\lambda \neq \lambda_1, \lambda_2 \dots \lambda_n$ is an open nhd of x in X. Also, $\overline{U}$ = × $\{U_\lambda : \lambda \in \Lambda\}$ which is compact by Tychonoff theorem. Thus we have shown that each point x ∈ X has a nhd whose closure is compact. Hence X is locally compact.

PROJECTION MAPPINGS

Definition: *The mappings*

$$\pi_x : X \times Y \to X : \pi_x((x, y)) = x \;\forall\; (x, y) \in X \times Y$$

and $\pi_y : X \times Y \to Y : \pi_y((x, y)) = \forall\, v\,(x, y) \in X \times Y$

are called the projection of the product X × Y on X and Y respectively.

Theorem 1:

Let (x, $\mathfrak{I}$) and (Y. v) be two topological spaces and let (X × Y, W) be the product space of two spaces. Then the projection px are continuous and open mappings. Further the product topology W is the coarsest topology for which the projections are continuous.

Proof:

Recall that π_x is a mapping of $X \times Y$ into X defined by $\pi_x((x, y)) = x$ for every $(x, y) \in X \times Y$. Let G be any $\mathfrak{I}$-open set. Then it is evident from the definition of π_x that $\pi_x^{-1}[G] = G \times Y$ which is a basic w-open subset of $X \times Y$. [$\because G \in \mathfrak{I}, Y \in v \Rightarrow G \times Y \in$ where E is the base for W]. Hence π_x is a W-$\mathfrak{I}$ continuous mapping. Similarly π_y is W-v continuous mapping. Now let U be any w-open subset of $X \times Y$. Then by the definition of the base E for W, we have

$$U = \cup \{G \times H : G \in \mathfrak{I}, H \in v \text{ and } G \times H \in E \subset\}$$

$$\begin{aligned} \text{Hence } \pi[U] &= \pi_x[\cup \{G \times H : G \in \mathfrak{I}, H \in v \text{ and } G \times G\ E\} \\ &= \cup \{\pi_x \{G \times H\} : G \in \mathfrak{I}, H \in v \text{ and } G \times G \in E\} \\ &= \cup \{G : G \in \mathfrak{I} \text{ and } G \times H \in E\} \end{aligned}$$

[By the definition of π_x] $\in \mathfrak{I}$.

It follows that π_x is an open mapping. Finally let W be any topology for $X \times Y$ which the projection are continuous and let U be any W-open set. Then

$$\begin{aligned} U &= \cup \{G \times H : G \in \mathfrak{I}, H \in v \text{ and } G \times H \in E'\} \text{ where } E' \subset E, \\ &= \cup \{(G \cap X) \times (Y \cap H) : G \in \mathfrak{I}, H \in S \text{ and } G \times H \in E'\} \end{aligned}$$

$= \cup \{(G \times Y) \cap (X \times H) : G \in \mathfrak{I}, H \in v \text{ and } G \times H \in E'\}$

$= \cup \{\pi_x^{-1} [G] \cap \pi_y^{-1}[H] : G \in \mathfrak{I}, H \in v \text{ and } G \subseteq H \in E'\}$

$\because$ π_x is W-v continuous $\Rightarrow \pi_y^{-1} [G] \in W$ and

π_y is W-$\mathfrak{I}$ continuous $\Rightarrow \pi_y^{-1} [H] \in W$ etc.

Thus every W-open set is W open and so W is coarser than W. Since W is any topology for $X \times Y$ for which the projections are continuous, it follows that W is the coarsest topology for which projections are continuous.

Theorem 2:

Let y_0 be a fixed element of Y and let $A = X \times \{y_0\}$. Then the restriction of π_x to A is a homeomorphism of the subspace A of $X \times Y$ onto X. Similarly the restriction of π_y to $B = \{\pi_0\} \times Y$, $\pi_0 \in X$ is a homeomorphism.

Proof:

Let g_x denote the restriction of π_x to A, that is, let

$$g_x : \rightarrow X : g_x ((x, y_0) = x \vee (x, y_0) \in A. \text{ Then}$$

$$g_x ((x_1, y_0)) = g_x ((x_2, y_0) \Rightarrow x_1 = x_2$$

$$\Rightarrow (x_1, y_0) = (x_2, y_0)$$

$$\Rightarrow g_x \text{ is one-one.}$$

g_x is evidently onto. Since by the preceding theorem π_x is continuous, it follows that g_x, is also continuous. We now show that g_x is open. Let V be any open subset of A. Then $V = A \cap U$ for some open subset U of $X \times Y$. But

$$U = \cup \{G \times H : G \in \mathfrak{I}, H \in v \text{ and } G \times H \in E'\} \text{ where } E' \in E$$

We then have

$g_x [V] = g_x [A \cap U]$

$= g_x [A \cap [\cup\{G \times H : G \in \mathfrak{I}, H \in v \text{ and } G \times H \in E'\}]$

$= g_x \{\cup\{A \cup (G \times H) : G \in \mathfrak{I}, H \in v \text{ and } G \times H \in E'\}]$

[Distributive law]

$= \cup \{g_x [(X \times \{y_0\}) \cap (G \times H)] : G \in \mathfrak{I}, H \in v \text{ and } G \times H \in E'\}$

$= \cup \{g_x [X \cap G) \times (\{y_0\} \cap H)\} : G \in \mathfrak{I}, H \in v$

and $G \times H \in E'\}$

$= \cup \{g_x[\times (\{y_0\} \cap H] : G \in \mathfrak{I}, H \in v \text{ and } G \times H \in E'\}$ ...(1)

If $y_0 \notin H$, then it is easy to see from (1) that $g_x [V] = \varnothing$. If $y_0 \in H$, then (1) gives

$g_x [V] = \cup \{g_x [G \times \{y_0\}] : G \in \mathfrak{I}, H \in v$

and $G \times H = \in E'\}$

$= \cup \{G : G \in \mathfrak{I} \text{ and } G \times H \in E'\} \in \mathfrak{I}.$

This implies that g_x is an open mapping as well. Thus we have shown that g_x is one-one, onto, continuous and an open mapping and consequently it is a homeomorphism.

QUOTIENT MAP

Definition : *Let X be non-empty set, and D be a decomposition of X. Then the mapping p from X onto D such that p (x) is the unique members of D to which x belongs is called the quotient map (or the projection map of canonical map). X/R (called the quotient set of X modulo R) so that X/ R is the family of all equivalence.*

DIAGONAL MAP

Definition : *Let Λ be any non-empty set. Then the mapping $d : X \to X^{\Lambda}$ defined in such a manner that for every $x \in X$, $(d(x))(\lambda) = x$ for every $\lambda \in \Lambda$ is called the diagonal map of a set X into the Λth power X^{Λ} of X.*

Since d (x) is an element of X^{Λ}, it follows from the definition of product that d (x) is a mapping of Λ into X so that $(d(x))(\lambda) \in X$ for all $\lambda \in \Lambda$. In the definition of d, we take this element $(d(x))(\lambda)$ of X equal to the pre-assigned element x of X for all $\lambda \in \Lambda$. Thus d (x) is a constant functions of Λ into X.

Theorem:

The diagonal map d is one-one.

Proof :

$d(x) = d(y) \Rightarrow d$ is one-one.

The subset d [X] of X^{3} is called the diagonal of X^{Λ}. The diagonal of X^{Λ} is the set of all constant functions of Λ into X. Also d is a one-one mapping of X onto d [X] and for each $\lambda \in \Lambda$, the restriction of the projection π_λ of X^{Λ} into X to d [X] is the inverse of d.

QUOTIENT SPACE

Definition : *Let $(X, \mathfrak{I})$ be a topological space and R an equivalence relation on X. Let f be the quotient map of X onto the quotient set X/R of X over R so that p (x) = [x] is the equivalence class to which x belongs.. Then the quotient space is the family X/R with quotient topology (relative to p).*

Remark : (i) The quotient topology for X/R is the strongest topology for X/R rendering the quotient map p is continuous on X. The quotient topology for X/R consists of all subsets G of X/R such that p^{-1} [G] is open in X.

(ii) If A is subset of X, then the set of all points which are R relative of points of A shall be denoted by [A] of R [A]. Thus R [A] [y : x, y) $\in$ R for some x $\in$ A}

In other words.

$$R [A] = \cup [D : D \subset X/R \text{ and } D \cap A \neq \emptyset].$$

If x is a point of X, then we abbreviate R [{x}] as R [x] or [{x}] as [x]. The set R [x] or [x] is the equivalence class to which s belongs and if p is the projection of X onto the decomposition, then p (x) = R [x].

(iii) If A $\subset$ X/R, then p^{-1} [A] = $\cup$ {A : A $\in \Im$} and therefore Á is open (closed) relative to the quotient topology iff $\cup$ {A : A $\in \Im$} is open (respectively closed) in X.

Theorem 1:

Let p be the quotient map of the topological space X onto the quotient space X/R. Then the following statements are equivalent.

(a) *p is open (closed) in mapping*

(b) *If A is open (closed) in X, then R [A] is open (respectively closed).*

(c) *If B is a closed (open) subset of X, then the union of all members of X/R which are subsets of B is closed (respectively open).*

Proof :

(a) ⇔ (b) : First observe that if A is a subset of X, then the set R [A] = π^{-1} [[A]].

Now let π be open map and let A be any open subset of X, Then by definition of an open map π [[A]] is open subset of X/R. Since π is continuous, π^{-1} [π [A]] is open in X and so by (1), R [A] is an open subset $2\pi^{-1}$ [p [A]] is open in X. Then π [A] is open in X/R, and consequently p is open mapping.

(b) ⇔ (c) : We first notice that the union of all these members of X/R which are subsets of B is X – R [X – B]. Now assume then R [A] be open for each open subset A of X and let B be any closed subset of X so that X – B is open in X. Hence by hypothesis R [X – B] is open X and consequently X – R [X – B] is closed in X, that is , the union of all those members of X/R which are subsets of B is closed. This proves (b) ⇒ (c). To prove (c) ⇒ (b), assume that for each closed subset B of X, the union of all those members of Y/R which are subsets of B is closed, that is assume that X − R [X − B] is closed. Let A be any open subset of X so that X − A is closed in X. Then and consequently R [A] is open.

If we interchange the words "open" and "closed" through, we get the proof of the dual proposition.

Theorem 2:

Let (X, $\mathfrak{I}$) be a topological space. If X is compact, connected locally connected separable or Lindelof, the so is X/R.

Proof:

Since compactness, connectedness, separability and Lindelofness are preserved under continuous mapping and since X/R is a continuous image of X under the quotient map π, it follows that X/R also has these properties. To prove local connectedness of X/R, we shall use the fact that a topological space X is locally connected iff every component of an open set is open in X. So let A be an open set of an open set of X/R and let B be a component of A. Then, the set G = π^{-1} [A] is an open subset of X containing the set C = π^{-1} [B]. Let x be any point of C and let C, be the component of C containing x.

Now $$x \in C \Rightarrow \pi(x) \in \pi[C] \Rightarrow \pi(x) \in \pi[\pi^{-1}(B)]$$

$$\Rightarrow \pi(x) \in B \text{ and } x \in C_x \Rightarrow \pi(x) \in \pi[C_x].$$

By definition of component, C_x is connected and so its continuous image π [C] is also connected. Also B, being a component, is connected. Since π (x) ∈ B and π (x) ∈ π [C_x], we have B ∪ π [C_x] ≠ ∅. It follows from theorem (2.24), ch. 5 that their union B ∪ π [C_x] is connected. Since π {C_x] is connected subset of A, it follows by the maximality of components that π [C_x] ⊂ B, and hence

$$\pi^{-1}[\pi[C_x]] \subset n^{-1}[B]$$

that is, C_x ⊂ C. Since X is locally connected, each component C_x must be open and hence C is a neighbourhood of x. Since x was an arbitrary point of C, we conclude that C is an open set in X. Finally, since C = π^{-1} [B].B is an open subset of X/R and consequently X/R is locally connected.

Example:

Give an example of a Hausdorff space X which has a non-Hausdorff quotient space.

Solution:

Consider the space (R, U) where R denotes the set of all real numbers and U is usual topology for R. We know that this space is Hausdorff. Now consider the relation R on R consisting of all pairs (x, y) relation on R. Also it is evident that the quotient space R/R in this case is indiscrete and consequently non-Hausdorff

Theorem 3:

Let X be a topological space and let the quotient space X/R be Hausdorff. Then R is a closed subset of the product space $X \times X$.

Proof:

We shall show that no point of $X \times X - R$ can be limit point of R. Let $(x, y) \in X \times X - R$. Then $(x, y) \notin R$ and hence $p(x) \neq \pi(y)$. Since X/R is Hausdorff, there exists open sets G and H inX/R such that

$$\pi(x) \in G, \pi(y) \in H$$

and $$G \cap H = \emptyset.$$

$$[\pi^{-1}[G]] = G \text{ and } \pi[\pi^{-1}(H)] = H$$

are disjoint, no member of $\pi^{-1}[G]$ can be R-related to a member of $\pi^{-1}[H]$. Hence π^{-1} Hence $\pi^{-1}[G] \times \pi^{-1}[H]$ is a neighbourhood of (x, y) which does not contain any point of R and consequently (x, y) is not a limit point of R. It follows that R is closed..

Theorem 4:

Let X be a topological space and let X/R be the quotient space. If the quotient map p is an open map and R is closed in $X \times X$, then X/R is a Hausdorff space.

Proof:

Let $\pi(x)$ and π y) be any two members of X/R such that $\pi(x) \neq \pi(y)$. Then x is no R-related to y so that $(x, y) \notin R$. Since R is closed, (x, y) can not be a limit point of R.

Hence there exists a basic open subset $G \times H$ of the product space $X \times X$ containing (x, y) and disjoint from R. This implies that G and H are open neighbourhoods of x and y respectively such that no point of G is R-related to a point of H.

Hence p [G] and p [H] are disjoint. Since p is given to be open map, it follows that $\pi[G]$ and $\pi[H]$ are open neighbourhoods of $\pi(x)$ and $\pi(y)$ respectively and consequently X/R is Hausdorff space.

UPPER SEMI-CONTINUOUS DECOMPOSITION

Definition: *Let X be a topological space. Then a decomposition D of X is said to be upper semi-continuous iff for every $D \in D$ and every open* *set G containing D there exists an open set H such that $D \subset H \subset G$ and H is the union of members of D.*

Theorem 1:

Let (X, $\mathfrak{I}$) be a topological space and let D be an upper semi-continuous decomposition of X and let D have the quotient topology v. If G is a $\mathfrak{I}$-open set containing a member A of D, then p [G] is a v-neighbourhood of A where p is the projection (or quotient) map.

Proof:

Since G is a $\mathfrak{I}$-open set containing $A \in D$, by the definition of upper semi-continuity, there exists a $\mathfrak{I}$-open set H which is a union of members of D such that

$$A \subset H \subset G.$$

Hence $\quad \pi [A] \subset \pi [H] \subset \pi [G]. \qquad ...(1)$

Since $A \in D$, by the definition of π, we have $\pi [A] = \{A\}$, and consequently

$$A \in \pi [A]. \qquad ...(2)$$

It will now be shown that π [H] is V-open. By definition of H, we have

$$H = \cup \{A_i : A_i \in D' \subset D\}$$

Then
$$\begin{aligned} \pi [H] &= \pi [\cup \{A_i : A_i \in D' \subset D\}] \\ &= \cup \{\pi [A_i] : A_i \in D' \subset D\} \\ &= \cup \{\{A_i\} : A_i \in D' \subset D\} \end{aligned}$$

Therefore
$$\begin{aligned} \pi^{-1} [\pi [H]] &= \pi^{-1} [\cup \{A_i : A_i \in D' \subset D\}] \\ &= \cup \{\pi^{-1} [\{A_i\}] : A_i \in D' \subset D\} \\ &= \cup \{A_i : A_i \in D' \subset D\} = H. \end{aligned}$$

Since H is $\mathfrak{I}$-open, it follows from the definition of quotient topology that p [H] is a v-open set. Hence it follows from (1) and (2) that π [G] is a v-neighbourhood of A.

Theorem 2:

Let (X , $\mathfrak{I}$) be a topological space and let D be an upper semi continuous decomposition of X such that every member of D is a compact subset of X and let D have the quotient topology v. If X is :

(i) Hausdorff,

(ii) Regular,

(iii) Locally compact or

(iv) Satisfies the second axiom of countability, then D also has the corresponding properties.

Proof:

(i) Let X be a Hausdorff space and let A and B be distinct members of D. Since A and B are compact subsets of the Hausdorff space X, there exists disjoint $\mathfrak{I}$-nhds N and M of A and B respectively. Then π [N] and p [M] are disjoint V-nhds of A and B respectively and consequently D is a Hausdorff space.

(ii) Let X be regular. Let A be any point of D and let N be a V-nhd of A. Then the union G of members of N is a $\mathfrak{I}$-nhd of A. Since Y is regular, for each $x \in A$, there exists an open nhd U_x of x such that $U_x \subset G$. Now the collection $\{U_x : x \in A\}$ is an open cover of A. Since A is compact there exists a limit subcover $U_1, U_2,... U_n$ of A of A. *Such* that $\overline{U_i} \subset G$ for each i so that $U_i \subset \cup \{A_i : i = 1, 2, ... n\}$. Let $V = \cup \{\overline{U} : i = 1, 2, ...n\}$. Then V is a $\mathfrak{I}$-closed neighbourhood of A such that $V \subset G$. It follows that π [V] is a V-closed neighbourhood of A such that $\pi [V] \subset \pi [G] = N$. Hence D is regular.

(iii) Let X be locally compact and A any member of D. Then by definition of local compactness, to each $x \in A$, there exists a $\mathfrak{I}$-open compact nhd U_x of x. Now the collection $\{U_x : x \in A\}$ is an open cover of A. Since A is compact, there exists a finite subcover $U_1, U_2,..., U_n$ so that

$A \subset \cup \{U_i : i = 1, 2. 3... n)$. Let $V = \cup \{U_i : i = 1, 2, 3,... n\}$. Also since V is the union of a finite member of compact neighbourhood of A. Since π is continuous, π [V] is compact. Also π [V] is a V-neighbourhood, and hence by definition of local compactness D is locally compact.

(iv) Let X be second countable and let B be a countable base for $\mathfrak{I}$. The family H, let H of unions of finite subfamilies of B is countable. For every member H of H, let H* denote the union of all those members of D which are subsets of H and M be the family of all sets H* corresponding to every $H \in H$. Then it is easy to see that π [H*] is open in D for every H* = M. We will show that the countable collection $\{\pi [H^*] : H^* \in M\}$ is a base for the quotient topology. For this purpose, it is sufficient to prove that for each $A \in D$ and each neighbourhood N of A, there exists some $H^* \in M$ such that $A \subset H^* \subset N$.

Since A is compact it may be covered by a finite number of members of B such that the union H of these members, which is members of H, is contained in N. Then $H^* \in M$ is such that

$$A \subset H^* \subset N.$$

This proves the theorem.

Theorem 3:

Let X be a topological space and let D be a decomposition of X. Then D is upper semi-continuous iff the projection p of X onto D is closed.

Proof:

We know that π is a closed map iff for each open subset G of X, the union H of all members of D which are subsets of G is an open set. Now let π be a closed map. Let $D \in D$ and C be an open set containing D. If H is the union of all those members of D which are subsets of G, then H is an open set such that $D \subset H \subset G$. Hence D is upper semi-continuous. To prove the converse, let D be upper semi-continuous. Let G be an open subset of X and let H be the union of all members of D which are subsets of G. We shall show that H is open. Let $x \in H$. Then by definition of union, $x \in D \subset G$ for some $D \in D$. Hence by upper semi-continuity, there exists an open set K which is the union of members of D such that $D \subset K \subset G$. It is follows that $K \subset H$. Thus we have shown that to each $x \in H$, there exists an open set K such that $x \in K \subset H$. Hence H is an open subset of X since it is a neighbourhood of each of its points. It follows that π is a closed map.

EVALUATION MAP

Definition : *For each λ in L let f_λ be a mapping of a set X into a set Y_λ. Then the mapping $e : X \to \times \{Y_\lambda : \lambda \in \Lambda\}$ defined in such a manner that for every $x \in X$ $(e, (x)) (\lambda) = f_\lambda (x)$ for all $\lambda \in \Lambda$ is called the evaluation up of X into the cartesian product $\times \{Y_\lambda : \lambda \in \Lambda\}$ determined by the family $\{f_\lambda : \lambda \in \Lambda\}$.*

Observe that e (x) is an element of the cartesian product $\times \{X_\lambda : \lambda \in \Lambda\}$ and so by the definition e (x) is itself a mapping of L into $\cup \{\Psi\lambda : \Lambda\}$ such that $(e (x) (l)) \in Y_\lambda$ for all $\lambda \in \Lambda$. Also f_λ is a pre-assigned mapping of X into Y_λ so that $f_\lambda (x) \in Y_\lambda$. It is this pre-assigned element $f_\lambda (x)$ of Y_λ which we take to be equal to (e, (x)) (l). In short, the lth coordinate of e (x) is taken to be $f_\lambda (x)$. That is why the map e is said to be determined by the family $\{f_\lambda : \lambda \in \Lambda\}$.

Thus $\quad e (x) = \{f_\lambda (x) : \lambda \in \Lambda\}$ for all $x \in X$ ...(1)

and $(\pi_\lambda \circ e) (x) = \pi_\lambda (e (x) = (e (x)) (\lambda) = f_\lambda (x)$ so that

$$\pi_\lambda \circ e = f_\lambda \text{ fol all } \lambda \in \Lambda. \qquad ...(2)$$

Remark : *If e' is a mapping of X into $\times \{Y_\lambda : \lambda \in \Lambda\}$ such that $\pi_\lambda \circ e'$ for all $\lambda \in \Lambda$, then $e' = e$.*

Proof:

Suppose, if possible, that e ≠ e. Then e' (x) ≠ e (x) for at least one x ∈ X. Let. e′ (x) = {y_λ λ : ∈ Λ} where $y_\lambda \in Y_\lambda$. Also by definition of e, we have e (x) = {(f_λ (x) : λ ∈ Λ}.

Since e′ (x) ≠ e (x) there is at lest one λ ∈ Λ; say λ_0, such that y_{λ_0} ≠ (x). Hence for this particular λ_0, we have

$$(\pi_{\lambda_0} \text{ o } e')(x) = \pi_{\lambda_0} (e'(x)) = \pi_{\lambda_0} ((y_\lambda : \lambda \in \Im\}\} = \pi_{\lambda_0} \neq (x)$$

which is a contradiction since π_λ o e′ = f_λ for all λ ∈ Λ.

Theorem 1:

f is a mapping X into × {Yl : L ∈ L}, then f is the evaluation of X into × {Y_λ : λ ∈ ℑ} determined by they family (π_λ 0 f : λ ∈ Λ}.

Proof:

Since f (x) is an element of the product × {Y_λ : λ ∈ Λ}, it follows by definition that f (x) is a mapping of ℑ into ∪ {Y_λ : ∈ Λ} such that (f (x)) (λ) ∈ Y_λ for all λ ∈ Λ. Let

$$(f(x))(\lambda) = y_\lambda \text{ where } y_\lambda \in Y_\lambda. \quad ...(1)$$

We may then write f (x) = {y_λ : λ ∈ Λ}.

Now $\quad (\pi_\lambda \text{ o } f)(x) = \pi_\lambda (f(x)) = (\{y_\lambda : \lambda \in \Lambda\}) = y_\lambda \quad ...(2)$

From (1) and (2), we have (f (x)) (λ) = (π_λ : o f) (x).

Hence f is the evaluation of X into × {Y_λ : λ ∈ Λ}, determined by the family {π_λ o f : l ∈ Λ}.

Definition (a) : *The family F = {f_λ : λ ∈ ℑ} defined, is said to distinguish points or to separate points of X iff, for every pair of distinct elements x_1 and x_2 of X, there exists a λ ∈ L such that $f_\lambda (x_1) \neq f_\lambda (x_2)$.*

(b) *F is said to* **Distinguish Points and Closed Sets** *iff for every closed subset H of X and each point x ∈ X – H,, there exists some f_λ in F such that $f_\lambda (x) \notin (f_\lambda [H])$.*

Theorem 2:

The family {f_λ : λ ∈ Λ} distinguishes points of X iff the evaluation map of X into × {Y_λ : λ ∈ Λ} determined by {f_λ : λ ∈ Λ} is one-one.

Proof :

If {f_λ : λ ∈ Λ} distinguishes points of X. Then $x_\lambda \neq x_2 \Rightarrow f_\lambda (x_\lambda)$ ' f_λ (x_2) for some λ ∈ Λ ⇒ e (x_1) ≠ e (x_2).

[∵ e (x_1) and e (x_2) differ at least in one coordinate].

Hence e is one- one

Conversely, let e be one-one. Then

$$x_1 \neq x_2 \Rightarrow e(x_1) \neq e(x_2)$$
$$\Rightarrow f_\lambda(x_1) \neq f_\lambda f(x_2) \text{ for some } \lambda \in \Im$$
$$\Rightarrow \{f_\lambda : \lambda \in \Lambda\} \text{ distinguishes points of X.}$$

SOLVED EXAMPLES

Example 1:

Describe open spheres for the usual metric on R^2.

Solution:

Let d be the usual metric on R^2 so tŀ t d is given by

$$d(x, y) = [(x_1 - x_1)^2 + (y_1 - y_2)^2]^{1/2}$$

where $x = (x_1, y_1) \in R^2$, $y = (x_2, y_2) \in R^2$.

Let $z_0 = (x_0, y_0) \in R^2$ and $r > 0$ be any real number

$$S_r(z_0) = \{(x, y) \in R^2 : z = (x, y),\ d(z, z_0) < r\}$$
$$= \{(x, y) \in R^2 : z = \sqrt{[(x - x_0)^2 + (y - y_0)^2]} < r\}$$
$$= \{(x, y) \in R^2 : (x - x_0)^2 + (y - y_0)^2 < r^2\}$$

This proves that open sphere $S_r(z_0)$ is the interior of the circle $(x - x_0)^2 + (y - y_0)^2 = r^2$, where $z_0 = (x_0, y_0)$.

Example 2:

Describe open sphere of unit radius having the centre at (0, 0) for each of the following metric on R.

(i) $d_1(z_1, z_2) = |x - x_2| + |y_1 - y_2|$

(ii) $d_2(z_1, z_2) = max.\ \{|x_1 - x_2|, |y_1 - y_2|\}$

where $z_1 = (x_1, y_1)$, $z_2 = (x_2, y_2) \in R^2$.

Solution:

Let $z_0 = (0, 0)$, $z = (x, y) \in R^2$.

To determine $S_1(z_0)$.

Step (i) $S_1(z_0) = \{(x, y) \in R^2 : d_1(z, z_0) < 1\}$

$$= \{(x, y) \in R^2 : |x - 0| + |y - 0| < 1\}$$
$$= \{(x, y) \in R^2 : |x| + |y| < 1\}$$

This shows that the open sphere $S_1(z_0)$ is the interior of the square bounded by the lines

$$x + y = 1, -x - y = 1, -x + y = 1, x - y = 1.$$

Step (ii) $\quad S_1(z_0) = \{(x, y) \in R^2 : d_1 (z, z_0) < 1\}$

$$= \{(x, y) \in R^2 : \max. \{|x - 0|, |y - 0|\} < 1\}$$

$$= \{(x, y) \in R^2 : \max. \{|x|, |y|\} < 1\}$$

This proves that the open sphere $S_1(z_0)$ is the interior of the square bounded by the lines

$x = 1, -x = 1, y = 1. -y = 1.$

Example 3:

Show that C the set of complex numbers is metric space under the metric d defined as follows:

$$d(z_1, z_2) = \frac{|z_1 - z_2|}{\left[1 + |z_1|^2\right]^{1/2}\left[1 + |z_2|^2\right]^{1/2}}$$

Solution:

Let z_1, z_2, z_3 be arbitrary complex numbers . It is given that

$$d(z_1, z_2) = \frac{|z_1 - z_2|}{\left[1 + |z_1|^2\right]^{1/2}\left[1 + |z_2|^2\right]^{1/2}}$$

To prove that (C, d) is a metric space.

We know that $|z| \geq 0$ for every complex number z.

If view of this, (1) says that $d(z_1, z_2) \geq 0$. ...(1)

Since $|z_1 - z_2| = |z_2 - z_1|$. In view of this, (1) $\Rightarrow$

$$d (z_1, z_2) = d (z_2, z_1) \quad ...(2)$$

$$d(z_1, z_2) = 0 \Leftrightarrow \frac{|z_1 - z_2|}{\left[1 + |z_1|^2\right]^{1/2}\left[1 + |z_2|^2\right]^{1/2}} = 0$$

$$\Leftrightarrow |z_1 - z_2| = 0 \Leftrightarrow z_1 = z_2$$

$\therefore \quad d (z_1, z_2) = 0 \Leftrightarrow z_1 = z_2$ ...(3)

$d (z_1, z_2) + d (z_2, z_3)$

$$= \frac{|z_1 - z_2|}{\left[1 + |z_1|^2\right]^{1/2}\left[1 + |z_2|^2\right]^{1/2}} + \frac{|z_2 - z_3|}{\left[1 + |z_2|^2\right]^{1/2}\left[1 + |z_3|^2\right]^{1/2}}$$

$$\geq \frac{|z_1 - z_2|}{\left[1+|z_1|^2\right]^{1/2}\left[1+|z_2|^2\right]^{1/2}\left[1+|z_3|^2\right]^{1/2}}$$

$$+ \frac{|z_2 - z_3|}{\left[1+|z_1|^2\right]^{1/2}\left[1+|z_2|^2\right]^{1/2}\left[1+|z_3|^2\right]^{1/2}}$$

$$= \frac{|z_1 - z_2| + |z_2 - z_3|}{\left[1+|z_1|^2\right]^{1/2}\left[1+|z_2|^2\right]^{1/2}\left[1+|z_3|^2\right]^{1/2}}$$

$$\geq \frac{|z_1 - z_3|}{\left[1+|z_1|^2\right]^{1/2}\left[1+|z_2|^2\right]^{1/2}\left[1+|z_3|^2\right]^{1/2}}$$

$$\geq \frac{|z_1 - z_3|}{\left[1+|z_1|^2\right]^{1/2}\left[1+|z_3|^2\right]^{1/2}} = d\left(z_{1,}z_3\right)$$

or $d(z_1, z_2) + d(z_2, z_3) \geq d(z_1, z_3)$ ...(4)

From (1), (2), (3) and (4), it follows that (C, d) is a metric space.

Example 4(a):

Show that the function d defined by d (p, d) = $\|p - q\|$, where p, q are vectors in a normal space V, is a metric on V.

Solution:

If p, q, r $\in$ V and K is any scalar, then by def. of normed vector space,

(N_1) $\|p - q\| \geq 0$, (N_2) $\|p - q\| = 0 \Leftrightarrow p = 0$

(N_3) $\|kp\| = |k| \|p\|$, (N_4) $\|p + q\| \leq \|p\| + \|q\|$

We have $d(p, q) = \|p - q\|$ ...(1)

Using (N_1) in (1), we find that $d(p, q) \geq 0$ ...(M_1)

Using (N_2) in (1), we see that

$$d(p, q) = 0 \Leftrightarrow \|p - q\| = 0 \Leftrightarrow p = q = 0$$

$$\Leftrightarrow p = q \quad ...(M_2)$$

or $\quad d(p, q) = 0 \Leftrightarrow p = q$ ...(M_3)

Using (N_3) in (1), we see that

$$d(q,p) = \|q - p\| = \|(-1)(p - q)\| = \|p - q\| = d(p,q)$$

or $\quad d(q, p) = d(p, q)$ ...(M_3)

Using (N_4) in (1), we see that

$$d(p, r) = \|p - r\| = \|(p - q) + (q, r)\| \leq \|p - q\| + \|q - r\|$$

$$d(p, q) + d(q, r)$$

or $\quad d(p, r) \leq d(p, q) + d(q, r)$

From (M_1), (M_2), (M_3) and (M_4), it follows that (C, d) is a metric space.

Example 4(b):

In a metric space (X, d), $S = S(x_0, r)$ is an open sphere and p is any point of it. Prove that $\exists$ an open sphere $T = S(p, \varepsilon)$ s.t. $T \subset S$.

Solution:

$p \in S(x_0, r) \Rightarrow d(p, x_0) < r \Rightarrow r - d(p, x_0) = r_1 > 0.$

Take $\varepsilon < r_1$, then $x \in S\,p\,\varepsilon \Rightarrow d(x, p) < \varepsilon < r_1 \Rightarrow d(x, p) < r_1$

or $\quad d(x, p) < r - d(p, x_0)$

or $\quad d(x, p) + d(p, x_0) < r$

or $\quad d(x, x_0) \leq d(x, p) + d(p, x_0) < r$

or $\quad d(p, x_0) < r$

or $\quad x \in S(x_0, r)$ ...(M_4)

Thus $\quad x \in S(p, \varepsilon) \Rightarrow x \in S(x_0, r).$

Hence $\quad S(p, \varepsilon) \subset S(x_0, r)$ or $T \subset S$.

Example 5:

If G_1, G_2 are open sets in metric space (X_1, d_1) and (X_2, d_2) then $G_1 \times G_2$ is open in the product metric space (X, d).

Solution:

Let $x = (x_1, x_2)$, $y\,(y_1, y_2) \in X = X_1 \times X_2$, then $x_1, y_1 \in X_1$ and $x_2\, y_2 \in X_2$. Let $x \in G_1 \times G_2$, then $x_1 \in G_1$, $x_2 \in G_2$.

By def. of open set,

$x \in G_1 \Rightarrow \exists\, r_1 > 0$ s.t. $S(x_1, r_1) \subset G_1$ ...(1)

Similarly, $S(x_2, r_2) \subset G_2$ ...(2)

By (1) and (2),

$S(x_1, r_1) \times S(x_2, r_2) \subset G_1 \times G_1$ ...(3)

Take $r = \min\{r_1, r_2\}$, then $r \leq r_1$, $r \leq r_2$

$\therefore\ S(x_1, r) \times S(x_2, r) \subset S(x_1, r_1) \times S(x_2, r_2) \subset G_1 \times G_2$

or $S(x, r) \subset G_1 \times G_2$, for any $x \in G_1 \times G_2$

By def., this proves that $G_1 \times G_2$ is open.

Example 6:

In a metric space, prove that $d(A) = d(\overline{A})$.

Solution:

Evidently $A \subset B \Rightarrow d(A) \leq d(B)$.

But $A \subset \overline{A}$ and so that $d(A) \leq d(\overline{A})$. ...(1)

Let $x, y \in \overline{A}$, then \$ $u, v \in A$ s.t.

$$d(x, u) < r,\ d(u, v) < r \text{ where } r > 0.$$

But $$d(x, y) \leq d(x, u) + d(u, v) + d(v, y)$$
$$< d(u, v) + 2r \leq d(A) + 2r$$

or $$d(x, y) \leq d(A) + 2r.$$

For $$d(A) = \sup d(x, y) \text{ where } x, y \in A.$$

This $$\Rightarrow d(\overline{A})' \leq d(A). \quad ...(2)$$

Combining (1) and (2), $d(A) = d(\overline{A})$.

Example 7(a):

Let X be a non-empty set and let d be a real valued function of ordered pairs of elements of X which satisfies the following conditions:

(i) $d(x, y) = 0 \Leftrightarrow x = y$

(ii) $d(x, y) \leq d(x, z) + d(y, z)$

Then prove that d is a metric on X.

Solution:

(1) To prove $d(x, y) \geq 0$.

By (ii), $d(x, x) \leq d(x, z) + d(x, z)$, on putting $x = y$.

or, $$0 \leq 2d(x, z) \text{ by (i)}$$

or, $$d(x, z) \geq 0$$

or, $$d(x, y) \geq 0$$

(2) To prove $d(x, y) = d(y, x)$.

putting $z = x$ in (ii),

$$d(x, y) \leq 0 + d(y, x)$$

or, $$d(x, y) \leq d(y, x)$$

Interchanging x and y,

d (y, x) $\leq$ d (x, y)

Combining the last two, d (x, y) = d (y, x).

(3) d (x, y) = 0 $\Leftrightarrow$ x = y by (i)

(4) To prove d (x, z) + d (z, y) $\geq$ d (x, y)

Using (2) in (ii), d (x, y) $\leq$ d (x, z) + d (z, y)

or, d (x, z) + d (z, y) $\geq$ d (x, y).

Example 8:

Every set in a discrete metric space is open.

Solution:

Let (X,d) be a discrete metric space. Let x, y $\in$ X be arbitrary.

By def. of discrete metric, d (x, y) = $\begin{cases} 1 \text{ if } x \neq y \\ 0 \text{ if } x = y \end{cases}$.

Let r be any positive real number s.t. r $\leq$ 1.

Then $$Sr(x) = \{y \in X : d(y, x) < r \leq 1\}$$
$$= \{y \in X : d(y, x) < 1\}$$
$$= \{y \in X : d(y, x) = 0, \text{ by def. of d.}$$
$$= \{y \in X : y = x\} = \{0\}.$$

or, $S_{r(x)} = \{0\}$.

But every open sphere is an open set.

$\therefore$ {x} is an open set in X $\forall$ x $\in$ X.

If A = $\{x_1, x_2, ..., x_n\}$ = finite set $\subset$ X, then

$A = \bigcup_{r=1}^{n} \{x_r\}$ = finite union of open sets.

= open set.

Hence every finite sub set of X is open set. ...(1)

If B = $\{x_1, x_2, x_3, ...\} \subset X$, then

B is an infinite subset of X.

Now $$B = \bigcup_{r=1}^{\infty} \{x_r\}$$

= Arbitrary union of open sets

= open set, by Theorem 8.

∴ B is an open set ...(2)

From (1) and (2), it follows that every subset (finite or infinite) is an open set in X.

Example 9:

Let ρ_1 and ρ_2 be metrics on the same set X, show that there exists a metric ρ on X such that

$$\rho(x, y) \geq \rho_i(x, y) \text{ for } i = 1, 2$$

and $\forall x, y \in X.$

Solution:

Let ρ_1 and ρ_2 be metrics on the same set X, Let x, y, z ∈ X be arbitrary.

To prove that ∃ a metric ρ on X s.t.

$$\rho(x, y) \geq \rho_1(x, y) \text{ for } i = 1, 2$$

Write $\rho(x, y) = \rho_1(x, y) + \rho_2(x, y).$ ...(1)

We claim ρ is metric on X.

(i) $\rho(x, y) \geq 0$

∵ $\rho_1(x, y) \geq 0,\ \rho_2(x, y) \geq 0.$

(ii) $\rho(x, y) = 0$ iff $x = y$

For $\rho(x, y) = 0 \Leftrightarrow \rho_1(x, y) + \rho_2(x, y) = 0.$

$$\Leftrightarrow \rho_1(x, y) = 0 = \rho_2(x, y).$$

Since ρ_1 and ρ_2 both are non-negative.

$$\Leftrightarrow x = y$$

(iii) $\rho(y, x) = \rho(x, y)$

For $\rho(y, x) = \rho_1(y, x) + \rho_2(y, x) = \rho_1(x, y) + \rho_2(x, y)$

$$\rho(x, y)$$

∵ $\rho_1(x, y) = \rho_1(y, x),\ \rho_2(x, y) = \rho_2(y, x)$

(iv) $\rho(x, y) + \rho(y, z) \geq \rho(x, z)$

For $\rho(x, y) + \rho(y, z) = \rho_1(x, y) + \rho_2(x, y) + \rho_1(y, z) + \rho_2(y, z)$

$$[\rho_1(x, y) + \rho_1(y, z)] + [\rho_2(x, y) + \rho_1(y, z)]$$

$$^3\ \rho_1(x, z) + \rho_2(x, z) = \rho(x, z).$$

From what has been done it follows that ρ is a metric on X.

Moreover (1) ⇒ $\rho(x, y) \geq \rho_1(x, y)$ for i = 1, 2.

Hence ρ is the required metric.

Example 10(a)

Show that the map $d : R^2 \times R^2 \to R$

define by $d(x, y) = max. \{|x_1 - y_1|, |x_2 - y_2|\}$.

where $x = (x_1, x_2), y = (y_1, y_2) \in R_2$

is a metric on R^2.

Solution:

Let $x = (x_1, x_2)$, $y = (y_1, y_2)$, $z = (z_1, z_2)$ be arbitrary elements or R^2. Also let

$d(x, y) = \max \{|x_1 - y_1|, |x_2 - y_2|\}$

(i) $d(x, y) \geq 0$.

Since modulus of any quantity ;is non-negative.

(ii) $d(x, y) = 0$ iff $x = y$.

For $d(x, y) = 0 \Leftrightarrow \max.\{|x_1 - y_1|, |x_2 - y_2|\} = 0$

$$\Leftrightarrow |x_1 - y_1| = 0 = |x_2 - y_2|$$

$$\Leftrightarrow x_1 = y_1, |x_2 = y_2$$

$$\Leftrightarrow (x_1, x_2) = y_2, y_2)$$

$$\Leftrightarrow x = y$$

(iii) $d(x, y) = d(y, x)$.

For $|a - b| = |b - a|$ and so

$\max.\{|x_1 - y_1|, |x_2 - y_2|\} = \max \{|y_1 - x_1|, |y_2 - x_2|\}$

(iv) $d(x, y) + d(y, z) \geq d(x, z)$

For $d(x, y) = \max \{|x_1 - z_1|, |x_2 - z_2|\}$

$$= \max. \{|x_1 - y_1 + y_1 - z_1|, |x_2 - y_2 + y_2 - z_2|\}$$

$$\leq \max. \{|x_1 - y_1| + |y_1 - z_1|, |x_2 - y_2| + |y_2 - z_2|\}$$

$$\leq \max. \{|x_1 - y_1|, |x_2 - y_2|\} + \max. \{|y_1 - z_1|, |y_2 - z_2|\}.$$

$$= d(x, y) + d(y, z).$$

The above facts lead to the conclusion that d is a metric on R^2.

Example 11:

In every metric space, all finite sets are closed

Solution:

Let (X, d) be a metric space.

To prove that every finite subset of X is closed.

For any subset A of X, we know that

$\overline{A} = \{x \in X : d(x, A) = 0\}$

Let p $\in$ X be arbitrary.

Then $\overline{\{p\}} = \{x \in X : d(x, \{p\}) = 0)$

$= \{p\}$

$\therefore$ $d(x, \{p\}) = 0 \Rightarrow x \in \{p\} \Rightarrow x = p$

$\{p\} = \overline{\{p\}}$,

Showing thereby, {p} is closed in X.

Hence every singleton set in X is closed. Moreover a finite union of closed sets is closed. A finite set is a finite union of singleton sets in which every singleton set is closed.

Example 12:

Let (X_1, d_1), (X_2, d_2) be two metric spaces.

Let $$d(x, y) = \left[d_1^2(x_1, y_1) + d_2^2(x_2, y_2)\right]^{1/2}$$

where $x = (x_1, y_1)$, $y = (y_1, y_2)$ *in* $X_1 \times X_2$

Prove that d is a metric on $X_1 \times X_2$

Solution:

Do your self.

Example 13(a):

Contor's set contains exactly as many points as the interval [0, 1].

Solution:

To prove $|E| = |[0,1]|$.

Also $|[0,1]| = C$

E is a closed set.

Since an infinite disjoint union of open sets is open, Hence the family of removed open intervals form an open set.

E = $[0,1]$ – family of removed open intervals.

= $[0,1]$ – open set = closed set.

For complement of open set is a closed set.

Example 13(b):

Given an example of two closed subset A and B of the real line r such that d (A, B) = 0, but A/B ∅.

Solution:

Let d be the usual metric on R. Write

$$N_1 = \{ 2, 3, 4, 5 \ldots \} = \{n + 1 : n \in N\}$$

$$A = \{2, 3, 4, 5 \ldots\},$$

$$B = \left\{2\frac{1}{2}, 3\frac{1}{3}, 4\frac{1}{4}, 5\frac{1}{5}\right\}$$

Then A and B are expressible as

$$A = \{n : n \in N_1\},$$

$$B = \left\{n + \frac{1}{n} : n \in N_1\right\}$$

If n ∈ A, then n + 1/n ∈ B and the distance between n + 1/n and n is 1/n, which → 0 as n → ∞ .

$$\begin{aligned} d (A, B) &= \inf \{d (x, y) : x \in A, y \in B\} \\ &= \inf \{|x - y| : x \in A, y \in B\} \\ &= \inf \{1/n : n \in A, n + 1/n \in B\} = 0. \end{aligned}$$

Thus, d (A, B) = 0. Also $A \text{ç} B = \varnothing$.

Therefore A and B are the required sets.

Example 14:

Every set in a discrete metric space is open

Solution:

Let (X,d) be a discrete metric space. Let x, y ∈ X be arbitrary.

By def. of discrete metric, $d (x, y) = \begin{cases} 1 \text{ if } x \neq y \\ 0 \text{ if } x = y \end{cases}$.

Let r be any positive real number s.t. r ≤ 1.

Then Sr(x) = {y ∈ X : d (y, x) < r ≤ 1}
= {y ∈ X : d (y, x) < 1}
= {y ∈ X : d (y, x) = 0, by def. of d.
= {y ∈ X : y= x} = {0}.

or, $S_{r(x)} = \{0\}$.

But every open sphere is an open set.

$\therefore$ {x} is an open set in X $\forall$ x $\in$ X.

If A = $\{x_1, x_2, ..., x_n\}$ = finite set $\subset$ X, then

$A = \bigcup_{r=1}^{n} \{x_r\}$ = finite union of open sets.

= open set.

Hence every finite sub set of X is open set. ...(1)

If B = $\{x_1, x_2, x_3, ...\} \subset$ X, then

B is an infinite subset of X.

Now $B = \bigcup_{r=1}^{\infty} \{x_r\}$

= Arbitrary union of open sets

= open set, by Theorem 8.

$\therefore$ B is an open set, ...(2)

From (1) and (2), it follows that every subset (finite or infinite) is an open set in X.

Corollary : In a discrete metric space, every subset of X is closed.

Solution: We know that G is open

$\Leftrightarrow$ X − G = G' is closed set.

Let A be a subset of a discrete metric space (X, d), then A is open.

Also A' = X − A is again sub set of X and so A' is open. Consequently A is closed.

Thus every subset of X is open as well as closed.

Example 15:

A finite set in any metric space has no limit point.

Solution:

Let A be a finite subset of a metric space (X, d). We know that "x $\in$ X is a limit point of any set B if every open sphere $S_{r(x)}$ contains an infinite number of points of B other than x."

This condition can not be satisfied here as A is finite set.

Hence A has no limit point.

Example 16(a)

Let C be the set of all real valued bounded continuous functions defined on [0, 1].

If d is defined by

d (f, g0 = sup {|f (x) – g (x) | : x ∈ [0, 1]}

then d is a metric on C, where f, g∈ C.

Solution:

Let f, g, h ∈ C and x ∈ [0, 1] be arbitrary. Then

(1) d (f, g) ≥ 0. For mod of any thing ≥ 0.

(2) d (f, g) = d (g, f). For |f (x) – g (x)| = |g (x) – f (x) |.

(3) d (f, g) = 0 ⇔ f = g

For sup {|f (x) – g (x) | : x ∈ [0, 1]} = 0 ⇔ |f (x) – g (x) | = 0

⇔ f (x) – g (x) ∀ x

⇔ f = g

(4) To prove d (f, h) ≤ d (f, g) + d (g, h)

Since |f – h| = |f – g + g – h|

≤ |f – g| + |g – h|

This ⇒ sup |f (x) – h (x)| ≤ sup |f (x) – g (x)| + sup |g (x) – h(x)|

⇒ d (f, h) ≤ d (f, g) + d (g, h)

From what has been it follows that d is a metric on C.

Example 16(b)

Let X be a non-empty set and let d be a real valued function of ordered pairs of elements of X which satisfies the following conditions:

(i) d (x, y) = 0 ⇔ x = y

(ii) d (x, y) ≤ d (x, z) + d (y, z)

Then prove that d is a metric on X.

Solution:

(1) To prove d(x, y) ≥ 0.

By (ii), d (x, x) ≤ d (x, z) + d (x, z), on putting x = y.

or, 0 ≤ 2d (x, z) by (i)

or, d (x, z) ≥ 0

or, d (x, y) ≥ 0

(2) To prove d (x, y) = d (y, x).

putting z = x in (ii),

$$d(x, y) \le 0 + d(y, x)$$

or, $\quad d(x, y) \le d(y, x)$

Interchanging x and y,

$d(y, x) \le d(x, y)$

Combining the last two, $d(x, y) = d(y, x)$.

(3) $d(x, y) = 0 \Leftrightarrow x = y$ by (i)

(4) To prove $d(x, z) + d(z, y) \ge d(x, y)$

Using (2) in (ii), $d(x, y) \le d(x, z) + d(z, y)$

or, $\quad d(x, z) + d(z, y) \ge d(x, y)$

Remark : This gives an alternative definition of metric.

Example 17:

In any metric space (X, d), show that $A \subset X$ is closed iff $A = \overline{A}$

Solution:

Let (X, d) be metric space and $A \subset X$. Firstly we shall establish the following lemmas :

(i) $\overline{A}$ is closed and $A \subset \overline{A}$.
(ii) $F \supset A$, F is closed $\Rightarrow F \supset C$

By definition, $\overline{A} = \cap\{F \subset X : F \text{ is closed}, F \supset A\}$, ...(1)
(i) Since an arbitrary; intersection of closed sets is closed and hence $\overline{A}$ is closed.

By (1), $\overline{A} \supset A$
(ii) Suppose $F \subset X$ is closed and $F \supset A$.

$\therefore$ F is one of those members whose intersection is $\overline{A}$ and hence $F \supset \overline{A}$. .E.D. lemma (ii)

Now we come to the proof of actual theorem.

Let $A = \overline{A}$. We claim A is closed. By (i). $\overline{A}$ is closed.

$\overline{A}$ is closed, $A = \overline{A} \Rightarrow$ is closed.

Conversely let A be closed. To prove that $A = \overline{A}$.

By the lemma (i), $A \subset \overline{A}$. ...(2)

By the lemma (ii), f is closed, $F \supset A \Rightarrow F \supset \overline{A}$. ...(3)
Since $A \supset A$ for every set A.

In particular (3) gives A is closed, $A \supset A \Rightarrow A \supset \overline{A}$ ...(4)

From (2) and (4), $A = \overline{A}$.

This concludes the problem.

Example 18:

In any metric space (X, d), prove that A is open

$\Leftrightarrow A^\circ = A$.

Solution:

Let A be a subset of a metric space (X, d). By def. of interior,

$$A^\circ = \cup \{S_{r(x)} : S_{r(x)} \subset A\} \quad ...(1)$$

Since every open sphere is an open set and arbitrary union of open sets is open. Consequently,

A° is an open set. ...(2)

By (1), it is clear that $A^\circ \subset A$ ...(3)

and A° is largest open subset of A. ...(4)

(i) Given $A = A^\circ$ ...(5)

Aim. A is an open set

(2) and (5) $\Rightarrow$ A is an open set.

(ii) Given A is an open set ...(6)

Aim, $A = A^\circ$

(4) and (6) $\Rightarrow = A\ A^\circ$

Example 19:

Show that a subset F of a metric space X is closed iff

$\{x \subset X : d(x, F) - 0\} \subset F$

Solution:

Let (X, d) be a metric space.

Let $F \subset X$ be closed.

To prove that

$\{x \in X : d(x, F) = 0\} \subset F$.

We know that $\bar{F} = \{x \in X : d(x, F) = 0\}$.

F is closed $\Rightarrow \bar{F} = F \Rightarrow \bar{F} \subset F$

$\Rightarrow \{x \in X : d(x, F) = 0\} \subset F$. on using (1)

Conversely suppose that F is a subset of a metric space (X, d) with the property that

$\{x \in X : d(x, F) = 0\} \subset F$ ...(2)

To prove that F is closed, it suffices to show that $F = \overline{F}$.

But $\overline{F} = \{x \in X : d(x, F) = 0\}$ for any set $F \subset X$. ...(3)

Now (2) and (3) $\Rightarrow \overline{F} \subset F$ ...(4)

But $F \subset \overline{F} \ \forall F$ ...(5)

Combining (4) and (5), $\overline{F} = F$.

Example 20:

In every metric space, all finite sets are closed

Solution:

Let (X, d) be a metric space.

To prove that every finite subset of X is closed.

For any subset A of X, we know that

$\overline{A} = \{x \in X : d(x, A) = 0\}$

Let $p \in X$ be arbitrary.

Then $\{\overline{p}\} = \{x \in X : d(x, \{p\}) = 0)$

$= \{p\}$

$\therefore$ $d(x, \{p\}) = 0 \Rightarrow x \in \{p\} \Rightarrow x = p$

$\{p\} = \{\overline{p}\}$,

Showing thereby, {p} is closed in X.

Hence every singleton set in X is closed. Moreover a finite union of closed sets is closed. A finite set is a finite union of singleton sets in which every singleton set is closed. Accordingly every finite set in a metric space is closed.

Example 21(a):

If d is a metric on X, prove that the metric ρ defined by

$$\rho(a,b) = \frac{d(a,b)}{1 + d(a,b)}$$

is also a metric on X.

Solution:

Do yourself.

Example 21(b)

Let R be the set of all real numbers and let

$$d(x,y) = \frac{|x-y|}{1+|x-y|} \textit{ for all } x, y, \in R.$$

Prove that d is a metric for R.

Solution:

Let x, y, z $\in$ R be arbitrary.

Let $\rho(x,y) = |x-y|$.

Then (1) $\Rightarrow$ $\rho(x, y) \geq 0$.

For modulus of any quantity is always non-negative.

(2) $\rho(x, y) = 0$ iff $x = y$.

For $\rho(x, y) = 0$

$\Rightarrow |x-y| = 0 \Rightarrow x = y.$

(3) $\rho(x, y) = \rho(y, x)$.

For $|x-y| = |y-x|$

(4) $\rho(x, y) + \rho(y, z) \geq \rho(x, z)$.

For $\rho(x, y) + \rho(y, z) = |x-y| + |y-z|$.

$\geq |x-y+y-z| = |x-z| = \rho(x, z)$.

i.e., $\rho(x, y) + \rho(y, z) \geq \rho(x, z)$.

The above arguments prove that ρ is a metric on R.

By what is given $d(x, y) = \dfrac{|x-y|}{1+|x-y|}$.

This is expressible as

$$d(x, y) = \frac{\rho(x,y)}{1+\rho(x,y)}$$

where ρ is a metric on R. We have to prove that d is a metric on R.

Example 21(c)

Let ρ_1 and ρ_2 be metrics on the same set X, show that there exists a metric ρ on X such that

$$\rho(x, y) \geq \rho_i(x, y) \textit{ for } i = 1, 2$$

and $\forall x, y \in X$.

Solution:

Let ρ_1 and ρ_2 be metrics on the same set X, Let x, y, z $\in$ X be arbitrary.

To prove that $\exists$ a metric ρ on X s.t.

$$\rho(x, y) \geq \rho_i(x, y) \text{ for } i = 1, 2$$

Write $\rho(x, y) = \rho_1(x, y) + \rho_2(x, y)$. ...(1)

We claim ρ is metric on X.

(i) $\rho(x, y) \geq 0$

$\because$ $\rho_1(x, y) \geq 0$, $\rho_2(x, y) \geq 0$.

(ii) $\rho(x, y) = 0$ iff $x = y$

For $\rho(x, y) = 0 \Leftrightarrow \rho_1(x, y) + \rho_2(x, y) = 0.$

$\Leftrightarrow \rho_1(x, y) = 0 = \rho_2(x, y).$

Since ρ_1 and ρ_2 both are non-negative.

$\Leftrightarrow x = y$

(iii) $\rho(y, x) = \rho(x, y)$

For $\rho(y, x) = \rho_1(y, x) + \rho_2(y, x) = \rho_1(x, y) + \rho_2(x, y)$

$\rho(x, y)$

$\because$ $\rho_1(x, y) = \rho_1(y, x), \rho_2(x, y) = \rho_2(y, x)$

(iv) $\rho(x, y) + \rho(y, z) \geq \rho(x, z)$

For $\rho(x, y) + \rho(y, z) = \rho_1(x, y) + \rho_2(x, y) + \rho_1(y, z) + \rho_2(y, z)$

$[\rho_1(x, y) + \rho_1(y, z)] + [\rho_2(x, y) + \rho_1(y, z)]$

${}^3\, \rho_1(x, z) + \rho_2(x, z) = \rho(x, z).$

From what has been done it follows that ρ is a metric on X.

Moreover (1) $\Rightarrow \rho(x, y) \geq \rho_i(x, y)$ for i = 1, 2.

Hence ρ is the required metric.

Example 22:

Let l_∞ denote the set of all bounded sequences $\langle x_n \rangle$. If $x = \langle x_n \rangle$, $y = \langle y_n \rangle$ in l_∞, then we define

$$d(x, y) = \sup \{|x_n - y_n| : n \varepsilon N\}.$$

Show that d is a metric on l_∞.

Solution:

(i) $d(x, y) = \sup |x_n - y_n|$

$\leq \sup |x_n| + \sup |y_n| < \infty$

or d (x, y) is finite. For $<x_n>$ and $<y_n>$ both are bounded.

This $\Rightarrow$ d (x, y) is well defined map.

(2) Evidently $d(x, y) \geq 0$.

(3) $d(x, y) = 0$ iff $x = y$

For $x = y \Leftrightarrow x_n = y_n \ \forall\ n \Leftrightarrow |x_n - y_n| = 0 \ \forall\ n$

$\Leftrightarrow \sup |x_n - y_n| = 0 \Leftrightarrow d(x, y) = 0$

(4) $d(x, y) = d(y, x)$.

For $|x_n - y_n| = |y_n - x_n|$

(5) $d(x, y) \leq d(x, z) + d(z, y)$

For $d(x, y) = \sup |x_n - y_n| = \sup |x_n - z_n + z_n - y_n|$

$\leq \sup [|x_n - z_n| + |z_n - y_n|]$

$\leq \sup |x_n - z_n| + \sup |z_n - y_n|$

$= d(x, z) + d(z, y)$

$\therefore (l_\infty, d)$ is a metric space.

Example 23:

Every derived set in a metric space is a closed set.

Solution:

Let (X, d) be a metric space and $A \subset X$. Let x be a limit point of D (A) . If we show that $x \varepsilon D(A)$, the result will be proved.

x is limit point of D (A) $\Rightarrow (S_r(x) - \{x\} \cap D(A) \neq \varnothing$ ...(1)

for every open sphere $S_r(x)$.

Then $\exists\ y_r \neq x \in S_r(x) \cap D(A)$

i.e. $y_r \in D(A),\ y_r \in S_r(x)$

$\therefore$ $d(x, y) < r$.

Take $r - d(x, y) = r_1 > 0$.

Evidently$S_{r1}(y_r) \subset S_r(x)$ ...(2)

$$\left[\begin{array}{l} For\ z \in S_{r1}(y_r) \Rightarrow d(z, y_r) < r_1 = r - d(x, y_r) \\ \qquad\qquad\qquad \Rightarrow d(x, y_r) + d(y_r, z) < r \\ \qquad\qquad\qquad \Rightarrow d(x, z) < r \Rightarrow z \in S_r(x) \end{array}\right]$$

$y_r \in D(A) \Rightarrow S_{r1}(y_r) \cap A \neq \varnothing \ \forall\ S_{r1}(y_r)$

$\Rightarrow$ $S_r(x) \cap A \neq \varnothing$, by (2)

$\Rightarrow$ $\in D(A)$.

Example 24:

Prove that the mapping defined by $d : R \times R \to R$ such that

$$d(x,y) = \begin{cases} o \text{ if } x = y \\ 1 \text{ if } x \neq y \end{cases}$$

is a metric on R.

Solution:

Let x, y, z $\in$ R be arbitrary and let

$$d(x,y) = \begin{cases} o \text{ if } x = y \\ 1 \text{ if } x \neq y \end{cases}$$

Now (i) d (x, y) $\geq$ 0

For d (x, y) = 0 or 1.

(ii) d (x, y) = 0 or iff x = y

(iii) d (x, y) = d (y, x)

For if d (x, y) = 0. Then d (y, x) = 0 so that d (x, y) = d (y, x); and if d (x, y) = 1, then d (y, x) = 1 so that d (x, y) = d (y, x)

$\therefore$ In either case d (y, x) = d (x, y)

(iv) To prove d (x, y) + d (y, z) $\geq$ d (x, z).

If x = z, then d (x, z) = 0, so that

d (x, y) + d (y, z) $\geq$ 0 = d (x, z).

If x $\neq$ z, then d (x, z) = 1 and if y is any real number, then y is at least different from one of x and z ad hence at least one of d (y, x) and d(y, z) be equal to one. Consequently,

d (x, z) $\leq$ d (x, y) + d (y, z)

The above facts prove that d is a metric on R.

Example 25(a)

Prove that the function $d : R^2 \times R^2 \to R$ s.t.

$$d(x, y) = [(x_1 - y_y)^2 + (x_2 - y_y)^2]^{1/2},$$

where $x = (x_{1,} x_y) = (y_2 - y_y) \in R^2$

Is a metric on R.

[This is called unusual metric on R^2]

Solution:

Let x = $(x_{1,} x_2)$, y = $(y_{1,} y_2)$, z = (z_1, z_2) be arbitrary elements of R^2

(i) d (x, y) $\geq$ 0.

For $(x_1 - y_1)^2 \geq 0, (x_2 - y_2)^2 \geq 0$ and hence

$[(x_1 - y_1)^2 + (x_2 - y_2)^2]^{1/2} \geq 0$

(ii) $d(x, y) = 0$ iff $x = y$

$$d(x, y) = 0$$
$$\Leftrightarrow (x_1 - y_1)^2 + (x_2 - y_2)^2 = 0$$
$$\Leftrightarrow x_1 - y_1 = 0, x_2 - y_2 = 0$$
$$\Leftrightarrow x_1 = y_1, x_2 = y_2$$
$$\Leftrightarrow (x_1, x_2) = (y_1, y_2)$$
$$\Leftrightarrow x = y$$

(iii) $d(x, y) = d(y, x)$.

For $[(x_1 - y)^2 + (x_2 - y_2)^2]^{1/2} = [(y_1 - x_1)^2 + (y_2 - x_2)^2]^{1/2}$

(iv) $d(x, y) + d(y, z) \geq d(x, z)$

For $d(x, y) + d(y, z) \geq [(x_1 - y_1)^2 + (x_2 - y_2)^2]^{1/2}$

$+ [(y_1 - z_1)^2 + (y_2 - z_2)^2]^{1/2} \geq [(x_1 - y_1 + y_1 - z_1)^2$
$+ (x_2 - y_2 + y_2 - z_2)^2]^{1/2}$ be Minkowski inequality.

$= [(x_1 - z_1)^2 + (x_2 - z_2)^2]^{1/2} = d(x, z)$

$\therefore d(x, y) + d(y, z) \geq d(x, z)$.

Example 25(b)

Prove that the mapping $d : R^2 \times R^2 \to R$

defined by $d(x, y) = |x_1 - y_1| + |x_2 - y_2|$,

where $x = (x_1, x_2), y = (y_1, y_2) \in R^2$.

is a metric on R^2.

Solution:

Let $x = (x_1, x_2), y = (y_1, y_2), z = (z_1, z_2)$ be arbitrary elements of R^2. Also let

$$d(x, y) = |x_1 - y_1| + |x_2 - y_2|.$$

(i) $d(x, y) \geq 0$.

Since modulus of any quantity is always non-negative.

(ii) $d(x, y) = 0$ iff $x = y$.

For $d(x, y) = 0 \Leftrightarrow |x_1 - y_1| + |x_2 - y_2| = 0$

$\Leftrightarrow |x_1 - y_1| = 0 = |x_2 - y_2|$

$\Leftrightarrow x_1 = y_1, x_2 = y_2$

$\Leftrightarrow (x_1, x_2) = (y_1, y_2)$

$\Leftrightarrow x = y.$

(iii) d (x, y) = d (y, x).

For $|x_1 - y_1| + |x_2 - y_2| = |y_1 - x_1| + |y_2 - x_2|.$

(iv) d (x, y) + d (y, z) ≥ d (x, z).

For $d(x, z) = |x_1 - z_1| + |x_2 - z_2|$

$= |x_1 - y_1 + y_1 - z_1| + |x_2 - y_2 + y_2 - z_2|$

$\leq |x_1 - y_1| + |y_1 - z_1| + |x_2 - y_2| + |y_2 - z_2|$

$= d(x, y) + d(y, z).$

$\therefore d(x, z) \leq d(x, y) + d(y, z).$

Therefore d is a metric on R^2.

Example 25(c)

Show that the map $d : R^2 \times R^2 \rightarrow R$

define by $d(x, y) = max. \{|x_1 - y_1|, |x_2 - y_2|\}.$

where $x = (x_1, x_2), y = (y_1, y_2) \in R_2$

is a metric on R^2.

Solution:

Let $x = (x_1, x_2), y = (y_1, y_2), z = (z_1, z_2)$ be arbitrary elements or R^2. Also let

$d(x, y) = \max \{|x_1 - y_1|, |x_2 - y_2|\}$

(i) d (x, y) ≥ 0.

Since modulus of any quantity ;is non-negative.

(ii) d (x, y) = 0 iff x = y.

For $d(x, y) = 0 \Leftrightarrow \max.\{|x_1 - y_1|, |x_2 - y_2|\} = 0$

$\Leftrightarrow |x_1 - y_1| = 0 = |x_2 - y_2|$

$\Leftrightarrow x_1 = y_1, |x_2 = y_2$

$\Leftrightarrow (x_1, x_2) = y_2, y_2)$

$\Leftrightarrow x = y$

(iii) d (x, y) = d (y, x).

For $|a - b| = |b - a|$ and so

$\max.\{|x_1 - y_1|, |x_2 - y_2|\} = \max \{|y_1 - x_1|, |y_2 - x_2|\}$

(iv) d (x, y) + d (y, z) ≥ d (x, z)

For $d(x, y) = \max \{|x_1 - z_1|, |x_2 - z_2|\}$

$$= \max. \{|x_1 - y_1 + y_1 - z_1|, |x_2 - y_2 + y_2 - z_2|\}$$
$$\leq \max. \{|x_1 - y_1| + |y_1 - z_1|, |x_2 - y_2| + |y_2 - z_2|\}$$
$$\leq \max. \{|x_1 - y_1|, |x_2 - y_2|\} + \max. \{|y_1 - z_1|, |y_2 - z_2|\}.$$
$$= d(x, y) + d(y, z).$$

The above facts lead to the conclusion that d is a metric on R^2.

Example 26:

Let X be a non-empty set and define d by

$$d(x,y) = \begin{cases} 0 \text{ if } x = y \\ 1 \text{ if } x \neq y \end{cases}$$

Prove that d is metric on X.

Solution:

Let $x, y, z \in X$ be arbitrary.

Then

(1) $d(x, y) \geq 0$

This follows from the definition of d.

(2) $d(x, y) = 0$ iff $x = y$, by def. of d.

(3) $d(x, y) = d(y, x)$ as value of d is free from order or x and y.

(4) $d(x, y) + d(y, z) \geq d(x, z)$

For if $x \neq y \neq z$, then

$$d(x, y) + d(y, z) = 1 + 1 = 2 > 1 = d(x, z)$$

or, $$d(x, y) + d(y, z) > d(x, z) \qquad \text{...(a)}$$

Again if $x = y \neq z$, then

$$d(x, y) + d(y, z) = 0 + 1 = d = (x, z)$$

or, d (x, y) + d (y [illegible] d (x, z) ...(b)

Combining (a) and (b), we get result (4).

$\therefore$ d is a metric on X.

Example 27:

Given two different metrics for the set R of all real numbers.

Solution:

Consider the map

$$d : R \times R \rightarrow R \text{ s. t.}$$

$$d(x,y) = \begin{cases} 0 \, if \, x = y \\ 1 \, if \, x \neq y \end{cases}$$

where $(x, y) \in R^2$.

Then d is metric on R.

Next consider the map

$$d' : R \times R \rightarrow R$$

$$d'(x, y) \, |x - y|.$$

where $x, y \in R$

Then d' is metric on R.

Thus d and d' are two different metrics on R.

Example 28(a):

If A is any non-empty subset of a metric space (X, d), prove that

$$|d(x, A) - d(y, A)| \leq d(x, y)$$

for any points $x, y \in X$.

Solution:

By def. of metric,

$$d(x, a) \leq d(x, y) + d(y, a)$$

and $$d(y, a) \leq d(y, x) + d(x, a)$$

Taking infimum over $a \in A$, we obtain

$$d(x, A) \leq d(x, y) + d(y, A)$$

and $$d(y, A) \leq d(y, x) + d(x, A)$$

i.e., $$d(x, A) - d(y, A) \leq d(x, y) \qquad ...(1)$$

$$d(y, A) - d(x, A) \leq d(x, y)$$

The last gives

$$d(x, A) - d(y, A) \geq - d(x, y)$$

From (1) and (2),

$$- d(x, y) \leq d(x, A) - d(y, A) \leq d(x, y)$$

This $\Rightarrow |d(x, A) - d(y, A)| \leq d(x, y)$.

Example 28(b):

Let A, B be non-empty subsets of a metric space (X, d).

Then show that

$$d(A \cap B) \leq d(A) + d(B) + d(A, B)$$

Solution:

Let $x, y \in A \cap B$ be arbitrary. Then there arise four cases.

(i) $x, y \in A$

This $\Rightarrow d(x, y) \leq d(A)$

For $d(A) = \sup \{d(x, y) : x, y \in A\}$

(ii) $x, y \in B$.

This $\Rightarrow d(x, y) \leq d(B)$

(iii) $x \in A$ and $y \in B$

(iv) $x \in B$ and $y \in A$

In case (iii), we have

$$d(x, y) \leq d(x, a) + d(a, y), \quad a \in A$$
$$\leq d(x, a) + d(a, b) + d(b, y), \; b \in B$$

$$\therefore \quad d(x, y) \leq d(x, a) + d(a, b) + d(b, y).$$

Similarly in case (iv), we have

$$d(x, y) \leq d(x, b) + d(b, a) + d(a, y)$$

This $\Rightarrow$

$$d(A \cup B) \leq d(B) + d(B, A) + d(A)$$

or $$d(A \cup B) \leq d(A) + d(B) + d(A, B).$$

Example 29.

Describe open spheres for the discrete metric on a space X.

Solution:

Let (X,d) be a discrete metric space so that d is given by

$$d(x,y) = \begin{cases} 0 \text{ if } x = y \\ 1 \text{ if } x \neq y \end{cases}$$

Let $x_0 \in X$ and $r > 0$ be any real number.

$$S_r(x_0) = \{x \in X : (x, x_0) < r\}.$$

Case (i) : When $r > 1$.

By definition metric, any $x \in X, x \neq x_0$

$$\Rightarrow d(x, x_0) = 1 < r$$

$\Rightarrow d(x, x_0) = < r$

$\Rightarrow x \in S_r(x_0)$

Also $x_0 \in S_r(x_0)$

$\because \quad d(x_0, x_0) = 0 < r.$

Combining these two results, we find that

any $\quad x \in X \Rightarrow x \in S_r(x_0)$

This $\quad \Rightarrow X = S_r(x)$

Case (ii) : when $r \leq 1$.

any $\quad x \in X, x \neq x_0 \Rightarrow d(x, x_0) = 1 < r$

$\Rightarrow d(x, x_0) < r.$

$\Rightarrow x \notin S_r(x_0)$

$x_0 \in S_r(x_0)$

Thus, any $x \in X, x \neq x_0 \Rightarrow x \notin S_r(x_0)$

and $\quad x_0 \in S_r(x_0)$.

From this it follows that $S_r(x_0) = \{x_0\}$.

Finally, $S_r(x_o) = \begin{cases} X & \text{if } r . 1. \\ \{x_0 & \text{if } r \leq 1.\} \end{cases}$.

Example 30:

Describe open spheres for the usual metric on R, *i.e.*, open spheres on the real line are precisely the bounded open intervals.

Solution:

Let d be the usual metric on R; then

$$d(x, y) = |x - y| \ \forall \ x, y \in R$$

Let $x_0 \in R$ and $r > 0$ be any real number.

$$\begin{aligned} S_r(x_0) &= \{x \in R : d(x, x_0) < r\} \\ &= \{x \in R : |x - x_0| < r\} \\ &= \{x \in R : -r < x - x_0 < r\} \\ &= \{x \in R : x_0 - r < x < x_0 + r\} \\ &= (x_0 - r, x_0 + r). \end{aligned}$$

This proves that the open spheres on the real line R are open intervals.

Example 31:

In any metric space, show that

$$x - \overline{A} = (X - A)^o$$

or, $$(\overline{A})' = (A')^o$$

Solution:

$(\overline{A})' = X - \overline{A}$

= X – Intersection of all closed super sets of A.

$= X - \bigcap_i F_i$ where F_i is closed and $F_i \supset A$.

$= \bigcup_i (X, F_i)$ where $X - F_i$ is open

and $X - F_i \subset X - A$

= Union of open subsets of $X - A = A'$

$= (A')^o$

Example 32(a):

Given an example of Pseudo metric which is not a metric.

Solution:

Suppose C is the family of all Riemann integral functions f; $[0, 1] \to R$.

Let f, g, h, $\in$ C be arbitrary. Define a map

$d : C \times C \to R$ s.t. $d(f, g) = \int_0^1 |f - g|(x)\, dx$.

To prove d is a Pseudo metric but; not metric on C.

d (f, g) $\geq$ For $|f - g|$ (x) ≥ 0 and so.

$\int_0^1 |f - g|(x)\, dx \geq 0$

(1) d (f, g) = d (g, f).

This follows form the fact that $|f - g| = |g - f|$

(2) d (f, g) + d (g, f).

$= \int_0^1 |f - g|(x)\, dx + \int_0^1 |g - h|(x)\, dx$

$= \int_0^1 [|f - g| + |g - h|](x)\, dx$

$\geq \int_0^1 |f - g + g - h|(x)\, dx$

$= \int_0^1 |f - h|(x)\, dx = d(f, h)$

(3) d (F, g) = 0 if f = g

If f = g, then

$$d(f, g) = \int_0^1 |f - g|(x)\, dx = \int_0^1 |g - g|(x)\, dx = \int_0^1 0\, dx = 0$$

(4) d (f, g) = 0 does not necessarily imply f = g.

That is to say, d (f, g) = 0

$\Rightarrow$ f $\neq$ g, in general.

Solution:

(IInd Part) Suppose $f(x) = \begin{cases} 4\, where\, x = 0 \\ 0\, when\, 0 < x \leq 1 \end{cases}$

and $g(x) = \begin{cases} 9\, when\, x = 0 \\ 0\, when\, 0 < x \leq 1 \end{cases}$

The f(x) $\neq$ g (x), so that f $\neq$ g.

Also $|f - g|(x) = \begin{cases} 9 - 4 = \; when\, x = 0 \\ 0 \quad when\, 0 < x \leq 1 \end{cases}$

Evidently f – g is Riemann integrable,

x = o is the only point of discontinuity.

$$d(f, g) = \int_0^1 |f - g|(x)\, dx = \int_0^1 0\, dx = 0$$

Thus, d (f, g) = 0 but f $\neq$ g.

This proves that d is a Pseudo metric but metric on C.

Similar Ex : *Let C be the class of bounded continuous functions f : [0, 1 $\rightarrow$ R. Define a map d : C $\times$ C $\rightarrow$ R such that*

$$d(f, g) = \int_0^1 |f - g|(x)\, dx \;\forall f, g \in C = \|f - g\|$$

Then d is a metric on C.

Example 32(b):

Let d be a Pseudo metric on X. Define a relation on X by setting X~y iff d (x, y) = 0. Show that this relation is an equivalence relation whose corresponding equivalence class can be made into a metric space in a natural way.

Solution: Let d be a Pseudo metric on X so that

$\forall$ x, y, z, $\in$ X we have

(1) d (x, y) ³ 0

(2) d (x, y) = d (y, x)

(3) d (x, x) = 0

(4) d (x, y) + d (y, z) ³ d (x, z)

Define a relation ~ on X s.t.

$$x \sim y \text{ iff } d(x, y) = 0$$

To prove that ~ is an equivalence relation

(i) ~ is reflexive

For $d(x, x) = 0$

$\Rightarrow$ $x \sim x$, by (3)

(ii) ~ is symmetric.

For $x \sim y \Rightarrow d(x, y) = 0$

$\Rightarrow$ $d(y, x) = 0$, by (2)

$\Rightarrow y \sim x$

(iii) ~ is transitive.

For $x \sim y, y \sim z$

$\Rightarrow$ $d(x, y) = 0 = d(y, z)$

$\Rightarrow$ $d(x, z) \leq (x, y) + d(y, z) = 0 + 0 = 0$

$\Rightarrow$ $d(x, z) \leq 0$

But $d(x, z) \geq 0$, by (1)

This $\Rightarrow d(x, z) = 0 \Rightarrow x \sim z$

Thus we have proved that ~ is an equivalence relation of X.

Hence the relation ~ decomposes X into family of disjoint subsets of X, known as equivalence classes. Let [X] denote the equivalence class determined by $x \in X$ and let C be the set of all equivalence classes,

i.e., $C = \{[x] : x \in X\}$

Define a map $f : C \times C \to R$ s.t.

$$f([x_1], [x_2]) = d(x_1, x_2)$$

where $[x_1], [x_2] \in C$

Then f is a metric for C on account of the fact that d is a pseudometric on X. The metric f defined as above is called natural metric for C.

Example 33

Let X be a non-empty set and let d be a real volued function of ordered pairs of elements of X which satisfies the following conditions:

(i) $d(x, y) = 0 \Leftrightarrow x = y$

(ii) $d(x, y) \leq d(x, z) + d(y, z)$

Then prove that d is a metric on X.

Solution:

(1) To prove $d(x, y) \geq 0$.

By (ii), $d(x, x) \leq d(x, z) + d(x, z)$, on putting $x = y$.

or, $0 \leq 2d(x, z)$ by (i)

or, $d(x, z) \geq 0$

or, $d(x, y) \geq 0$

(2) To prove $d(x, y) = d(y, x)$.

putting $z = x$ in (ii),

$d(x, y) \leq 0 + d(y, x)$

or, $d(x, y) \leq d(y, x)$

Interchanging x and y,

$d(y, x) \leq d(x, y)$

Combining the last two, $d(x, y) = d(y, x)$.

(3) $d(x, y) = 0 \Leftrightarrow x = y$ by (i)

(4) To prove $d(x, z) + d(z, y) \geq d(x, y)$

Using (2) in (ii), $d(x, y) \leq d(x, z) + d(z, y)$

or, $d(x, z) + d(z, y) \geq d(x, y)$

Remark : This gives an alternative definition of metric.

Example 34:

In any metric space (X, d), show that $A \subset X$ is closed iff $A = \overline{A}$

Solution:

Let (X, d) be metric space and $A \subset X$. Firstly we shall establish the following lemmas :

(i) $\overline{A}$ is closed and $A \subset \overline{A}$.

(ii) $F \supset A$, F is closed $\Rightarrow F \supset C$

By definition, $\overline{A} = \cap\{F \subset X : F \text{ is closed}, F \supset A\}$, ...(1)

(i) Since an arbitrary; intersection of closed sets is closed and hence $\overline{A}$ is closed.

By (1), $\overline{A} \supset A$

(ii) Suppose $F \subset X$ is closed and $F \supset A$.

$\therefore$ F is one of those members whose intersection is $\overline{A}$ and hence $F \supset \overline{A}$. .E.D. lemma (ii)

Now we come to the proof of actual theorem.

Let A = $\overline{A}$. We claim A is closed. By (i). $\overline{A}$ is closed.

$\overline{A}$ is closed, A = $\overline{A}$ $\Rightarrow$ is closed.

Conversely let A be closed. To prove that A = $\overline{A}$.

By the lemma (i), $A \subset \overline{A}$. ...(2)

By the lemma (ii), f is closed, $F \supset A \Rightarrow F \supset \overline{A}$. ...(3)

Since $A \supset A$ for every set A.

In particular (3) gives A is closed, $A \supset A \Rightarrow A \supset \overline{A}$ ...(4)

From (2) and (4), A= $\overline{A}$.

This concludes the problem.

EXERCISES

1. Let a be a subset of a metric space (X, d). Then prove that "a is dense $\Leftrightarrow$ the only closed super set of A is X $\Leftrightarrow$ the only open set disjoint from A is ϕ $\Leftrightarrow$ A intersects every sphere.
2. Prove that Cauchy sequence is convergent $\Leftrightarrow$ it has a convergent subsequence.
3. Prove that the set of all real continuous functions defined on [0, 1] is a metric space if we define

$$d(f, g) = \sup_{0 < t \leq 1} |f(t) - g(t)|$$

as the distance between the functions f and g.

How is the metric convergence related to uniform convergence.

4. Let A and B be two subsets of a metric space X.

Prove that :

(i) Int A $\cup$ Int (B) $\subseteq$ Int (A $\cup$ B)

(ii) Int A $\cap$ Int (B) = Int (A $\cup$ B).

Give an example on the real line to show that inclusion in (i) above can be strict.

5. Define convergence of a sequence in a metric space (X, d). Prove that, for $A \subset X$, $x_0 \in X$, the point x_0 is a limit point of the set A iff there is a sequence of distinct points of A converging to x_0.
6. State and prove Minkowski's inequality.
7. (a) Define a space and give one example.

Let ρ be a metric on X. Show that ρ defined by

$$\sigma(x, y) = \frac{\rho(x,y)}{1+\rho(x,y)} \text{ is a metric on X.}$$

(b) Define an open ball; and an open set in a metric space. Prove that a finite intersection of open sets is open. Give an example to show that an infinite intersection of open sets need not be open.

8. Define pointwise convergence and uniform convergence of a sequence of functions on a metric space X into a metric space y.

9. (a) Define a metric space. Prove that the set of all bounded continuous real functions defined on the closed unit interval has a metric defined by

$$d(f, g) = \sup \{|f(x) - g(x) \in [0, 1]\}$$

(b) When is a subset of a metric space called open ?

Prove that in a metric space :

(i) any union of open sets is open.

(ii) any finite intersection of open sets is open.

10. (a) Prove that in any metric space, each closed sphere is a closed set.

(b) State and prove Cantor's intersection theorem.

11. (a) Define a metric on a set X. Prove that on the complex plan function d defined by

$$d(z_1, z_2) = |z_1 - z_2|$$

is a metric

(b) Define an open set in a metric space (X, d) and prove that

(i) union of open sets in (X, d) is open.

(c) Let $S_r(x)$ be an open sphere in a metric space X whose centre is x and radius r. If A is a subset of X with diameter less than r which intersects then prove that $A \subset S_{2r}(x)$.

12. If d is a metric on X, prove that ρ defined by

$$\rho(a, b) = \frac{d(a, b)}{1 + d(a, b)}$$

is also a metric on X.

13. Let X be a non-empty set, and let d be a real function of ordered pairs of elements of X which satisfies the following two conditions: $d(x, y) = 0 \Leftrightarrow x = y$, and $d(x, y) \le d(x, z) + d(y, z)$. Show that d is a metric on X.

14. Let X be a non-empty set, and let d be a real function of ordered pairs of elements of X which satisfies the following three conditions: $d(x, y) \geq 0$, and $x = y \Rightarrow d(x, y) = 0$; $d(x, y) = d(y, x)$; and $d(x, y) \leq d(x, z) + d(z, y)$. A function d with these properties is called a *pseudo-metric* on X. A metric is obviously a pseudo-metric. Give an example of a pseudo-metric which is not a metric. Let d be a pseudo-metric on X, define a relation ~ in X by means of

$$x \sim y \Leftrightarrow d(x, y) = 0,$$

and show that this is an equivalence relation whose corresponding class of equivalence sets can be made into a metric space in a natural way.

15. Let I be a subset of the real line. Show that I is an interval $\Leftrightarrow$ it is non-empty and contains each point between any two of its points (in the sense that if x and z are in I and $x \leq y \leq z$, then y is in I). If $\{I_i\}$ is a non-empty class of intervals on the real line such that $\cap_i I_i$ is non-empty, show that $\cup_i I_i$ is an interval.

16. Let X be a metric space with metric d. If x is a point of X and A a subset of X, show the following: if A is non-empty, d(x, A) is a non-negative real number; and $d(x, A) = +\infty \Leftrightarrow$ A is empty.

17. Let X be a metric space with metric d and A a subset of X. Show the following: if A is non-empty, d(A) is a non-negative extended real number; $d(A) = -\infty \Leftrightarrow$ A is empty; and if A is bounded, it is non-empty.

18. Let $(X_1 ; \Im_1)$, $(X_2, \Im_2)$ and $(X_3, \Im_3)$ be topological spaces. Show that a mapping $f : X_3 \to X_1 \times X_2$ is continuous iff π_{x_1} of $: X_3 \to X_1$ and π_{x_2} of $: X_3 \to X_2$ are continuous.

19. (a) Let $\{X_i : i \in I\}$ be an indexed family of topological spaces and let X be its topological product. Prove that the projection function $\pi_i : X \to X_i$ is open for all $i \in I$.

    (b) Prove that a sequence $\{x_n\}$ in X above converges to a point x in X if an only if for every $i \in I$, $\{\pi_i (x_n)\}$ converges to $\pi_i (x)$ in X_i.

20. Prove that the topological product is, upto a homeomorphism

    (i) a well defined, (ii) commutative, and, (iii) associative operation.

21. Let $(X, \Im)$ and (Y, v) be topological spaces and L and M be sub-bases for $\Im$ and v respectively. Show that the collection A of all subset of the from $\Lambda \times Y$ and $X \times M$ where $\Lambda \in \Lambda$ and $M \in M$ is a subbase for the product topology on $X \subseteq Y$.

22. Let X be a. metric space and $S_r(x)$ the open sphere in X with center x and radius r. Let A be a subset of X with diameter less than r which intersects $S_r(x)$. Prove that $A \subseteq S_{2r}(x)$.

33. Let X be a metric space. Show that every subset of X is open ⇔ each subset of X which consists of a single point is open.

34. Let X be a metric space with metric d, and let d_1 be the metric defined. Show that the two metric spaces (X, d) and (X, d_1) have precisely the same open sets. (*Hint:* show that they have the same open spheres with one exception. What is this exception?)

35. If $X = X_1 \times X_2 \times ... \times X_n$ is the product, and if d and 3 are the metrics on X defined in that problem, show that the two metric spaces (X, d) and (X, d) have precisely the same open sets. Observe that in this case the spaces do not have the same open spheres.

36. Let Y be a subspace of a metric space X, and let A be a subset of the metric space Y. Show that A is open as a subset of Y ⇔ it is the intersection with Y of a set which is open in X.

37. Describe the interior of each of the following subsets of the real line: the set of all integers; the set of all rationale; the set of all irrationals; (0, 1); [0, 1]; [0, 1) ∪ {1, 2}. Do the same for each of the following subsets of the complex plane: $\{z:|z| < 1\}$; $|z:|z| \leq 1\}$; $\{z:I(z) = 0\}$; $[z:R(z)$ is rational$\}$.

38. Let A and B be two subsets of a metric space X, and prove the following:

    (a) Int (A) ∪ Int (B) ⊆ Int (A ∪ B);

    (b) Int (A) ∩ Int (B) = Int (A ∩ B).

    Give an example of two subsets A and B of the real line such that Int (A) ∪ Int (B) ≠ Int (A ∪ B).

39. Let X_1, X_2 be two topological spaces and let X be the product space. Let π be the projective defined by $\pi(x_1, x_2) = x$. Which of the following is a wrong statement:

    (a) p is onto, (b) p is continuous,

    (b) p is open, (d) p is closed.

40. Let L be a lattice. If x, y, and z are elements of L, prove the following:

$$x \wedge x = x, x \vee x = x,$$

$$x \wedge y = y \wedge x, x \wedge y = y \vee x, x \vee (y \wedge z) = (x \wedge y) \wedge z,$$

$$x \vee (y \vee z) = (x \vee y) \vee z, (x \wedge y) \vee x = x,$$

$(x \vee y) \wedge x = x$.

If X has the weak topology induced by a collection

$\{f_\lambda\ \lambda \in \Lambda\}$

of functions $f_\lambda : X \to X_\lambda$, then $f : Y \to X$ is continuous iff f_λ of is continuous for each $\lambda \in \Lambda$.

41. Let f : X → Y continuous and let Y be. Hausdorff. Show that f is a closed subset of X × Y.

42. Let X be any topological space and let Y be a compact space. Show that the projection mapping π_x is a closed mapping of X × Y into Y.

43. Show that each projection map π_{x_i} of the product space is a continuous map onto X_i.

44. Show that the product space of two completely regular spaces is a completely regular space.

45. Let $(x_i\ \Im_i)$, i = 1, 2, ..., n be n topological spaces. Show that the collection $B = \{G_1 \times G_2 \times ... \times G_n : G_i \in \Im_1, i = 1, 2 ..., n\}$ is a base for a topology for $X_1 \times X_2 \times ... \times X_n$,

    [The topology W generated by B in exercise 10 above is called the product topology for the set $X_1 \times X_2 \times ... X_n = X$, say and (X, W) is product space of the given spaces. The mapping $\pi_{x_i} : X \to X_i$: $\pi_{x_i}(x) = x_i \ \forall\ x \in X$ where $X = X_1 \times X_2 ... \times X_n$ is called the i-th projection mapping or the i-th projection].

46. Show that the product space of two separable spaces is a separable space.

47. Suppose X and Y are compact spaces and v is an open cover of X × Y. Prove that for every y ∈ Y there exists an open set V_y in Y containing Y such that, X × V_y is covered by finitely many members of U. Hence show that X × Y is compact.

48. Prove that X is Hausdroff if and only if the diagonal in X is a closed subset of X × X.

49. Prove that the product of two locally connected spaces is locally connected.

50. Show that the product topology W is the smallest topology for the set $X_1 \times X_2 \times ... \times X_n$ for which each projection π_1 is continuous.

51. Let f : X Y be onto, where Y is a topological space. If X is given the weak topology induce by f, show that f becomes a quotient map.

52. Show that each coordinate space X of a product space X × $\{X_\lambda : \lambda \in L\}$ has a quotient topology induced by the projection π_λ.

53. Let X and Y be topological spaces and if X/R is the quotient space of X, show that any continuous function $f : X \xrightarrow{onto} Y$ induces a continuous functions g from X/R onto Y, provided $(x, y) \in \to f(x) = f(y)$, *i.e.*, $R \subset \{(x, y) : f(x) = f(y)\}$.

54. A function g of a quotient space X/R into a space Y is continuous if and only if the compositive function g o π, where π denotes the quotient map is continuous on the space X.

55. Let f : g be maps from a set X onto two topological spaces Y_f and Y_g, and let X have the weak topology induced by {f, g}. Show that g need not be the quotient map.

56. Show that a space X is. Hausdorff if and only if its diagonal $L = \{x, x) : x \in X\}$ is a closed set in the product space X × X.

57. If F_λ is a proper closed set in X_λ, is it true that $F = \pi F_\lambda$ is closed in $X = \pi X_\lambda$ where λ is an infinite set ?

$\lambda \in \Im$

58. Let $X = \times \{X_\lambda : \lambda \in \Lambda\}$ be a product space. Prove that $\times \{A_\lambda : \lambda \in \Lambda\}$ is dense in X iff A_λ is dense in X_λ for every $\lambda \in \Lambda$.

59. Let $X = \times \{X_\lambda : \in \Lambda\}$ be a product space. If $A_\lambda \subset X_\lambda$ for each $\lambda \in \Lambda$, show that $\times \{A_\lambda : \lambda \in \Lambda\}$ (where each A_λ has the subspace topology has the same topology as it is considered as a subspace of X.

60. Assuming that the product of two connected spaces is connected, prove that the product of any non-empty family of connected spaces is cornected.

61. Show that the product space $X = \times \{X_\lambda : \lambda \in \Lambda\}$ is T_0-space iff each coordinate space X_λ is T_0.

62. Prove that a non-empty, product space of a countable number of separable spaces is separable.

63. Show that a non-empty subset A of a Banach space is bounded ⇔ there exists a real number K such that $\|f\| \le K$ for every x in A.

64. Let X and Y be metric spaces with metrics d_1 and d_2, and $\{f_n\}$ $\{f_n\}$ be a sequence of mappings of X into Y which converges pointwise to a mapping f of X into Y, in the sense that $f_n(x) \to f(x)$ for each x in X. Define what ought to be meant by the statement that f, converges uniformly to f_n and prove that under this assumption f is continuous if each f_n is continuous.

65. Let X and Y be metric spaces and f a mapping of X into Y. If f is a constant mapping, show that f is continuous. Use this to show that a continuous mapping need not have the property that the image of every open set is open.

66. Let $X \neq f$ and let d be a real function of ordered pairs of X which satisfies the following tow conditions :

$$d(x, y) = 0 \Leftrightarrow x = y$$

and $\quad d(x, y) \leq d(x, z) + D(z, y)$.

Show that d is a metric on X.

67. Let d be a pseudo metric on X. Define a relation ~ on X by means of $x \sim y \Leftrightarrow d(x, y) = 0$.

Show that it is an equivalence relation whose corresponding class of equivalence sets can be made into a metric space in a natural way.

68. Suppose (X, d) is a metric space. Show that any two distinct points of X can be separated by open spheres in the following sense : if x, y are distinct points in X, then there is a pair of open spheres each of which is centred on one of the points.

69. If {x} be a subset of a metric space (X, d) show that its complement {x}' is open. More generally, show that A' is open if A is a finite subset of X.

70. Find the boundary of the following subsets of R :

(i) Integers, (ii) Rationals,

(iii) [0, 1], (iv) (0, 1),

(v) $\{z \in C : |z| \leq 1\}$, (vi) $\{z \in C : |z| \leq 1\}$,

(vii) $\{z \in C : I(z) \leq 1\}$.

71. Let X be a metric space. Show that every subset of X is open $\Leftrightarrow$ each subset of X which consists of single point is open.

72. In a metric space prove that

(i) $(\overline{A}) = \text{Int}(A')$

(ii) $\overline{A} = \{x : d(x, A) = 0\}$.

73. Let $S_r(x)$ be an open sphere in a metric space (X, d) and $A \subset X$ with diameter < r which intersects $S_r(x)$. Prove that

$$A \subset S_{2r}(x).$$

74. Let X be a metric space with metric d, and let x_0 be a fixed point in X. Show that the real function f_{x_0} defined on X by $f_{x_0}(x) = d(x, x_0)$ is continuous. Is it uniformly continuous?

75. Let X and Y be metric spaces and A a non-empty subset of X. If f and g are continuous mappings of X into Y such that f(x) = g(x) for every x in A, show that f(x) = g(x) for every x in $\overline{A}$.

76. Let X and Y be metric spaces and f a mapping of X into Y. Show that f is continuous $\Rightarrow f^{-1}(F)$ is closed in X whenever F is closed in Y $\Leftrightarrow f(\overline{A}) \subseteq \overline{f(A)}$ for every subset A of X.

77. Show that any mapping of the metric space any other metric space is continuous.

78. Consider the real function f defined on the real line R by $f(x) = x^2$. If b is a given positive real number, show that the restriction of f to the closed interval [0, b] is uniformly continuous by starting with an $\epsilon > 0$ and exhibiting a $\delta > 0$ which satisfies the requirement of the definition.

79. Determine which of the following functions are uniformly continuous on the open unit interval (0, 1) : $1/(1-x)$; $1/(2-x)$; sin x; sin (1/x); $x^{1/2}$, x^3. Which are uniformly continuous on the open interval $(0, +\infty)$?

80. In the proof of Theorem D we used the following fact: the image of a Cauchy sequence under a uniformly continuous mapping is again a Cauchy sequence. Give the details of the proof.

81. Let f be a continuous real function defined on R which satisfies the functional equation $f(x + y) = f(x) + f(y)$. Show that this function must have the form f(x) = mx for some real number m.

82. Let X be a metric space, proving the following statements:

    (a) any point and disjoint closed set in X can be separated by open sets, in the sense that if a: is a point and F a closed set which does not contain x, then there exists a disjoint pair of open sets G_1 and G_2 such that $x \in G_1$ and $F \subseteq G_2$;

    (b) any disjoint pair of closed sets in X can be separated by open sets, in the sense that if Fi and Fi are disjoint closed sets, then there exists a disjoint pair of open sets Gi and Gi such that $F_1 \subseteq G_1$ and $F_2 \subseteq G_2$.

83. Let X be a metric space, and let A be a subset of X. If a: is a limit point of A, show that each open sphere centered on x contains an

infinite number of distinct points of A. Use this result to show that a finite subset of X is closed.

84. Show that a subset of a metric space is bounded <=> it is non-empty and is contained in some closed sphere.

85. Give an example of an infinite class of closed sets whose union is not closed. Give an example of a set which :

 (a) is both open and closed;

 (b) is neither open nor closed;

 (c) contains a point which is not a limit point of the set; and

 (d) contains no point which is not a limit point of the set.

86. Let X be a metric space and A a subset of X. Prove the following facts:

 (a) $\overline{A}$ = Int (A');

 (b) $\overline{A}$ = {x:d(x, A) = 0}.

87. Describe the closure of each of the following subsets of the real line: the integers; the rationals; the Cantor set; (0, + ∞); (–1, 0) ∪ (0, 1). Do the same for each of the following subsets of the complex plane: {z:|z| is rational}; {z:l/R(z) is an integer}; {z:\z\<1 and I(z) < 0}.

88. Let X be a metric space, let a; be a point of X, and let r be a positive real number. One is inclined to believe that the closure of $S_r(x)$ must equal $S_r[x]$. Give an example to show that this is not necessarily true.

89. Let X be a metric space, and let G be an open set in X. Prove that G is disjoint from a set A ⇔ G is disjoint from A.

90. Doooribe the boundary of each of the following subsets of the real line: the integers; the rationals; [0, 1]; (0, 1). Do the same for each of the following subsets of the complex plane: [z:|z| < I}; {z:|z| ≤ l}; [z:I(z) > 0}.

91. Let X be a metric space and A a subset of X. A is said to be dense (or *everywhere dense*) if $\overline{A}$ = X. Prove that A is dense ⇔ the only closed superset of A is X ⇔ the only open set disjoint from A is ϕ ⇔ A intersects every non-empty open set ⇔ A intersects every open sphere.

92. Let X be a metric space. If $\{x_n\}$ and $\{y_n\}$ are sequences in X such that $x_n \to x$ and $y_n \to y$, show that $d(x_n, y_n) \to d(x, y)$.

93. Show that a Cauchy sequence is convergent ⇔ it has a convergent subsequence.

94. If $X = X_1 \times X_2 \times ... \times X_n$ is the product, and if each of the coordin spaces $X_1, X_2, ..., X_n$ is complete, show that X is complete w respect to each of the metrics d and $\bar{d}$ defined in that problem.

95. Let X be any non-empty set. The set of all bounded real functio defined on X is a metric space with respect to the metric induced the norm defined in that problem. Show that this metric space complete.

96. The following functions f_n defined on [0, 1] form a Cauchy sequer in this space which is not convergent:

$$f_n(x) = 1 \text{ if } 0 \le x \le 1/2,$$

$$f_n(x) = -2^n\left\{x - \frac{1}{2}\right\} + 1$$

if $\frac{1}{2} \le x \le \frac{1}{2} + \left(\frac{1}{2}\right)^n$,

and $f_n(x) = 0$ if $\frac{1}{2} + \left(\frac{1}{2}\right)^n \le x \le 1$.

97. Give an example to show that the set F in Cantor's intersecti theorem may be empty if the hypothesis $d(F_n) \to 0$ is dropped.